D0152397

ACS SYMPOSIUM SERIES **489**

Viscoelasticity of Biomaterials

Wolfgang G. Glasser, EDITOR
Virginia Polytechnic Institute and State University

Hyoe Hatakeyama, EDITOR
Industrial Products Research Institute

Developed from a symposium sponsored
by the Division of Cellulose, Textile, and Paper Chemistry
at the199th National Meeting of the American Chemical Society,
Boston, Massachusetts,
April 22–27, 1990

American Chemical Society, Washington, DC 1992

Library of Congress Cataloging-in-Publication Data

Viscoelasticity of biomaterials: developed from a symposium sponsored by the Division of Cellulose, Paper, and Textile Chemistry at the 199th National Meeting of the American Chemical Society, Boston, Massachusetts, April 22–27, 1990 / Wolfgang G. Glasser, editor, Hyoe Hatakeyama, editor.

p. cm.—(ACS Symposium Series, 0097–6156; 489).
Includes bibliographical references and index.
ISBN 0–8412–2221–5

1. Biomedical materials—Congresses. 2. Viscoelasticity—Congresses. 3. Biopolymers—Congresses. I. Glasser, Wolfgang G., 1941– . II. Hatakeyama, Hyoe, 1939– . III. American Chemical Society. Cellulose, paper, and Textile Division. IV. American Chemical Society. Meeting (199th: 1990: Boston, Mass.). V. Series.

R857.M3V57 1992
610'.28—dc20 92–10653
CIP

The paper used in this publication meets the minimum requirements of American National Standard for Information Sciences—Permanence of Paper for Printed Library Materials, ANSI Z39.48–1984. ♾

PRINTED IN THE UNITED STATES OF AMERICA

ACS Symposium Series

Foreword

THE ACS SYMPOSIUM SERIES was founded in 1974 to provide a medium for publishing symposia quickly in book form. The format of the Series parallels that of the continuing ADVANCES IN CHEMISTRY SERIES except that, in order to save time, the papers are not typeset, but are reproduced as they are submitted by the authors in camera-ready form. Papers are reviewed under the supervision of the editors with the assistance of the Advisory Board and are selected to maintain the integrity of the symposia. Both reviews and reports of research are acceptable, because symposia may embrace both types of presentation. However, verbatim reproductions of previously published papers are not accepted.

Contents

BIOGELS AND GELATION

RELAXATION PHENOMENA

Preface

NATURE CONSTRUCTS A VARIETY OF MATERIALS to use in performing complex and multifaceted tasks. These molecular assemblies are made up of a virtually unlimited number of combinations of a few seemingly simple molecules that themselves consist of even simpler repeating subunits. Nature's skill in arranging molecules into complicated biomaterials might be compared to that of a virtuoso's musical mastery in arranging musical phrases and chords into complex melodies and symphonies.

Virtually none of nature's biological materials are perfectly elastic, and none are completely inelastic. Biomaterials span the range from elastic solids to viscous liquids. The complexity of this material characteristic of elasticity vs. plasticity reflects the eons of "natural selection process" it has taken to reach this level. The complexity indigenous to biomaterials is based on the intricacies of their molecular architecture. That complexity is gradually yielding to scientific advances, particularly in the fields of nuclear magnetic resonance spectroscopy, thermal and mechanical analysis, and electron microscopy.

This book brings together recent advances in the understanding of solid and liquid biomaterials and of viscoelastic solids and liquids. These advances are the result of a vast range of scientific activities. Some are focused on material properties of wood and of textile fibers. Others seek new material applications related to human well being. These advances involve understanding the essence of controlling both the mechanical performance and the fluidity of biomaterials.

Viscoelasticity of Biomaterials is divided into three sections. The first offers a materials design lesson on the architectural arrangement of biopolymers in collagen. Included also are reviews on solution properties of polysaccharides, chiral and liquid crystalline solution characteristics of cellulose derivatives, and viscoelastic properties of wood and wood fiber reinforced thermoplastics. The second section, "Biogels and Gelation", discusses the molecular arrangements of highly hydrated biomaterials such as mucus, gums, skinlike tissue, and silk fibroin. The physical effects that result from the transition from a liquid to a solid state are the subject of the third section, which focuses on relaxation phenomena. Gel formation, the conformation of domain structures, and motional aspects of complex biomaterials are described in terms of recent experimental advances in various fields.

Collectively, the 26 chapters provide a perspective on the current state of knowledge of biomaterial science as it affects the medical, nutritional, and structural fields of biomaterial science.

We acknowledge with gratitude the efforts of the individual authors who contributed to this book. Additional thanks go to Kathryn Hollandsworth, who skillfully and proficiently prepared the final manuscript. A. Maureen Rouhi and Barbara C. Tansill are owed thanks for serving as never-tiring acquisition editors. Particular thanks are due to the donors of the Petroleum Research Fund, administered by the American Chemical Society, and to Aqualon Company, AKZO America, Inc., Monsanto Chemical Company, and Weyerhaeuser; their generous support has contributed in very real terms to the success of this symposium. It was a distinct privilege to be part of such an exciting endeavor.

WOLFGANG G. GLASSER
Virginia Polytechnic Institute and State University
Blacksburg, VA 24061

HYOE HATAKEYAMA
Industrial Products Research Institute
Tsukua, Ibaraki 305, Japan

February 10, 1992

STRUCTURE–PROPERTY RELATIONSHIPS

Chapter 1

Hierarchical Structure of Collagen Composite Systems

Lessons from Biology

Eric Baer, James J. Cassidy, and Anne Hiltner

Department of Macromolecular Science, Case Western Reserve University, Cleveland, OH 44106

Hierarchical structure in biocomposite systems such as in collagenous connective tissue have many scales or levels, have highly specific interactions between these levels, and have the architecture to accommodate a complex spectrum of property requirements. As examples, the hierarchical structure-property relationships are described in three soft connective tissues: tendon, intestine and intervertebral disc. In all instances, we observed numerous levels of organization with highly specific interconnectivity and with unique architectures that are designed to give the required spectrum of properties for each oriented composite system. From these lessons in biology, the laws of complex composite systems for functional macromolecular assemblies are considered.

Soft connective tissues, designed to serve specific functions in the bodies of man and animals, are among the most advanced composite materials known made of macromolecular building blocks. By utilizing the same basic macromolecular design and only by varying the hierarchical structure, a wide range of tissues possessing very different properties are synthesized by the cellular organism. This article reviews our recent work on the hierarchical structure of soft connective tissues from the animal kingdom in hopes that these "lessons from biology" may provide polymer and materials scientists with new ideas for the design of high performance composite polymeric materials. Only with an appreciation and understanding of the

---

Reprinted with permission from *Pure Appl. Chem.* **1991,** *63*(7), 961–973

unique structure-properties relationships in such biosystems can this be achieved.

All soft connective tissues have remarkably similar chemistry at the macromolecular and fibrillar levels of structure. This similitude extends through the collagen fibril which is the basic building block of all soft connective tissues. Differentiation in the hierarchical structure takes place when these fibrils are arranged in a particular architecture, thus constructing a particular tissue for a unique function (1). The hierarchical structure of a tissue reflects and depends upon the different stress states in which the tissue is required to function. Other requirements are also accommodated or coupled simultaneously such as the transport of water and the diffusion of the products of digestion. In each of these cases, the upper levels of the hierarchical structure or architecture are organized with specific mechanical and transport requirements in mind. Furthermore, the many discrete layers of the hierarchical structure must interact in such a way as to make the performance of these tasks possible (2).

The basic structural fiber in all connective tissues is collagen. It is often the most abundant protein in animals and is widely distributed in the structural elements of the body. The common elements in the structure of soft connective tissues begin at the molecular level with similarities in the amino acid sequences of this class of proteins. This similitude lays the groundwork for the development of a variety of tissues that share common chemical and physical properties. Collagens contain the amino acid hydroxyproline, which is not commonly found in other proteins. Along with proline and glycine, these three amino acids account for more than 50% of the total amino acid content in all collagen types. Addition of other amino acids or variations in the ratio of these amino acids differentiate between the different collagen types. Amino acids play a major role in determining the three-dimensional conformation of the precursor to collagen, the tropocollagen molecule. This molecule is a coiled coil of three helical polypeptides 290 nm in length (Figure 1). Five of these molecules align longitudinally with an overlap of approximately one-quarter the molecular length to form a microfibril of diameter 3.6 nm. This so-called quarter stagger combined with the gap between successive macromolecules is responsible for the characteristic 64 nm banding pattern observed in the electron microscope and by x-ray diffraction. The microfibrils are then assembled into collagen fibrils that may vary in thickness from 35 to 500 nm (Table I). These basic building blocks are combined, oriented, and laid up to form higher ordered structures with a particular morphology to suit the requirements of a tissue.

The collagen fibrils are surrounded by an extracellular matrix that maintains the integrity and architecture of the collagen. The primary component of this matrix is high molecular weight hyaluronic acid with a highly branched aggregate of proteoglycans. Proteoglycans consist of a core protein with numerous pendant mucopolysaccharide molecules. These mucopolysaccharides include chondroitin sulfates 4 and 6 and keratan sulfate. The amounts of these molecules and their ratios vary between connective tissues, with location within tissues, and with age. It is the ability of the

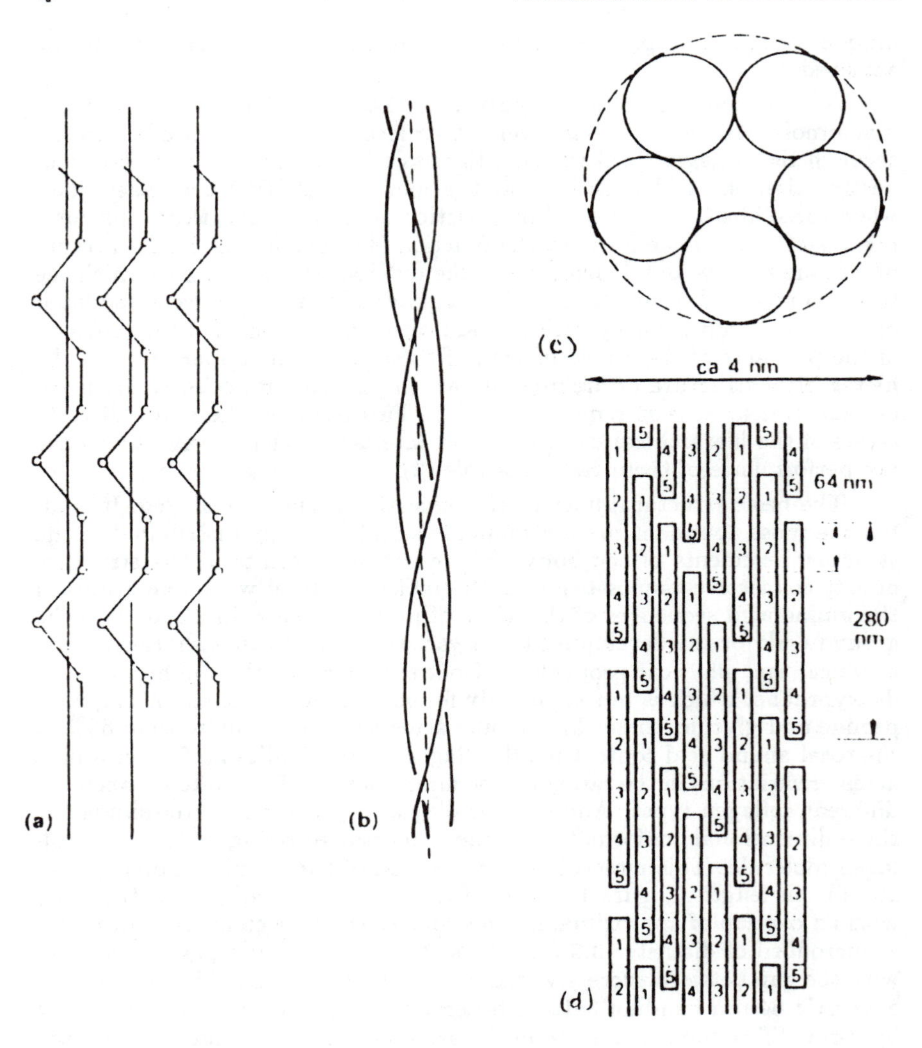

Figure 1. Building blocks of the collagen fibril. The association of amino acid molecules into the collagen fibril is illustrated schematically in these diagrams. (a) Three polypeptide chains, each coiled separately about a minor axis, form a coiled-coil (b) about a single central axis. This triple helical molecule is tropocollagen. (c) Five tropocollagen molecules aggregate to form a microfibril with diameter approximately 4 nm. (d) In a longitudinal section, the quarter staggered arrangement of the tropocollagen helices along the length of the microfibril is shown. This quarter stagger and the gap between the ends of the molecules give rise to the 64 nm banding pattern observed in collagen fibrils. **(Reproduced with permission from reference 10. Copyright 1968 Macmillan Magazines, Ltd.)**

Table I. Collagen fiber size and distribution in various tissues

	Diameter (nm)	Distribution
Rat tail tendon		
fetal	30	Normal (Broad)
adult	450	Broad (Bimodal)
Human periodontal ligament		
fetal	70	Normal (Narrow)
adult	40	Normal (Narrow)
Human intervertebral disk		
fetal	31	Normal
adult	40, 100–150	Bimodal
Human heart valve leaflet		
adult	30–50	Normal
Rat intestine (2 years)	400	Normal (Narrow)
Rabbit cornea	20–25	Normal (Narrow)

proteoglycans to imbibe water which swells the matrix and supports the collagen fibrils. The mechanical properties of the matrix are regulated by its water content which, in turn, affects the properties of the composite tissue as a whole.

The overwhelming consideration in the arrangement of collagen fibrils to form connective tissues is its function in the body. This will be illustrated by three tissue types studied in our laboratories at Case Western Reserve University.

Tendons

Tendons connect muscle to bone around a joint thereby converting muscle contraction into joint motion. As such, the tendon is subjected almost exclusively to uniaxial tensile stresses oriented along its length. This situation requires that the tendon be elastomeric yet sufficiently stiff to efficiently transmit the force generated by the muscle. At the same time, it must be capable of absorbing large amounts of energy without fracturing; for example, absorbing the force generated about the knee joint in a fall. It accomplishes this through a unique hierarchical structure in which all levels of organization from the molecular through the macroscopic are oriented to maximize the reversible and irreversible tensile properties in the longitudinal direction without fracture.

In the tendon, collagen fibrils are organized into larger fibers. These fibrils are arrayed parallel to one another and oriented longitudinally between the muscle and bone. When the fibers are observed in between crossed polarizers in the optical microscope, they have an undulating appearance (Figure 2). Further examination reveals the waveform to be a planar zig-zag or crimp rather than a helix (3). That is, the macroscopic

Figure 2. The morphology of the collagen fibril in tendon. Viewed between crossed polarizers in the optical microscope, the collagen fibrils that make up the tendon have a wavy appearance. Upon further examination this waveform is characterized as a planar zig-zag. Adjacent fibrils are all crimped in register. It is this wavy conformation of the fibrils on the microscopic level of the structure that imparts a high degree of elasticity to the tendon, enabling it to be stretched repeatedly longitudinally without damaging the underlying structure on the nano- and molecular levels. (Reproduced with permission from reference 11. Copyright 1986 Scientific American.)

structure does not reflect the helical conformation of the collagen macromolecules. When the tendon is stretched along its length, this crimp waveform is gradually straightened (Figure 3). It is the magnitude of this waveform that determines the reversible elastic properties of the tendon. As the tendon is pulled further, all the crimp in the collagen fibers is eventually pulled out. The waviness observed in collagen fibers in the rat tail tendon is also found in tendons from other species and in other connective tissue types (Table II) (4). This generality across species and tissue lines indicates the ubiquitousness of this crimp morphology and its importance in determining the mechanical response of all soft connective tissues. As the upper levels of structural architecture are varied to meet the mechanical and other environmental requirements of a particular tissue, so are the parameters of the crimp waveform altered to adjust the mechanical response of that tissue.

Table II. Crimp parameters for various tissues

Source and age	Period (*mum*)	Angle (deg)
Tendons:		
Rat tail (14 months)	200	12
Human diaphragm (51 years)	120	12
Kangaroo tail (11.7 years)	150	8–9
Human achilles (46 years)	40–100	6–8
Other tissues:		
Human periodontal ligament (8 years)	32	25
Human intervertebral disc (31 and 36 years)	12–16	20–42[a]
Heart valve leaflet	20	28–30
Rat intestine (3 months)	20	30–56
Rabbit cornea (adult)	14	(sine wave)

[a] The disc is a gradient structure and crimp parameters change with radial distance through the annulus fibrosus.

Further examination of the structure of tendon for structural hierarchy reveals both the size distribution and the composite nature of the collagen fibrils. By cutting the tendon transversely, one observes that in young animals, the fibrils are of small diameter and they are all essentially the same size (Figure 4) (5). As the animal matures and ages, the fibrils get larger and the range of fibril sizes broadens considerably (Table I). Interspersed between the fibrils is the matrix material discussed earlier. A longitudinal section through a tendon that has been extended to remove the crimp shows the 64 nm banding pattern that is characteristic of collagen and results from the staggered arrangement of the microfibrils. It is apparent that these soft connective tissues are a hierarchical structure constructed of many collagen fibrils that are themselves composite materials (Figure 5) (6).

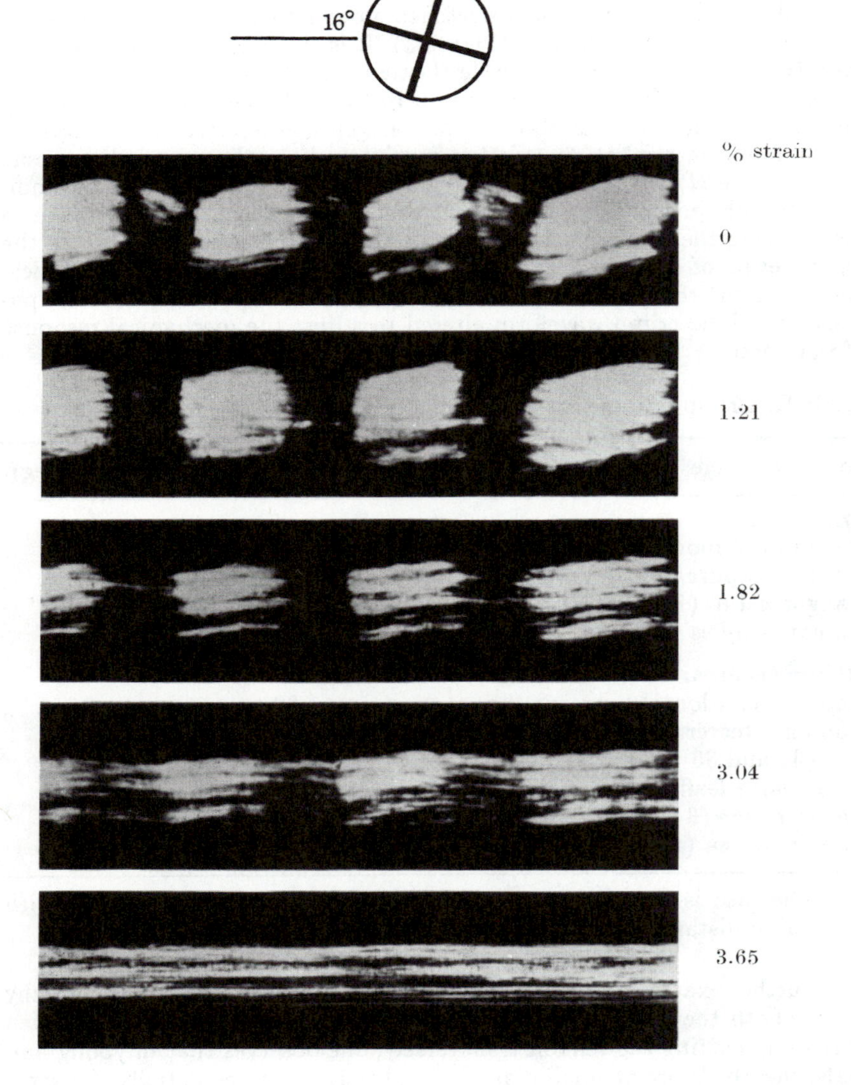

Figure 3. Crimp straightening in the tendon. The performance of the tendon in the toe region of the stress-strain curve is directly attributable to the wavy conformation of the collagen fibrils. In this series of photomicrographs, the progressive straightening of the planar zig-zag with increasing tensile strain is shown. The physiological loading range of the tendon is in this regime, so most loads are borne by this crimp waveform. (Reproduced with permission from reference 12. Copyright 1972 The Royal Society.)

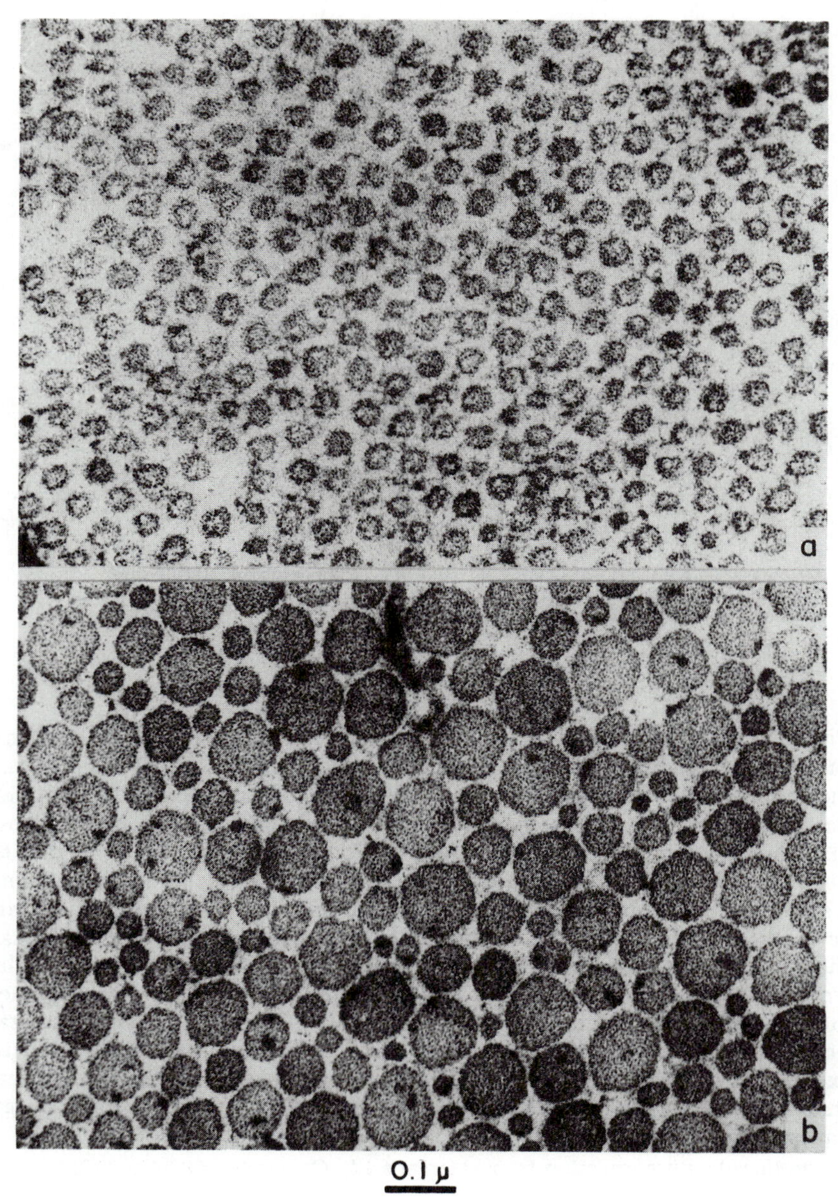

Figure 4. Electron microscopy of collagen fibers in tendon. Transverse sections of undeformed rat tail tendon show the change in fibril size and distribution with age. (a) In the newborn rat, the fibrils are of uniform size, 30 nm in diameter. (b) At 30 months, there is a distribution of fibril sizes up to 350 nm in diameter in this section.

(Reproduced with permission from reference 13. Copyright 1974.)

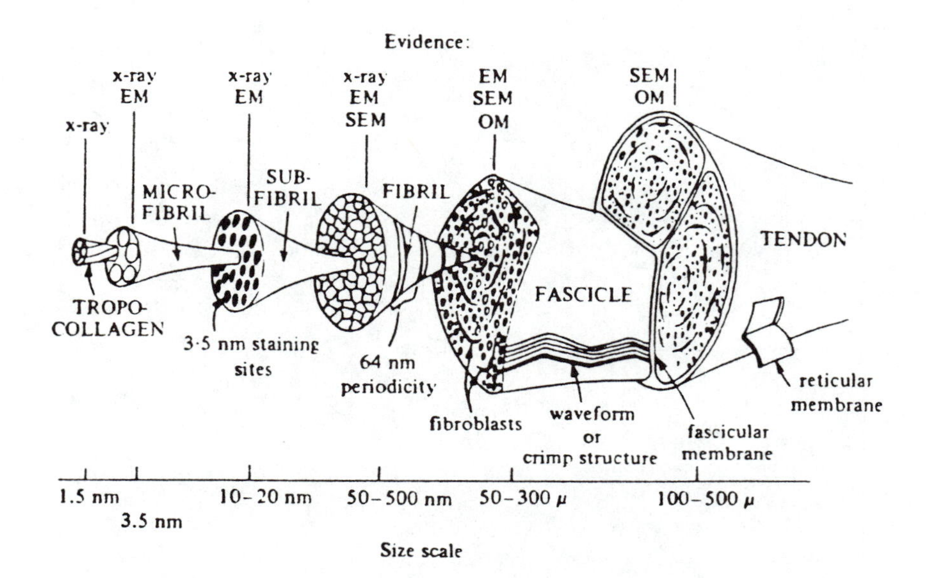

Figure 5. Hierarchical structure of the tendon. The hierarchical organization of connective tissues is illustrated in the tendon. Beginning at the molecular level with tropocollagen, progressively larger and more complex structures are built up on the nano- and microscopic scales. At the most fundamental level is the tropocollagen helix. These molecules aggregate to form microfibrils which, in turn, are packed into a lattice structure forming a subfibril. The subfibrils are then joined to form fibrils in which the characteristic 64 nm banding pattern is evident. It is these basic building blocks that, in the tendon, form a unit called a fascicle. At the fascicular level the wavy nature of the collagen fibrils is evident. Two or three fascicles together form the structure referred to as a tendon. It is this multi-level organization that imparts toughness to the tendon. If the tendon is subjected to excessive stresses, individual elements at different levels of the hierarchical structure can fail independently. In this way, the elements absorb energy and protect the tendon as a whole from catastrophic failure.

This response of the elements of the hierarchical structure of the tendon is reflected in the shape of the stress-strain curve (Figure 6). At small tensile deformations, the stress-strain curve is non-linear, which is the case for all soft connective tissue. As the tendon is stretched further, the stress-strain curve becomes linear as a result of progressive straightening of the collagen fibers. This is referred to as the toe region of the curve. All normal physiological loads are confined to this non-linear toe region. When all the fibers are straight, the modulus is constant since the structure is uncrimped. In the linear region, the fully straightened collagen fibers are further pulled elastically. If the load is released, the tendon will entirely recover to its initial crimped morphology. At high strains, the tendon shows yielding and irreversible damage is imparted to the structure since the collagen fibrils begin to disassociate into sub- and microfibrils. Localized slippage and voiding between hierarchical levels account for the yielding observed at the macroscopic level. Thus, once the tendon yields, it cannot fully recover to its initial state. The hierarchical design distributes the remote stress locally by imparting damage efficiently throughout the different levels of structure, thereby minimizing damage concentrations that could precipitate failure and fracture. This allows the damaged tendon to continue to function in a nearly normal way. The failure of these small structural elements also absorbs a tremendous amount of energy, thereby preventing catastrophic failure. Ligaments behave in essentially the same manner as tendons, except that they connect bone to bone around a joint. As such, their function is to provide a stabilizing force on the joint and prevent unnatural motions from occurring. Examined between crossed polarizers in the optical microscope, collagen fibers in ligaments have the same wavy morphology as in tendons.

Intervertebral Discs

The intervertebral discs are interspersed between the vertebral bones in the spinal column. In this location, the disc is subjected to compressive loads generated on the spine by the weight of the body as well as torsion and bending loads during movement. Another design consideration is impact loading superimposed upon the body weight during activities such as walking or jumping. The disc is composed of two parts: a gelatinous nucleus pulposus containing a matrix of fine collagen fibers, hydrophilic proteoglycan molecules, and up to 88% water; and the annulus fibrosus, having concentric cylindrical layers of fibrous collagen arrayed around the nucleus like the layers of an onion skin (Figure 7). The collagen fibers that comprise the annulus are inclined with respect to the spinal axis by the interlamellar angle. In successive lamellae, this angle alternates like a layered layup in man-made composite materials (7). This angle varies with radial distance through the annulus from ±62 degrees at the periphery to ±45 degrees in the vicinity of the nucleus, thus imparting a structurally graduated architecture to the disc. The collagen fibers that make up these layers have the same planar crimped morphology as seen in the tendon (Figure 8). The crimp parameters vary with radial distance through the annulus

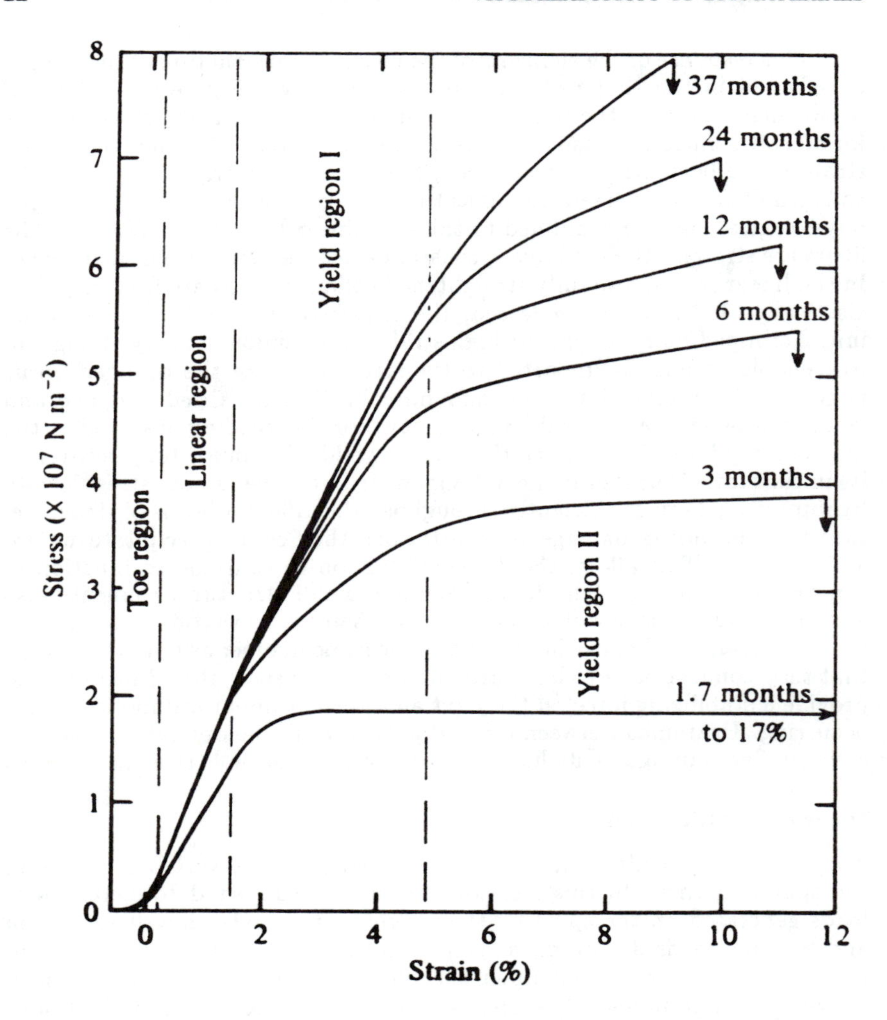

Figure 6. Stress-strain behavior of rat tail tendon. Stress response of the tendon is derived from the underlying architecture of the collagenous elements in the tissue. The stress-strain curve in tension is divided into three regions. In the toe region, the slope of the stress-strain curve (the modulus of the tendon) gradually increases with increasing length. This corresponds to a progressive straightening of the waviness in the collagen fibrils. Once all the fibers are fully straightened, the elements in the structural hierarchy are stretched elastically, giving rise to a region of constant modulus on the stress-strain curve. This behavior continues until individual elements at the sub- and microfibrillar levels begin to break or slip in relation to one another. The modulus then decreases with increasing strain until the tendon fails catastrophically. Differences in the hierarchical structure related to the age of the animal are manifested in all these regions.

Figure 7. Radial anatomy of the intervertebral disc. A slice through the anterior of the disc cut parallel to the axis of the spinal column and photographed through the optical microscope in crossed polarized light reveals the complex collagenous architecture. The relationship between the annulus fibrosus, nucleus pulposus, and the cartilage endplates is shown. While the nucleus and cartilage exhibit little optical activity, the lamellae of the annulus are highly active. These lamellae are made up of large collagen fibrils, the cut ends of which are seen in this view.
(Reproduced with permission from reference 15. Copyright 1989 Gordon & Breach.)

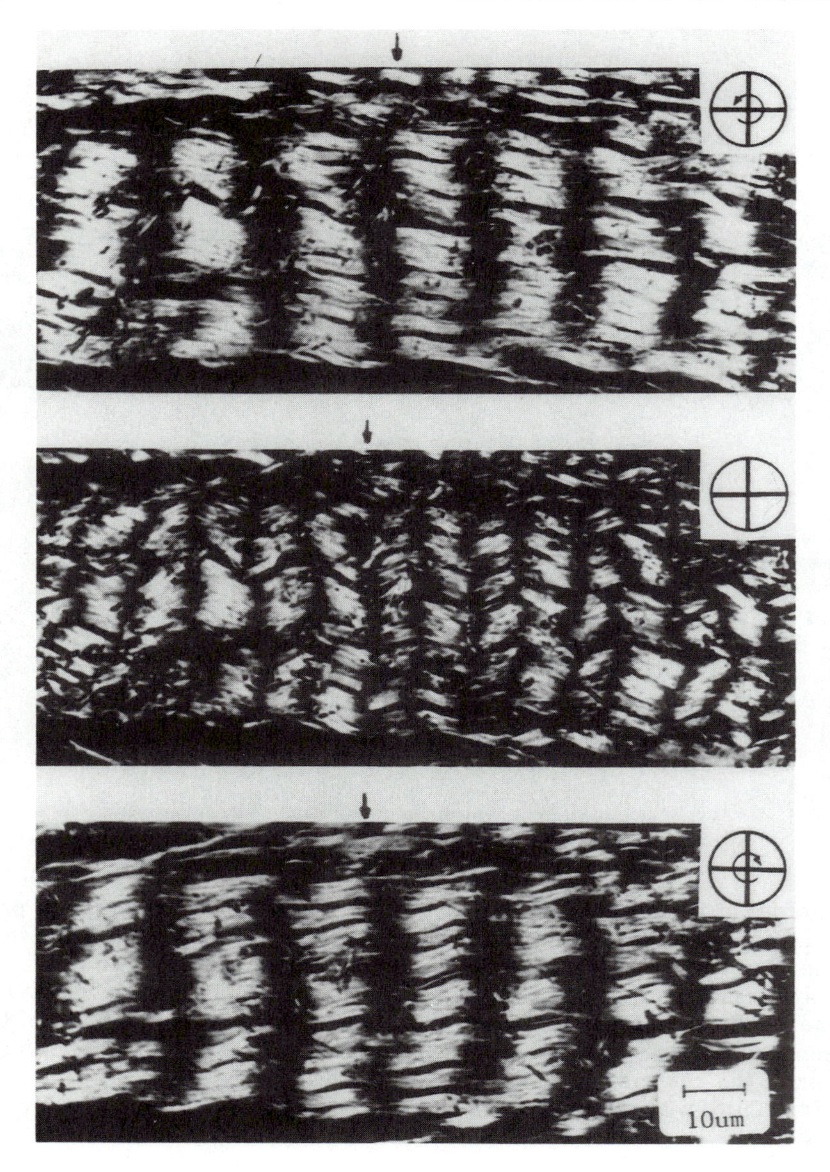

Figure 8. Collagen fibril morphology in the disc. The collagen fibrils that comprise a single lamellae are shown in crossed polarized light at high magnification in the optical microscope. Extinction bands are observed evenly spaced along their length. Upon rotation of the microscope stage, these bands approach one another and eventually merge. The behavior of these fibrils is identical to that of fibrils taken from the tendon and is characteristic of the planar zig-zag waveform observed in those tissues. (Reproduced with permission from reference 15. Copyright 1989 Gordon & Breach.)

as well, with the crimp angle increasing from 22 degrees at the periphery to 42 degrees near the nucleus. The unique hierarchical organization of the disc utilizes the superior tensile properties of the collagen fibers of the annulus in resisting compressive forces oriented along the spinal axis as well as torsional and bending forces (Figure 9). Examination of fixed discs in compression reveals that as the disc is compressed, the nucleus pulposus spreads outward laterally, causing the lamellae of the annulus to bulge outward. Since the ends of the fibers are anchored to the vertebral bone, they are stretched in tension. This elongation of the crimped fibers again accounts for the observed shape of the stress-strain curve. The stress-strain curve of the disc in compression is similar to that of the tendon in tension, containing toe, linear, and yield regions.

While it has not been demonstrated experimentally, one can imagine that the toe, linear, and yield regions of the stress-strain curve for a disc in compression correspond to the same morphological requirements and changes in the collagen fibers that are present in tendons. The biaxial layup of the fibers in the annulus also stabilizes the disc in torsion and bending. In torsion, the collagen fibers in the lamellae oriented in the direction of twist are stretched while the balance are unloaded. In bending, the biaxial layup of the fibers prevents the disc from buckling on the flexion side while resisting excessive bulging on the extension side.

The high water content of the disc plays a vital role in determining the time-dependent mechanical response of the disc in compression (8). If the disc is compressed and held, as it is when the body weight presses the spine during stance, water is squeezed out of the disc. This water loss and the concomitant decrease in disc height is responsible for the "shrinkage" in height that people experience during the course of a normal day's activities. This water is transported out of the disc through cartilage endplates separating the disc from the vertebral bones and into the general circulation. When the load is released, the water is reabsorbed by the disc in the same manner. The stress relaxation and creep responses of the disc in compression are the result of this transport process, rather than the molecular reorganization normally associated with time-dependent mechanical phenomena in materials.

Intestines

The intestine is a semipermeable flexible tube in which mechanical breakup of food is accomplished and through which the products of digestion diffuse into the circulation. As this process takes place, the walls of the intestine are stretched in both the longitudinal and the radial or hoop directions. While this requires a design with adequate tensile strength in these two directions, movement of nutrients across the wall of the intestine demands that efficient routes exist for diffusion to take place. The intestine fulfills both these requirements through an innovative hierarchical structure. The intestinal wall is a layered structure with four distinct layers. Two layers contain smooth muscle cells oriented in the longitudinal and circumferential directions to macerate food particles; another is a covering that isolates the intestine from the other organs in the body cavity.

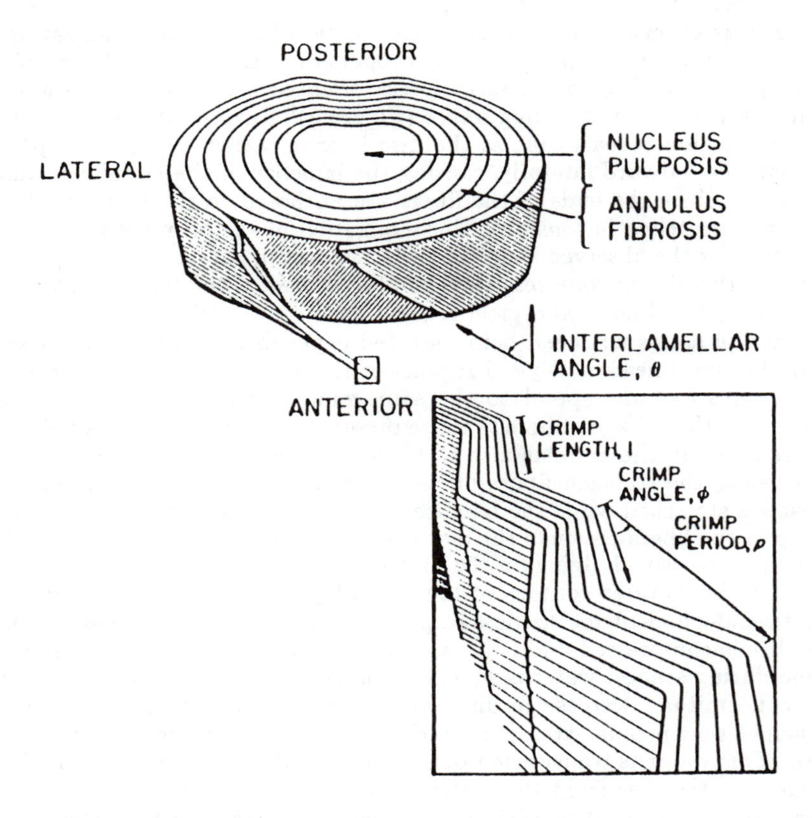

Figure 9. Hierarchical structure of the intervertebral disc. In the intervertebral disc, collagen fibrils are organized into lamellar sheets in the annulus fibrosus which surround a gelatinous and highly hydrated nucleus pulposus. The thickness of lamellae vary with location and are thicker at the anterior and lateral aspects of the disc than at the posterior. Within lamellae, fibrils are parallel and inclined with respect to the axis of the spinal column by an interlamellar angle which alternates in successive lamellae. This angle decreases from the edge of the disc inward. At higher magnification, the fibrils have a planar zig-zag waveform. The crimp angle is largest in fibrils close in to the nucleus and decreases toward the periphery. The orientation of the collagen fibrils in the annulus gives the disc strength and stability in tension, bending, and torsional motions. Based upon optical microscope observations of the morphology of the collagen fibrils, the levels of structural hierarchy below the fibrils are assumed to be identical to that of the tendon and intestine
(Reproduced with permission from reference 15. Copyright 1989 Gordon & Breach.)

The predominant structural member of the intestine is the submucosa, which contains the majority of the collagen fibrils. Fibrils are arranged in layers oriented biaxially around the lumen. In successive layers the fibrils are oriented at ±60 degrees to one another forming helices of opposite sense arrayed at ±30 degrees to the longitudinal axis winding around the intestine (Figure 10). This biaxial winding of the collagen fibrils prevents buckling of the intestine as it winds through the body cavity while the multiple layers of helically wound fibers gives maximum resistance to internal bursting pressures. Within each layer the collagen fibrils are planar crimped as in the tendon. The hierarchical structure of the intestine below the level of the fibrils is identical to that of the tendon (Figure 11).

The crimped morphology in conjunction with the biaxial orientation of the collagen fibers gives rise to the shape of the stress-strain curve in tension and to the anisotropic nature of the mechanical response. At low tensile strains, a similar toe region exists as in the tendon. However, in the intestine, there are two deformation mechanisms at work due to the requirement of simultaneous biaxial expansion. A limited reorientation of the fibers, as evidenced by a decrease in the angle between fibers in adjacent layers, takes place along with the progressive straightening of the crimp in the fibers. The biaxial layup of the fibers also causes an anisotropic mechanical response in tension. If the intestine is stretched along different orientations with respect to the longitudinal axis, different stress-strain responses are observed (Figure 12). Intestine is the stiffest and least extensible at ±30 degrees to longitudinal, in alignment with the fibers in one set of layers, while the greatest extensibility is at 90 degrees to the longitudinal axis. This is suited to the function of the intestine, which necessitates changes in the diameter of the intestine as food passes rather than changes in length. Once fully straightened, the collagen fibers are strong and inextensible, effectively resisting bursting.

Virtually all connective tissues, whether soft or hard, have hierarchical structural designs arranged at discrete levels of structure. For example, the cornea has mechanical requirements similar to that of the intestine. Collagen fibers must contain the pressures generated internally in the eye by the vitreous. However, its distinguishing characteristic is its transparency to light in the visible region of the spectrum. Both these functions are accomplished using a layered structure as in the intestine. The principal structural layer is the stroma which makes up 90 percent of the corneal thickness. The stroma is a layered structure with approximately 200 lamellae between 2 and 3 μm thick. Within these lamellae, the collagen fibrils are parallel to the surface of the cornea and extend entirely across it. Within lamellae, the fibrils are parallel to one another while in successive lamellae, fibrils make large angles to one another (Figure 13). This lamellar structure and cross-ply orientation provide enhanced reinforcement for the eye when compared to an isotropic arrangement of collagen fibrils. The arrow distribution of fibril thicknesses, uniform packing scheme, and the highly ordered arrangement of the fibrils in the lamellae permit the stroma to perform its mechanical function while passing visible light undisturbed.

Figure 10. The biaxial crimped layered structure of collagen fibers in the intestine. Collagen fibers in the intestine have the same planar crimped morphology observed in the tendon. The parameters of the waveforms (crimp angles, period) are different, however, with the greater extensibility required in the intestine being reflected in a large crimp angle. (Reproduced with permission from reference 17. Copyright 1983 Gordon & Breach.)

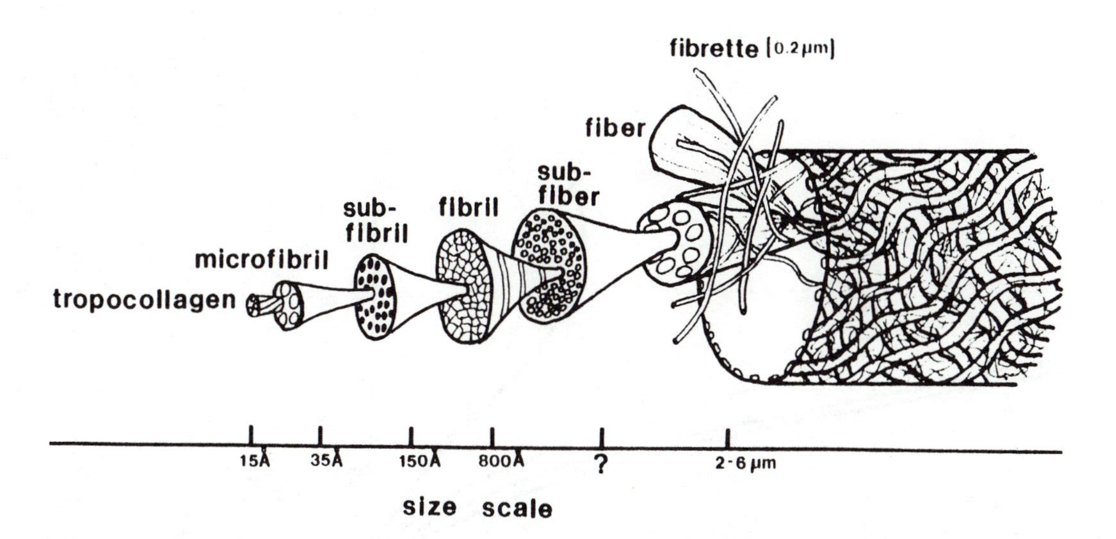

Figure 11. Hierarchical structure of the intestine. The hierarchical organization of the intestine is identical to that of the tendon from the molecular through the fibrillar level. The two tissues differ only in how the collagen fibrils are arranged into the highest levels of structure. Note that in the intestine no fascicles are present. Instead, the collagen fibrils aggregate into fibers which are then wound around the intestine in a helical fashion. (Reproduced with permission from reference 16. Copyright 1982 Gordon & Breach.)

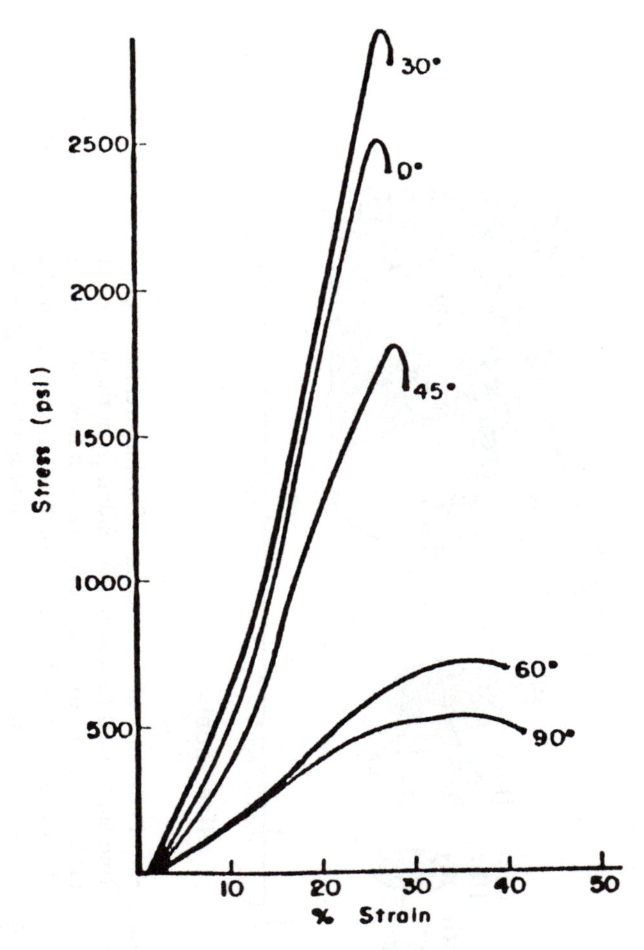

Figure 12. Stress-strain behavior of the intestine in tension. The tensile properties of the intestine vary with orientation, as would be expected in a tissue made up of biaxially oriented fibers. The greatest modulus and least extensibility occur when the intestine is stretched parallel to the fibers in the layers (30°). The properties along the longitudinal direction (0°) are only slightly less than in the fiber direction. The greatest extensibility and least modulus are found perpendicular to the long axis in the hoop direction (90°). These properties represent a structural adaptation to the function demands of the intestine. Digestion of food requires the ability to accommodate changes in the diameter of the intestine as food passes through and for rapid diffusion of nutrients through a permeable wall. At the same time, changes in the length of the intestine must be minimized in order for the peristaltic action of the smooth muscle to operate efficiently. The arrangement of collagen fibrils in the intestine enables these demands to be met. (Reproduced with permission from reference 18. Copyright 1981 Blackwell.)

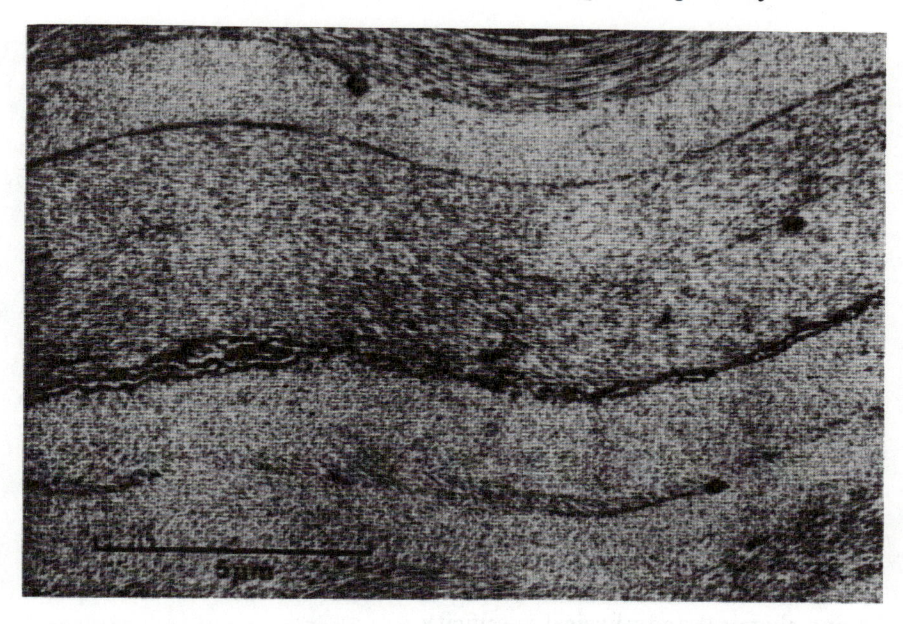

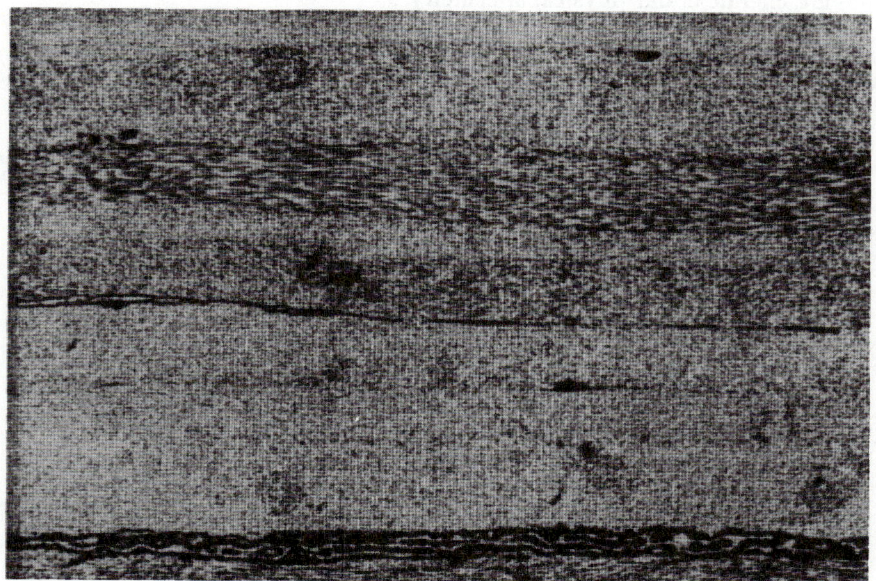

Figure 13. Collagen fiber layup in the cornea. Electron micrographs of the central rabbit cornea show the wavy morphology of the collagen fibers extending through the thickness of the stroma. Note also the layered structure of the stroma, the uniform fiber diameter, and the variation in the orientation of the fibers in each layer.
(Reproduced with permission from reference 19. Copyright 1983 Academic.)

General Comments

It is tempting to suggest that all connective tissues follow at least three "Laws for Complex Assemblies." Starting with a very similar macromolecular design (the fibrous protein, collagen, and the ground substance, mucopolysaccharide), the hierarchical structures are assembled into distinct and totally different systems both in their morphology and their function.

First, the macromolecules associate into discrete levels of organizations. Usually these are in the form of fibrils which are themselves composed of smaller subfibrils and microfibrils. The fibrils are subsequently arranged in layered structures reflecting the specific functional requirements of the overall composite system. The minimum number of discrete levels or scales observed thus far in biocomposites are four. That is, structural levels at the molecular, at the nano, at the micro, and at the macro scale are the minimum required components within an ordered hierarchical biocomposite system.

Second, these levels are held together by highly specific interactions between surfaces. Considerable evidence exists for surface-to-surface interactions due to intermolecular bonding at specific active sites, and for epitactic arrangements of a crystallographic nature. Whatever the nature of the interfacial mechanism, strong interfacial interactions are required having chemical and physical specificity.

Third, the highly interacting fibers and layers are organized into an oriented hierarchical composite system which is designed to meet a spectrum of property or function requirements. This law of architecture stipulates that as the complexity of the overall system and its uses increases, the system is capable of a higher degree of adaptation in difficult and unexpected environments. The so-called "intelligent composite system" results from a complex architectural arrangement which is designed to serve highly specific functions.

At the very least, the analysis of complex behavior in natural polymeric systems in terms of hierarchies is an approach that interrelates our understanding of structure at various scales. Such an approach may prove invaluable in the design of new advanced polymers. Structural hierarchy is more than a convenient vehicle for description and analysis. Important and difficult questions that remain to be addressed include the physical and chemical factors that give rise to relatively discrete levels of structure and the relations that govern their scaling (2).

Literature Cited

1. Baer, E. *Scient. Amer.* 1986, **254**, 179.
2. Baer, E.; Hiltner, A.; Keith, H. D. *Science* 1987, **235**, 1015.
3. Baer, E.; Diamant, J.; Litt, M.; Arridge, R. G. C.; Keller. *Proc. Royal Soc.* 1972, **B180**, 293.
4. Cassidy, J. J.; Hiltner, A.; Baer, E. *Ann. Rev. Mater. Sci.* 1985, **15**, 455.

5. Baer, E.; Torp, S.; Friedman, B. *Structure of Fibrous Biopolymers*; Atkin, E. D. T.; Keller, A., Eds.; Butterworth: London, 1975; p 223.
6. Baer, E.; Kastelic, J.; Galeski, A. *J. Connect. Tissue Res.* 1978, **6**, 11. Baer, E.; Kastelic, J.; Palley, I. *J. Biomech.* 1980, **13**, 887.
7. Hiltner, A.; Cassidy, J.; Baer, E. *J. Connect. Tissue Res.* 1989, **23**, 75.
8. Silverstein, M. S.; Cassidy, J.; Hiltner, A.; Baer, E. *J. Mater. Sci., Mater. in Medicine* 1990, in press.
9. Hiltner, A.; Orberg, J.; Baer, E. *J. Connect. Tissue Res.* 1983, **11**, 285.
10. Smith, J. W. *Nature* **1968,** *254,* 157.
11. Baer, E. *Sci. Am.* **1986,** *10,* 185.
12. Diamant, J.; Keller, A.; Baer, E.; Litt, M.; Arridge, R. G. C. *Proc. Roy. Soc. Lond. B.* **172,** *180,* plate 24.
13. Torp, S.; Baer, E., Friedman, B. *Proc. Colston Conf.* **1974,** 226.
14. Kastelic, J. Baer, E.; *Soc. Exp. Biol.* **1980,** 398.
15. Cassidy, J. J.; Hiltner, A.; Baer, E. *Connect. Tissue Res.* **1989,** *23,* 77, 85, 87.
16. Orberg, J. W.; Baer, E.; Hiltner, A. *Connect. Tissue Res.* **1982,** *9,* 187.
17. Orberg, J. W.; Baer, E.; Hiltner, A. *Connect. Tissue Res.* **1983,** *11,* 287.
18. Fackler, K.; Klein, L.; Hiltner, A. *J. Microsc.* **1981,** *124,* 305
19. McCally, R. L.; Bargeron, C. B.; Green, W. R.; Farrell, R. A. *Exp. Eye Res.* **1983,** *37,* 548.

RECEIVED December 16, 1991

Chapter 2

Physicochemical Properties of Polysaccharides in Relation to Their Molecular Structure

Marguerite Rinaudo

Centre de Recherches sur les Macromolécules Végétales, Centre National de la Recherche Scientifique, B.P. 53X, Grenoble, 38041 France

Several relationships between the chemical structure of some bacterial polysaccharides and their physical properties in aqueous solutions are reviewed. These native microbial polysaccharides adopt an ordered conformation in dilute solution which controls the stiffness of the molecule and thereby the viscoelastic behavior of their solutions. Two types of gelation observed with polysaccharides are described. Especially characteristic are cooperative interactions in thermoreversible gels that are influenced by the thermodynamic behavior of the solutions.

This review concerns some native polysaccharides produced by plants, animals, and microorganisms. Even though modified polysaccharides are very important for industrial applications, this class of polymers will not be considered here because chemical modifications usually induce heterogeneities in chemical structure that result in the disappearance of some characteristic such as stereoregularity.

The polysaccharide cellulose is one of the most important natural polymers; it is used as an insoluble material in the textile and paper industries. Different reviews have been published on its structure, biosynthesis and physical properties (1–6). It will not be covered here. Starch is also of great importance; in some conditions, it is soluble in an aqueous medium and forms gels by retrogradation. A large bibliography exists concerning starch, starch derivatives, and the morphology of starch granules (7–10); this compound will also not be considered in the following. Among the most important sources of polysaccharides is also the crab shell or other crustacea from which chitin is extracted (11–14). Alkaline deacetylation of chitin produces chitosan, which becomes soluble in acid; this polysaccharide has a specific chelating property related to the $-NH_2$ group which

0097–6156/92/0489–0024$06.00/0

is unusual in natural polysaccharides. In the fully deacetylated form, it is poly[β(1 $\rightarrow$ 4)-2-amino-2-deoxy-D-glucose] which becomes an important and interesting polymer for many applications.

Most of the water-soluble native polysaccharides are used as thickening polymers mainly for food applications but also for other industrial purposes (15).

This review will be devoted in large part to the microbial water-soluble polysaccharides. The aim will be to show how the chemical structure (monomeric unit, osidic linkage, ionic site, side groups, etc.) controls the physical properties in solution and during gelation.

Solution Properties

Xanthan Gum. The most important bacterial polysaccharide produced on large scale is the xanthan gum excreted by *Xanthomonas campestris* (Figure 1a). Many papers have been published on this polysaccharide, but there are still some questions under discussion.

The thermodynamic behavior (i.e., activity coefficients of counterions) suggests that xanthan is a single chain molecule which adopts a helical conformation under certain conditions (excess of external salt, low temperature) (16,17). Xanthan undergoes a reversible conformational transition (helix $\rightleftharpoons$ coil) as demonstrated by optical rotation-temperature experiments. A characteristic temperature, T_m, taken at half the transition, is assumed to characterize the conformational transition. As for other biopolymers, a linear dependence relates log C_T to ($1/T_m$) for monovalent counterions where C_T is the total ion concentration taking into account the external salt concentration C_s and the osmotic free counterions ΦC_p (C_p is the concentration of polymer ionic sites expressed in equivalents/liter, and Φ is the osmotic coefficient directly related to the charge parameter of the ionic polymer).

The change in viscosity from helical conformation to coil conformation when temperature increases is not important due to the intrinsic stiffness of the β(1 $\rightarrow$ 4)D-glucose chain and the presence of the three sugar units as a side chain. In dilute solution, in the helical conformation, xanthan behaves as a worm-like chain characterized by an intrinsic persistence length around 350Å for the preheated molecule (18); this value was determined using Yamakawa-Fujii and Odijk (19) treatments. This great stiffness of the xanthan molecule justifies the high viscosity for a given molecular weight and the small dependence of viscosity on the ionic strength; a large solvent draining effect exists even in infinite ionic strength in monovalent electrolyte (20,21). This insensitivity to an excess of salt was appreciated for tertiary oil recovery compared to the usual flexible ionic synthetic polymers.

In direct relation to the stiffness of the molecule, in concentrated solutions, cholesteric liquid crystals are formed (22); the same type of structure was also obtained with succinoglycan (23) and scleroglucan (24).

The most important physical properties of xanthan and related polysaccharides are the rheological properties in dilute and semi-dilute solutions. This point has been much investigated in the literature. Never-

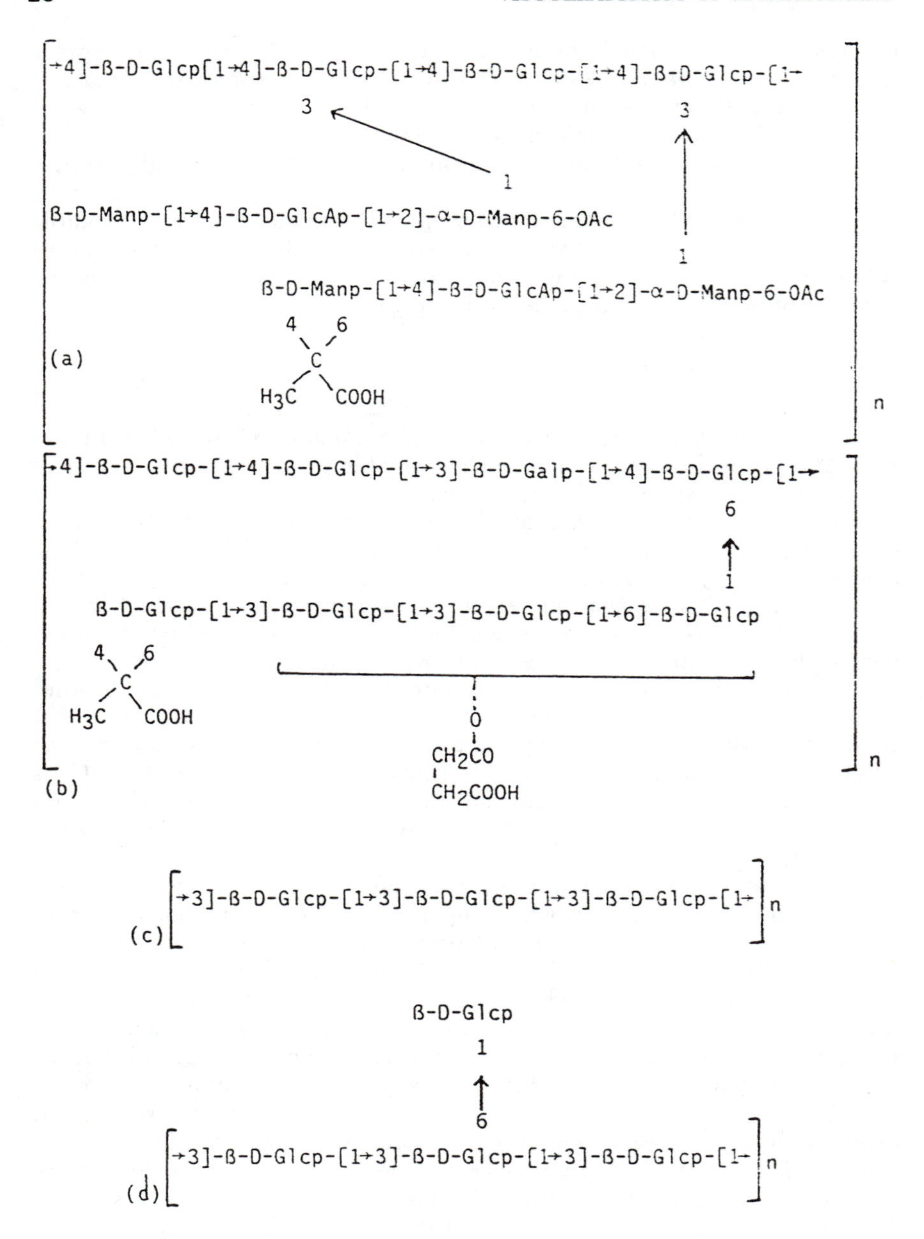

Figure 1. Repeating unit structure of some microbial polysaccharides. (a) xanthan; (b) succinoglycan (29); (c) curdlan (30); (d) scleroglucan (35).

theless, many of the results were obtained on commercial samples in which microgels and aggregates preexist due mainly to post-fermentation treatments. This allowed some authors to draw conclusions about the gel-like properties of xanthan solutions (25). In fact, xanthan directly isolated from the broth behaves as a single chain solution as recently discussed (26). In the dilute regime (i.e., polymer concentration lower than the overlap concentration, $C^* \sim [\eta]^{-1}$, where $[\eta]$ is the intrinsic viscosity), the viscosity in the Newtonian regime is directly related to the polymer concentration C and the molecular weight M or the intrinsic viscosity following the relation:

$$\eta_{sp} = C[\eta] + k'(C[\eta])^2$$

in which $[\eta] = KM^a$.

In helical conformation, in direct connection with the stiffness of the chain, the parameter a equals 1.14 in 0.1 M NaCl (27), and it approaches 1 for infinite ionic concentration (18). Such high values for the a coefficient are still not justified in the literature; they seem to correspond to the original behavior of worm-like chains in which the exponent a increases with the persistence length in the range of intermediate molecular weights even when the persistence length is considered in infinite salt concentration thereby eliminating the electrostatic contribution (see Yamakawa and Fujii treatment, ref. 19a, Figure 1).

The semi-dilute regime appears over C*; in this range, the non-Newtonian behavior (*i.e.*, decrease of the viscosity as a function of the shear rate) becomes more and more important when the polymer concentration and molecular weight increase.

A second critical concentration, C**, corresponds to a linear variation of log η_{sp} with $C[\eta]$ in the range of $C^{**}[\eta] \simeq$ 8-10; over C**, the viscosity follows the relation:

$$\eta_{sp} \propto (C[\eta])^{3.7} \propto C^{3.7}M^4$$

This dependence looks like that of synthetic polymers in the melt (28). Over C**, the viscoelastic properties of the fluid become important and a yield stress (τ_r) rapidly appears when the polymer concentration increases. The low values of $C[\eta]$ corresponding to the appearance of τ_r seem to be related to the stiffness of the xanthan molecule. This may be related to the conditions under which the screening length, ξ, in semi-dilute solution is of the same order as the persistence length (20). In the same range of polymer concentration, the free diffusion of the molecules becomes nearly independent of the polymer concentration (20).

The viscoelastic behavior of xanthan solutions has recently been examined in flow and dynamic experiments (26).

To conclude, xanthan gum behaves as a worm-like chain with a large local stiffness in its helical conformation. Some essential differences between xanthan gum and synthetic flexible polymers become apparent. The influence of the chemical structure (i.e., content of pyruvate or acetyl substituents depending on strain type and/or on post-fermentation treatment) on the main physical properties in solution seems to be negligible, but it may be important for polymer-polymer interactive processes (i.e., adsorption on solids, etc.).

Other Microbial Polysaccharides. More recently, the behavior of a succinoglycan excreted by *Pseudomonas* sp. NCIB 11592 and produced by Shell Company has been examined. The main chain is a $\beta(1 \rightarrow 4)\beta(1 \rightarrow 3)$ glycanic chain with a four sugar unit side chain attached to each fourth unit of the main chain (Figure 1b). The stiffness of this backbone is much lower than that of xanthan. In the disordered conformation, the viscosity is very low; by contrast, in the helical conformation induced by excess salt, the stiffness becomes greater than that of xanthan (29). The role of the interaction of the side chain with the backbone is essential for stabilizing the helical conformation, and this seems of prime importance for the coordination of the conformational transitions (30,31). In direct relation, a large isotopic effect is demonstrated when the melting temperature T_m is determined in H_2O vs. D_2O, and this indicates the role of H bonds in the helical conformation (23).

$\beta(1 \rightarrow 3)$ *Glucans.* Curdlan is a polysaccharide produced by *Alkaligenes faecalis* var. *myxogenes* 10 C3 and specially investigated by Harada in Japan (32). This $\beta(1 \rightarrow 3)$ linear glucan is a gel-forming polymer (33); it is necessary to heat the aqueous dispersion of curdlan to dissolve it (Figure 1c). A gel is formed upon cooling, and this process is thermally reversible. Curdlan forms firm gels with some syneresis; the mechanism of gelation was investigated by Fulton and Atkins (34).

Scleroglucan has the same backbone structure as curdlan; it is a $\beta(1 \rightarrow 3)$ glucan with a $\beta(1 \rightarrow 6)$ glucose as side chain on every third glucose unit of the main chain. Scleroglucan is produced by *Sclerotium rolfsii.* A very regular structure has been revealed by ^{13}C NMR (35). It was demonstrated that schizophyllan (same structure as scleroglucan) produced by *Schizophyllum commune* forms a stiff triple helix with a persistence length of 200 nm (36). It is a neutral polymer soluble in water but with some aggregation. Solubility increases in NaOH (0.01N), and this produces good solutions (37).

The triple helix of curdlan (38) (whose near-insolubility is reflected by the formation of strong and opaque gels) is rendered soluble by substitution of every third glucosyl sidechain, and this inhibits packing without affecting the local helical structure (36).

An irreversible conformational transition was demonstrated in DMSO and 0.2 M NaOH from the stiff triple helix conformation of scleroglucan to a single chain coil with a very low viscosity. Thermal melting was also observed in water around 135°C (36b). In the ordered conformation, the molecule is very stiff, and it forms liquid crystals (24).

The presence of the side group is essential for increasing the solubility of the polymer and decreasing the interchain interactions compared with curdlan. A phase transition is observed around 8°C leading to the formation of a gel. This behavior is due to interchain interactions with a large isotopic effect. This effect demonstrates the interaction between the side group and the main chain involving the surrounding water molecules (39,40).

Gellan-Like Polysaccharides. Gellan is a polysaccharide excreted by the bacteria *Pseudomonas elodea.* In the literature, it has been referred to as S-60 or PS60. It is produced by Kelco (USA) as a gel-forming polymer. The chemical structure of the native polymer was examined by different groups (41) (Figure 2). The substituents reduce the ability to gel, but they are readily hydrolyzed in alkaline conditions. The linear polymer thus obtained is called Gelrite (Kelco trade name) (Figure 3a), and this produces firm gels under certain thermodynamic conditions (i.e., excess of external salt, low temperature, etc.). The gels are capable of displacing agarose in some applications (42). In the solid state, Chandrasekaran *et al.* established a double helical structure in which the helices are parallel (43).

In dilute aqueous solution, gellan exists as a coil at high temperature, and forms a double helix when the temperature decreases or when an excess external salt is present. In the dilute regime, in the ordered conformation, gellan behaves as a worm-like chain with a persistence length of 72 nm. In the disordered conformation the length was determined to be 6.1 nm (42).

Gellan is a weakly charged polyelectrolyte even in double helix structure (charge parameter $\lambda = 0.75$). The activity of counterions in dilute solution confirms the double helical conformation which is more stable in the presence of divalent counterions (44).

Even with its low charge parameter, this polyelectrolyte displays a high ionic selectivity in the double helix form. With monovalent counterions, this ionic selectivity is in the order: $Li^+ > Na^+ > K^+ > TMA^+$ (TMA^+ = tetramethyl–ammonium ion). This selectivity also controls the degree of aggregation and the gel-forming ability (45).

Welan is a new polysaccharide produced by Alcaligenes AT CC 31555 and referred to as S-130 or Biozan, the trade name of Kelco. This polymer has the same backbone as that of gellan, but since it is branched (Figure 3b), it is a non-gelling polymer. It is advocated as a thickening and suspending agent just as rhamsan (S194—produced by Alcalignes AT CC 31961) (Figure 3c) or the polysaccharide S-657 excreted by Xanthomonas AT CC 53159 (Figure 3d) (46).

For welan, the following relation was found

$$[\eta] = 3.37 \times 10^{-7} M^{1.41} \qquad ([\eta] \text{ in } dl/g)$$

between the intrinsic viscosity $[\eta]$ and the molecular weight, M, in 0.1 M NaCl solution. The high exponent ($a = 1.41$) seems to indicate a very stiff conformation (47). Brant *et al.* seem to rule out a double helix conformation. Based on thermodynamic arguments, it was recently proven that welan forms a very stable double helix structure (48).

The physical properties of these three polysaccharides with comb-like structure were also investigated by Crescenzi *et al.* (49) and Brant (50) so as to discuss the role of the comb-like structure on the physical properties. As pointed out by Robinson (51), no conformational transition was observed in welan for different ionic strengths, and no detectable high resolution NMR

```
              L-Glyceric
                  1
                  ↓
                  2
[→3)-β-D-Glcp-(1→4)-β-D-GlcpA-(1→4)-β-D-Glcp-(1→4)-α-L-Rhap-(1→]n
                  6
                  ↑
               Ac0.5
```

Figure 2. Repeating unit structure of the native gellan (39a).

(a)

```
–3)-β-D-Glcp-(1–4)-β-D-GlcpA-(1–4)-β-D-Glcp-(1–4)-α-L-Rhap-(1–
```

(b)

```
–3)-β-D-Glcp-(1–4)-β-D-GlcpA-(1–4)-β-D-Glcp-(1–4)-α-L-Rhap-(1–
                                          3
                                          ↑
                                          1
                          α – L – Rhap or α – L-Manp.
```

(c)

```
–3)-β-D-Glcp-(1–4)-β-D-GlcpA-(1–4)-β-D-Glcp-(1–4)-α-L-Rhap-(1–
       6
       ↑
       1
   α-D-Glcp
       6
       ↑
       1
   β-D-Glcp
```

(d)

```
→3)-β-D-Glcp-(1→4)-β-D-GlcpA-(1→4)-β-D-Glcp-(1→4)-α-L-Rhap-(1→
                                        3
                                        ↑
                                        1
                          α-L-Rhap-(1→4)-α-L-Rhap
```

Figure 3. Repeating unit structure of gellan-like polysaccharides. (a) gellan (39); (b) welan (50a); (c) rhamsan (50b); (d) S-657 (39d).

could be recorded. The conclusion is that the ordered rigid structure is stable.

It is apparent again that side chains inhibit aggregation and thermoreversible gelation; and that gellan-like polysaccharides behave as worm-like chains with a high persistence length corresponding to a double helical structure.

Polysaccharide Gelation

Some polysaccharides are well known to form gels in given thermodynamic conditions; they form three-dimensional networks stabilized by non-covalent linkages. Two kinds of physical gels are formed depending on the chemical structure of the polysaccharides: thermoreversible gels with pectins in acid medium or agarose, carrageenans; and crosslinked gels in the presence of calcium counterions in alginate, pectinates, etc.

The mechanisms of gelation and the properties of these gels have been investigated for a long time. More recently, it appeared that synthetic polymers, such as atactic polystyrene, also form thermoreversible gels. This system was described by Keller and coworkers (51,52). They related gelation to a liquid-liquid phase segregation.

Thermoreversible Gels. The behavior of carrageenans was investigated in 1979 in our laboratory (53); κ and ι forms are gelling polymers in the presence of an excess of external salt and/or at low temperatures. The firmness of the gels depends directly on the chemical structure of the polysaccharides as pointed out recently (54) in a series of galactan gels.

The lower the charge density (lower yield of $-O-SO_3{}^-$ group per disaccharidic unit), the lower the solubility of the polysaccharides and the greater their ability to form gels. The degree of deswelling in non-solvents demonstrates the stability of the junction zones. Almost no deswelling of the agarose gel formed in water was observed in acetone. The properties are explained by the mechanism of gelation suggested previously (55) for κ-carrageenan.

κ-carrageenans in dilute regime and high temperature are coils which undergo a conformational transition when temperature decreases and/or ionic content increases as follows:

$$2 \text{ coiled molecules} \rightleftarrows 1 \text{ stiff double helix}$$

The conformational transition is characterized by T_m, the conformational change temperature depending directly on the total ionic content of the solution C_T in monovalent electrolytes (with $C_T = Cs + \overline{\gamma}\, Cp$, where Cs is the concentration of external electrolyte, Cp is the ionic concentration in polyelectrolyte, and $\overline{\gamma}$ is the average activity coefficient of the counterions averaged on the coiled and double helical structure values). These results are in agreement with the following relation:

$$\Delta H = -R(\Phi_c - \Phi_h)\frac{d\ \ln C_T}{d\ 1/T_m}$$

in which ΔH is the enthalpy of conformational change, Φ_c and Φ_h are the osmotic coefficients of the counterions under the coil and helical conformations respectively (56). When a gel is formed, a contribution of the junction zones modifies the ΔH values.

The mechanisms of gelation imply the aggregation of the stiff double helices; in such cases (lower temperatures and/or higher salt content) a hysteresis is observed and the temperature of conformational transition corresponding to the gelation temperature is lower than the temperature of melting (T_f) of the gel.

$$\text{Sol} \underset{T_f}{\overset{T_m}{\rightleftarrows}} \text{Gel} \qquad T_m < T_f$$

In the galactan series, the lower the charge density of the polysaccharide, the larger the hysteresis, *i.e.*, the difference $|T_f - T_m|$ for a polymer concentration in the same range.

While nearly no salt selectivity was observed for the neutral pure agarose, with K and i carrageenans a large ionic selectivity was observed. It was related to the ability to form ion pairs in dilute solution (57). The ion selectivity followed the order: Rb > K, Cs > Na, TMA > Li.

This order corresponds to the ability of counterions to induce the double helix but also the gelation. In fact, thermoreversible gelation is a two-step mechanism as demonstrated (58) in specific thermodynamic conditions. The two processes are directly related: ion pairs decrease the net charge, *i.e.*, the solubility, and favor aggregation.

More recently, the same mechanism of gelation was suggested for gellan, a microbial polysaccharide (42). Contrary to κ-carrageenans, it was demonstrated that the conformational transition is not sensitive to the nature of the monovalent counterions. This may be related to the lower charge density of the gellan ($\lambda = 0.37$) compared with the κ-carrageenan ($\lambda = 0.65$) in the single chain conformation.

The ionic selectivity found (43) is the following: K > Na > Li > TMA. The hysteresis is usually low and the divalent counterions are very effective to induce gelation due to their higher ionic valency.

The mechanical properties of the gels formed were investigated on κ-carrageenans (59) and gellan (42a).

The influence of molecular weight was established especially for κ-carrageenan gels. First, it was shown that the polymer concentration must be larger than C*, the overlap concentration (C* is estimated in the range of $[\eta]^{-1}$ with $[\eta]$ the intrinsic viscosity).

The elastic modulus E at a given ionic concentration varies as $E = KC^n$, with K depending on the ionic content and $n = 2$, which seems an original characteristic for physical gels compared to covalent gels (59).

At a given polymer concentration and ionic content, the elastic modulus increases when the molecular weight increases up to a limit around $\overline{M_w} = 180000$; on the contrary, the yield stress, F_m, increases linearly with $\overline{M_w}$ in the same conditions on κ-carrageenans (60).

In general, the elastic modulus is greater when the polymer solubility decreases in given conditions corresponding to larger junction zones. The junction zones should be characterized by the numbers of double helices aggregated and the length and degree of cooperativity of the interacting zones.

In cases of moderate ionic concentration but the same polymer concentration, gellan forms stronger gels than κ-carrageenan, and the counterions play an important role. For a given polymer concentration (13g/l) and electrolyte concentration (C_s = 0.1N), the following values were obtained (42a):

	K	Na	Li
$E \times 10^{-4}$ (N/m^2)	31.4	11.9	0.44
Fm (N)	23.1	13.3	1.40

Calcium-Induced Gelation

The ability of gelation with calcium is directly related to the configuration of the sugar units involved in the polysaccharide.

In pectins, depending on the degree of esterification, the crosslink density is related to the galacturonic acid content. This was described previously using pectins with different degrees of methylation (61–64).

The sol-gel diagram for calcium pectin systems was revised by Durand *et al.* (65) in terms of the cooperativity of the crosslink.

The actual discussion concerns the minimum number of calcium ions involved in a stable cooperative crosslinkage; this number was suggested to be around seven ions.

In alginate, considered as a copolymer formed with α-L-guluronic acid blocks, β-D-mannuronic acid blocks and alternated zones, the crosslink is based on the α-L-guluronic acid blocks (having a configuration looking like D-galacturonic acid units in pectins). The same mechanism of junction for alginates and pectins was proposed to be the egg box model (66). An ionic selectivity is also demonstrated in this case with the order Ba > Ca > Sr, corresponding to decreasing stability of the gel. For both polymers, no gel is ever formed in the presence of Mg^{2+}.

Applications of alginate are developed in different industrial fields and especially for cell immobilization (67). The mechanical properties of these gels were investigated by Bouffar-Roupe; in this work, the role of the guluronic block on the elastic modulus was clearly demonstrated (68). In the absence of divalent counterions, alginates have a moderate stiffness (persistence length around 100Å) (69), depending on their chemical structure (68).

Conclusion

In this study, the main results obtained on some polysaccharides were reviewed. The water-soluble polymers were chosen to demonstrate the role of chemical structure on their physical properties in aqueous solution.

The microbial polysaccharides form a series of new polysaccharides with a large variety of physical properties, and they hold great interest for investigating the fundamental structure-properties relationship. Xanthan and succinoglycan can be considered as single-chain polymers adopting a helical conformation in the presence of an excess of neutral salts. Their stiffness is great, and they behave as worm-like chains with persistence lengths of about 400Å.

The $\beta(1 \rightarrow 3)$ glucans form triple helical structures, such as curdlan, a well-known gel-forming polysaccharide. When side groups ($\beta(1 \rightarrow 6)$ glucose) exist, as in scleroglucan, the polysaccharide becomes soluble in water, and then it is a good thickening polymer. It forms a gel in water only at temperatures below 8°C.

The gellan-like polysaccharides are produced by bacteria. Gellan is a gelling polymer, but the comb-like polysaccharides (welan, rhamsan, S-657) having the same backbone also form a double helical structure but without gelation.

Many analogies exist among all these polysaccharides and are related to the local stiffness of the molecules, the low ionic content sensitivity, and large draining effect were explained.

Finally, some information concerning the gelling polysaccharides are recalled for the two series of systems usually recognized: the thermoreversible systems and the divalent crosslinked gels.

In light of various data obtained on these polysaccharides, it appears that general conclusions relating their chemical structure to their physical properties can now be drawn. Some points are still under discussion and will be investigated in the near future.

Literature Cited

1. Ott, E.; Spurlin, H. M.; Grafflin, M. W. In *High Polymers*; Interscience Publishers, 1963; Vol. V, Parts I, II, III.
2. Bikales, N. M.; Segal, L. In *High Polymers*; Interscience Publishers, 1971; Vol. V, Parts IV, V.
3. Mark, H. F.; Bikales, N. M.; Overberger, C. G.; Menges, G.; Kroschwitz, J. I., Eds. In *Encyclopedia of Polymer Science and Engineering*; John Wiley: New York, 1985; Vol. 3, pp 60-270.
4. Pigman, W.; Horton, D.; Herp, A., Eds. In *The Carbohydrates*; Academic: New York, 1970; Vol. II, Chs. 34–36.
5. Marchessault, R. H.; Sundarajan, P. R. In *Polysaccharides*; Aspinall, G. O., Ed.; Academic: New York, 1983; Vol. 2, p 11-95.
6. Arthur, J. C., Ed. *Cellulose Chemistry and Technology*; ACS Symposium Series No. 48; American Chemical Society: Washington, DC, 1983.

7. Guilbot, A.; Mercier, C. In *The Polysaccharides*; Aspinall, G. O., Ed.; Academic: New York, 1985; Vol. 3, pp 209-282.
8. Greenwood, C. T. In *The Carbohydrates*; Pigman, W.; Horton, D.; Herp, A., Eds.; Academic: New York, 1970; Vol. IIB, pp 471-513.
9. Imberty, A. Thesis, Grenoble, 1988.
10. Whistler, R. L.; BeMiller, J. N.; Paschall, E. F., Eds. *Starch: Chemistry and Technology*; Academic: New York, 1984.
11. Muzzarelli, R. A. A. *Chitin*; Pergamon Press, 1977.
12. Domard, A.; Rinaudo, M. *Int. J. Biol. Macromol.* 1983, **5**, 49-51.
13. Batista, I.; Roberts, G. A. F. *Makromol. Chem.* 1990, **191**, 429-434.
14. Rinaudo, M.; Domard, A. In *Chitin and Chitosan*; Skjak-Braek, G.; Anthonsen, T.; Sandford, P., Eds.; Elsevier, 1989; pp 71-86.
15. Blanshard, J. M. V.; Mitchell, J. R. *Polysaccharides in Food*; Butterworths, 1979.
16. Rinaudo, M.; Milas, M. *Biopolymers* 1978, **17**, 2663-2678.
17. Rinaudo, M.; Milas, M. *Carbohydr. Res.* 1979, **76**, 186-196.
18. Tinland, B. Thesis, Grenoble, 1988.

19a. Yamakawa, H.; Fujii, M. *Macromolecules* 1974, **7**, 128-135.

19b. T. Odijk. *Biopolymers* 1979, **18**, 3111-3113.

20. Tinland, B.; Maret, G.; Rinaudo, M. *Macromolecules* 1990, **23**, 596-602.
21. Tinland, B.; Rinaudo, M. *Macromolecules* 1989, **22**, 1863-1865.
22. Maret, G.; Milas, M.; Rinaudo, M. *Polymer Bull.* 1981, **4**, 291-297.
23. Gravanis, G. Thesis, Grenoble, 1985.
24. Kojima, T.; Itou, T.; Teramoto, A. *Polymer J.* 1987, **19**, 1225-1229.
25. Milas, M.; Rinaudo, M.; Knipper, M.; Schuppiser, J. L. *Macromolecules*, 1990, **23**, 2506-2511.

26a. Milas, M.; Rinaudo, M.; Tinland, B. *Polym. Bull.* 1985, **14**, 157-164.

26b. Kulicke, W. M.; Kniewske, R. *Rheo. Acta* 1984, **23**, 75-83.

27. Rinaudo, M.; Milas, M. in *Industrial Polysaccharides: Genetic Engineering, Structure/Property Relations, and Applications*; Yalpani, M., Ed.; Elsevier, 1987; pp 217-228.
28. Gravanis, G.; Milas, M.; Rinaudo, M.; Tinland, B. *Carbohydr. Polym.* 1987, **160**, 259-265.
29. Gravanis, G.; Milas, M.; Rinaudo, M.; Sturman, A. J. C. *Int. J. Mol. Biol.*, 1990, **12**, 195-200; 201-206.
30. Harada, T.; Masada, K.; Fujimori, K.; Maeda, I. *Agr. Biochem.* 1966, **30**, 196.
31. Kasai, N.; Harada, T. In *Fiber Diffraction Methods*; French, A. D.; Gardner, K. C. H., Eds.; ACS Symposium Series No. 141; American Chemical Society: Washington, DC, 1980; pp 363-383.
32. Fulton, W. S.; Atkins, E. D. T. In *Fiber Diffraction Methods*; French, A. D.; Gardner, K. C. H., Eds.; ACS Symposium Series No. 141; American Chemical Society: Washington, DC, 1980; pp 384-410.
33. Deslandes, Y.; Marchessault, R. H.; Sarko, A. *Macromolecules* 1980, **13**, 1466-1471.

34a. Yanaki, T.; Norisuye, T.; Fujita, H. *Macromolecules* 1980, **13**, 1462-1466.
34b. Yanaki, T. Thesis, Osaka, Japan, 1984.
35. Rinaudo, M.; Vincendon, M. *Carbohydr. Polym.* 1982, **2**, 135-144.
36. Bo, S.; Milas, M.; Rinaudo, M. *Int. J. Biol. Macromol.* 1987, **9**, 153-157.
37. Itou, T.; Teramoto, A.; Matsuo, T.; Suga, H. *Carbohydr. Res.* 1987, **160**, 243-257.
38. Itou, T.; Teramoto, A.; Matsuo, T.; Suga, H. *Macromolecules* 1986, **19**, 1234-1240.
39a. Kuo, M.; Mort, A. J.; Dell, A. *Carbohydr. Res.* 1986, **156**, 173-187.
39b. Jansson, B.; Lindberg, B.; Sandford, P. A. *Carbohydr. Res.* 1983, **124**, 135-139.
39c. O'Neill, N. A.; Selvendran, R. R.; Morris, V. J. *Carbohydr. Res.* 1983, **124**, 123-133.
39d. Chowhury, T. A.; Lindberg, B.; Lindquist, U.; Baird, J. *Carbohydr. Res.* 1987, **164**, 117-122.
40. Kang, K. S.; Velder, G. T.; Cottrell, I. W. *Progr. in Ind. Microbiol.* 1983, **18**, 231-253.
41a. Chandrasekaran, R.; Millane, R. P.; Arnott, S.; Atkins, E. D. T. *Carbohydr. Res.* 1988, **175**, 1-15.
41b. Attwool, P. T.; Atkins, E. D. T.; Miles, J.; Morris, V. J. *Carbohydr. Res.* 1986, **148**, C1-C4.
42a. Shi, X. Thesis, Grenoble, 1990.
42b. Milas, M.; Rinaudo, M.; Shi, X. *Polym. Prepr.* 1988, **29**, 629-630.
43. Milas, M.; Shi, X.; Rinaudo, M. *Biopolymers* 1990, **30**, 451-464.
44. Urbani, R.; Brant, D. A. *Carbohydr. Res.* 1989, **11**, 169-191.
45. Campana, S.; Andrade, C.; Milas, M.; Rinaudo, M. *Int. J. Biol. Macromol.* 1990, **12**, 379-384.
46. Crescenzi, V.; Dentini, M.; Dea, I. C. M. *Carbohydr. Res.* 1987, **160**, 283-302.
47. Crescenzi, V.; Dentini, M.; Coviello, T.; Rizzo, R. *Carbohydr. Res.* 1986, **149**, 425-432.
48. Talashek, T. A.; Brant, D. A. *Carbohydr. Res.* 1987, **160**, 303-316.
49. Robinson, G.; Manning, C. E.; Morris, E. R.; Dea, I. C. M. In *Gums and Stabilizers for Industry*; Phillips, G. O.; Wedlock, D. J.; Williams, P. A., Eds.; IRL Press, 1988; Vol. 4, pp 173-181.
50a. Jansson, P. E.; Lindberg, B.; Widmaim, G.; Sandford, P. A. *Carbohydr. Res.* 1985, **139**, 217-223.
50b. Jansson, P. E.; Lindberg, B.; Lindberg, J. Maekawa, E.; Sandford, P. A. *Carbohydr. Res.* 1986, **156**, 157-163.
51. Frank, F. C.; Keller, A. *Polymer Comm.* 1988, **29**, 186-189.
52. Hikmet, R. M.; Callister, S.; Keller, A. *Polymer* 1988, **29**, 1378-1388.
53. Rinaudo, M.; Karimian, A.; Milas, M. *Biopolymers* 1979, **29**, 1378-1388.
54. Rinaudo, M.; Landry, S. *Polymer Bull.* 1987, **17**, 563-565.
55. Rochas, C.; Rinaudo, M. *Biopolymers* 1984, **23**, 735-745.

56. Rochas, C.; Rinaudo, M. *Carbohydr. Res.* 1982, **105**, 227-236.
57. Rinaudo, M.; Rochas, C.; Michels, B. *J. Chim. Phys.* 1983, **80**, 305-308.
58. Rochas, C. Thesis, Grenoble, 1982.
59. Landry, S. Thesis, Grenoble, 1987.
60. Rochas, C.; Rinaudo, M.; Landry, S. *Carbohydr. Polym.* 1990, **12**, 255-266.
61. Thibault, J. F.; Rinaudo, M. *Biopolymers* 1985, **24**, 2131-2134.
62. Thibault, J. F.; Rinaudo, M. *Biopolymers* 1986, **25**, 455-468.
63. Thibault, J. F.; Rinaudo, M. *Brit. Polym. J.* 1985, **17**, 181-184.
64. Rinaudo, M. In *Biogenesis and Biodegradation of Plant Cell Wall Polymers*; Lewis, N. G.; Paice, M., Eds.; ACS Symposium Series No. 399; American Chemical Society: Washington, DC, 1989; pp 324-332.
65. Durand, D.; Bertrand, C.; Clark, A. H.; Lips, A. *Int. J. Biol. Macromol.* 1990, **12**, 14-18.
66. Liang, J. N.; Stevens, E. S.; Frangou, S. A.; Morris, E. R.; Rees, D. A. *Int. J. Biol. Macromol.* 1980, **2**, 204.
67. Smidsrod, D.; Skjak-Braek, G. *Tibtech* 1990, **8**, 71-78.
68. Bouffar-Roupe, C. Thesis, Grenoble, 1989.
69. Rinaudo, M.; Graebling, D. *Polym. Bull.* 1986, **15**, 253-256.

RECEIVED February 10, 1992

Chapter 3

Structure–Property Relationship of α- and β-Chitin

M. Takai[1], Y. Shimizu[1], J. Hayashi[1], S. Tokura[2], M. Ogawa[2], T. Kohriyama[3], M. Satake[3], and T. Fujita[3]

[1]Department of Applied Chemistry and [2]Department of Polymer Science, Hokkaido University, Sapporo 060, Japan
[3]Central Research Institute, Nippon Suisan Company, Ltd., Hachiohji 192, Japan

High resolution ^{13}C NMR spectra of α-chitin and β-chitin in solid state can be distinguished from each other as well as the x-ray patterns or IR spectra. α-chitin from crab or shrimp shell is known to be slightly soluble in general organic solvents and to have poor reactivity due to its rigid crystalline structure. β-chitin, on the other hand, from squid bone or Loligo pen, easily forms slurries when ground with water. This is responsible for the loose crystalline structure and the high hydrophilicity of Loligo pen chitin. Therefore it is much easier to prepare paper sheets from β-chitin than α-chitin without binder. Direct sheet preparation from Loligo pen gel is economically of greater advantage than nonwoven sheets from crab shell. The handmade chitin paper from Loligo pen shows a high bursting factor of 7.4 and breaking length of 6.9 Km, compared with those of α-chitin from crab shell, 1.0 and 3.0 Km. The Loligo pen paper is obviously softer than that from crab shell as to stiffness calculated from Young's modulus. Furthermore, the higher permeability of moisture together with water regain, a strong affinity for blood protein such as fibrinogen, albumin or γ-globulin, and slow biodegradation by lysozyme were observed.

Chitin is a linear mucopolysaccharide having a similar chemical structure to cellulose and occurring widely in nature; in crustaceans, insects, mushrooms, and in the cell walls of bacteria. This is composed of N-acetylglucosamine repeat units through β-1,4 glucosidic linkages, and it can be regarded as analogue of cellulose, with an aminoacetyl group instead of the C_2 hydroxyl group. However, the chemical properties of chitin are

0097–6156/92/0489–0038$06.00/0

completely different from those of cellulose in spite of their similar structures. In particular, chitin is very resistant to chemical reagents compared to cellulose. The difference between chitin and cellulose in their chemical properties is due to the strong hydrogen bonding between the aminoacetyl and hydroxyl groups. Thus, chitin is not so far regarded as a useful resource compared to other polysaccharides such as cellulose or starch. However, recently chitin and chitin derivatives have been reported to have some useful medical applications. For example, chitin administration to animals attacked by certain bacteria and fungi has proved very useful as a highly effective antigen (1). Chitin has been found suitable as a biodegradable pharmaceutical carrier (2), a blood anticoagulant (3), a wound-healing accelerator (4), and as a highly selective medium for isolation of actinomycetes (5).

Chitin exists in three different polymorphic forms, denoted as α-chitin, β-chitin, and γ-chitin, depending upon the crystalline structure. α-chitin is the most abundant and most stable natural polymorphic form. On the other hand, β-chitin is a metastable form and is transformed into α-form during the dissolving process. α-chitin requires a tedious process to remove calcium carbonate, but β-chitin hardly has calcium carbonate and can be easily obtained by deprotonating squid bone with dilute alkaline solution. Furthermore, β-chitin easily forms a gel when ground with water. Then direct sheet preparation from β-chitin is possible without binder. In this paper, relationships between the structure and mechanical and physical properties of α- and β-chitin are discussed in relation to handmade paper properties.

Materials and Methods

Chitin. Table 1 summarizes chitin, ash, and protein contents of Loligo pen, shrimp and crab shells obtained by element analysis. Protein content was calculated from the equation shown in Table 1. Various kinds of shrimp and crab shells contain large amounts of ash (mainly calcium carbonate) ranging from 30 to 40%, while that of Loligo pen is only about 0.5%. This is a remarkable difference between Loligo pen and shrimp or crab shells. Thus, α-chitin from shrimp or crab shells requires a tedious process of calcium carbonate removal. On the other hand, β-chitin from Loligo pen can easily be obtained by deprotonating squid bone with boiling 4% aqueous NaOH, and then exhaustively washing the residue with water and successively neutralizing the last traces of alkali with 1N HCl. These treatments were repeated twice. It was found that β-chitin is tightly bound to one mole of water per each N-acetylglucosamine residue. Analysis found: C, 42.52; H, 6.61; N, 6.31. Calculated for $C_8H_{13}NO_5.H_2O$: C, 43.46; H, 6.78; N, 6.33. Purified and powdered chitin of crab shell (Alaska King Crab) was obtained from Nippon Suisan Co., Ltd. It was also found by elemental analysis that the α-chitin is bound to half a mole of water per each residue. Analysis found: C, 45.91; H, 6.59; N, 6.73. Calculated for $C_8H_{13}NO_5.1/2H_2O$: C, 45.30; H, 6.60; N, 6.60.

Paper-Making Procedures for Handmade Chitin Paper. Three grams of

Table 1. Chitin, Ash, and Protein Components of Loligo Pen, Shrimp and Crab Shells

Sample	Content (%)		
	Chitin	Ash	Protein
Loligo Pen	33.7	0.5	65.5
Shrimp Shell			
Black Tiger	24.0	28.8	38.4
Prawn	23.5	33.6	35.5
Argentina Pink	21.0	33.5	48.4
Lobster	10.0	40.0	30.6
Antarctic Krill	7.9	33.6	48.4
Crab Shell(King Crab)	27.5	44.8	14.5

Protein = (total N % - chitin N %) x 6.25

purified Loligo pen chitin were suspended in 300 ml of water and ground to fine pieces using a homogenizer at room temperature. This was used as stock solution. Freeness of the homogenate was almost zero. Thus, the papermaking apparatus required a suction system. A handmade apparatus was designed in such a manner as to resemble a Buchner funnel of 185 mm diameter. The resulting chitin suspension was diluted to 1700 ml by water (final chitin consistency: 1.5 g/l) and subjected to the paper preparation. Basis weight of the chitin paper was controlled by the volume of chitin suspension charged to the funnel. α-chitin paper from purified crab shell was also prepared by the same method. Nonwoven sheet was commercially available from Unitika Co., Ltd., as "Beschitin" and used for a control sample.

^{13}C CP/MAS NMR. The ^{13}C CP/MAS high-resolution solid state NMR spectra of the chitin samples were obtained using a Brucker MSL-400 operating at room temperature and ^{13}C frequency of 100.630 MHz. The samples (ca. 100 mg) enclosed in the rotor were measured at a rotating frequency of 4000 cps. Contact times were typically 2 ms. The number of scans was 120 per spectrum, and the recycle delays were 5 s. ^{13}C chemical shifts in ppm were calibrated using the carbonyl carbon of glycine at 176.03 ppm as an external standard.

X-Ray and IR Analyses. X-ray powder patterns of the chitins were measured by the reflection method with nickel-filtered CuK_α radiation of Rigaku Denki x-ray diffractometer operated in the w-20 scanning mode between 5° and 30°. IR spectra were obtained with a JASCO A-202 IR spectrophotometer by the KBr pellet technique for powdered samples, and directly using the paper sheets.

Mechanical/Physical Properties Testing. Mechanical properties of the chitin papers and others were measured by TAPPI standard methods for usual wood pulp sheets (6). Dynamic Young's modulus was calculated from a vibrating reed method (7) and calculated from the real part by the following equation:

$$E' = \frac{48\pi^2 \bullet 1^4\delta}{a^4 d^2}[f_r{}^2 + \frac{1}{8}(\Delta f)^2]$$

where:

d = thickness of sample (cm)
l = length of sample (cm)
δ = density of sample (g/cm^3)
a = constant = 1.874
f_r = frequency for maximum vibration width (Hz)
Δf = frequency difference for half vibrating width (Hz)

Permeability of O_2 gas or water vapor was measured by JIS methods (8).

Results and Discussion

NMR Spectra. The ^{13}C CP/MAS NMR spectra of α- and β-chitins are shown in Figure 1 and their ^{13}C chemical shifts are summarized in Table 2. As can be seen in Figure 1C, β-chitin from Loligo pen shows a relatively broad and single peak at 74.7 ppm assigned to C_5 and C_3, while α-chitin from King crab shell or tendon shows very definite and sharp signals that display conspicuous splittings at 75.4 and 73.0 ppm for C_5 and C_3 (Figure 1A or 1B). The difference is probably due to a different configuration of C_5 and C_3 carbons by the hydrogen bonds. The x-ray diffraction patterns for α-chitin from crab tendon and crab shell give three sharp peaks at 9.50 ($2\theta = 9.3°$), 4.60 (19.3°), and 3.81 (23.3°)Å as shown in Figure 2. They are typical powder patterns of α-chitin crystal structure determined by Blackwell *et al.* (10). β-chitin from Loligo pen gives a broader pattern at 10.6($2\theta = 8.3°$) and 4.55(19.5°)Å (Figure 2). Both α- and β-chitin can also be distinguished from each other by their ^{13}C CP/MAS NMR spectra. According to the structural analysis by Blackwell *et al.*, the structure of β-chitin consists of an array of poly-N-acetyl-D-glucosamine chains all having the same sense (Figure 3), and these are linked together in sheets by N–H...O=C hydrogen bonds of the amide groups. In addition to O–3′–H...O–5 intramolecular hydrogen bonds, the CH_2OH side chain forms an intrasheet hydrogen bond to the carbonyl oxygen on the next chain (9). On the other hand, α-chitin has an antiparallel structure such that all the hydroxyl groups form hydrogen bonds (10): O–3′–H...O–5 intramolecular, O–6–H...O–6′ intermolecular, and O–6′–H...O–7=C intramolecular hydrogen bonds. As a result, the structure of α-chitin contains two types of amide groups, which differ in their hydrogen bonding, and account for the splitting of C_5 and C_3 signals in the NMR spectra. In addition, existence of the two kinds of C=O groups shows definite asymmetry consistent with the presence of an unresolved shoulder at the carbonyl carbon resonance of lower field (Figure 1A), while that of β-chitin is a fundamental singlet peak (Figure 1C).

Chitin fiber from Loligo pen was prepared by spinning the solution of Loligo pen dissolved in formic acid-dichloroacetic acid mixture into acetone and then EtOH–H_2O 50/50, and stretching it in H_2O immediately after coagulation. Finally, the fiber was washed in H_2O and alkaline EtOH to remove any trace of dichloroacetic acid or formic acid, and then dried in air. Usually, if β-chitin is dissolved in formic acid, then chains recrystallize on precipitation in the more stable α-chitin form with antiparallel chains. The ^{13}C NMR spectra of the chitin fibers as prepared above are shown in Figure 4. Figure 4A shows the spectrum of the chitin fibers before annealing. At C_5 and C_3 carbon signal, two peaks characteristic of partially resolved doublets were observed. On the other hand, Figure 4B after annealing by superheated steam at 170°C for 2 hr shows the sharper doublet for C_5 and C_3 carbons at 75.3 ppm and 73.3 ppm, respectively, and this is assigned to that of α-chitin (Figure 1A or 1B). The x-ray diffraction pattern of the chitin fiber is relatively broad, but it can be assigned to α-chitin at 9.50($2\theta = 9.3°$) and 4.60(19.3°)$\AA$ as shown in Figure 5A. Figure

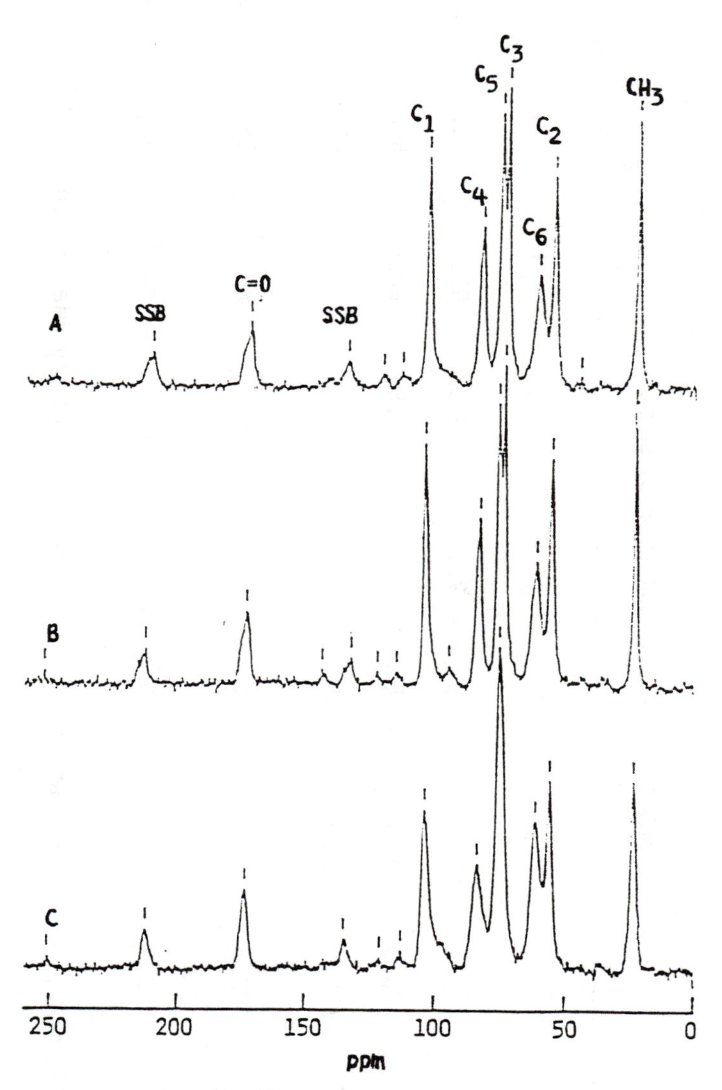

Figure 1. CP/MAS ^{13}C NMR spectra of chitin polymorphs: α-chitin from (A) crab shell; (B) crab tendon; (C) β-chitin from Loligo pen.

Table 2. ^{13}C CP/MAS NMR Chemical Shift of Chitin

Chitin	C=O	C_1	C_4	C_5	C_3	C_6	C_2	CH_3
Crab Shell	172.7	103.8	82.6	75.4	73.0	60.8	54.9	22.5
Crab Tendon	173.3	103.8	82.6	75.3	72.9	60.6	54.7	22.4
Loligo Pen	173.7	103.6	83.5	74.7		60.8	55.3	22.9
Loligo Pen Fiber (untreated)	173.3	103.6	83.2	74.8	74.0	60.9	55.1	22.9
Loligo Pen Fiber (annealed)	173.7	103.7	83.0	75.3	73.3	61.0	54.9	22.7

chemical shift in ppm, external standard ; carbonyl carbon of glycine, 176.0

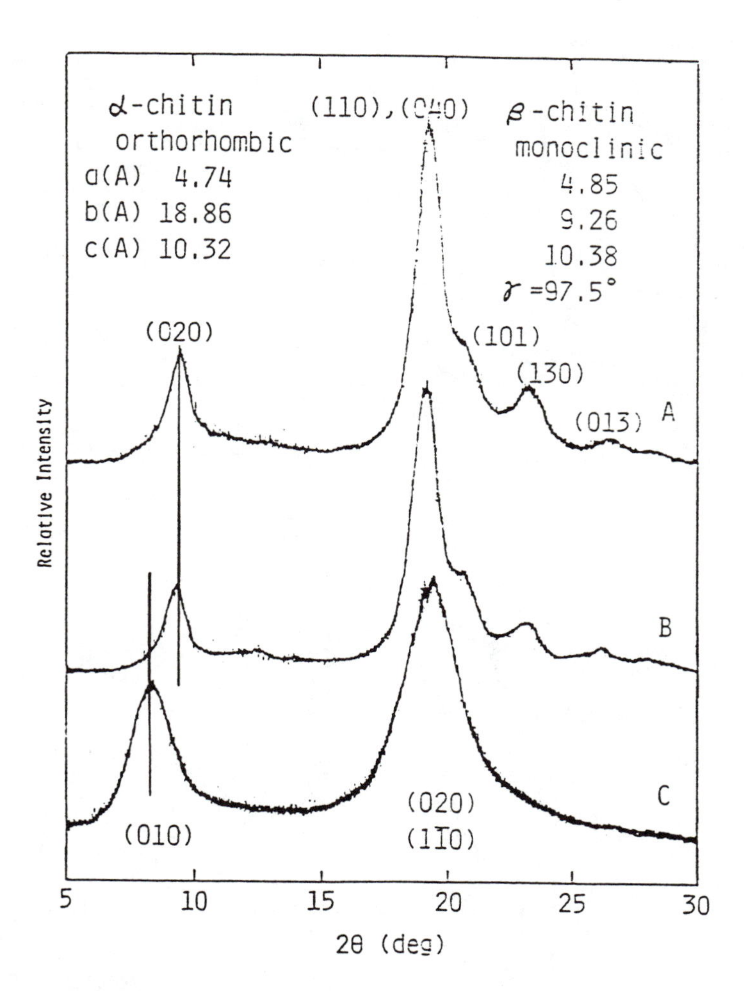

Figure 2. X-ray diffraction patterns of chitins from (A) crab tendon; (B) crab shell; and (C) Loligo pen.

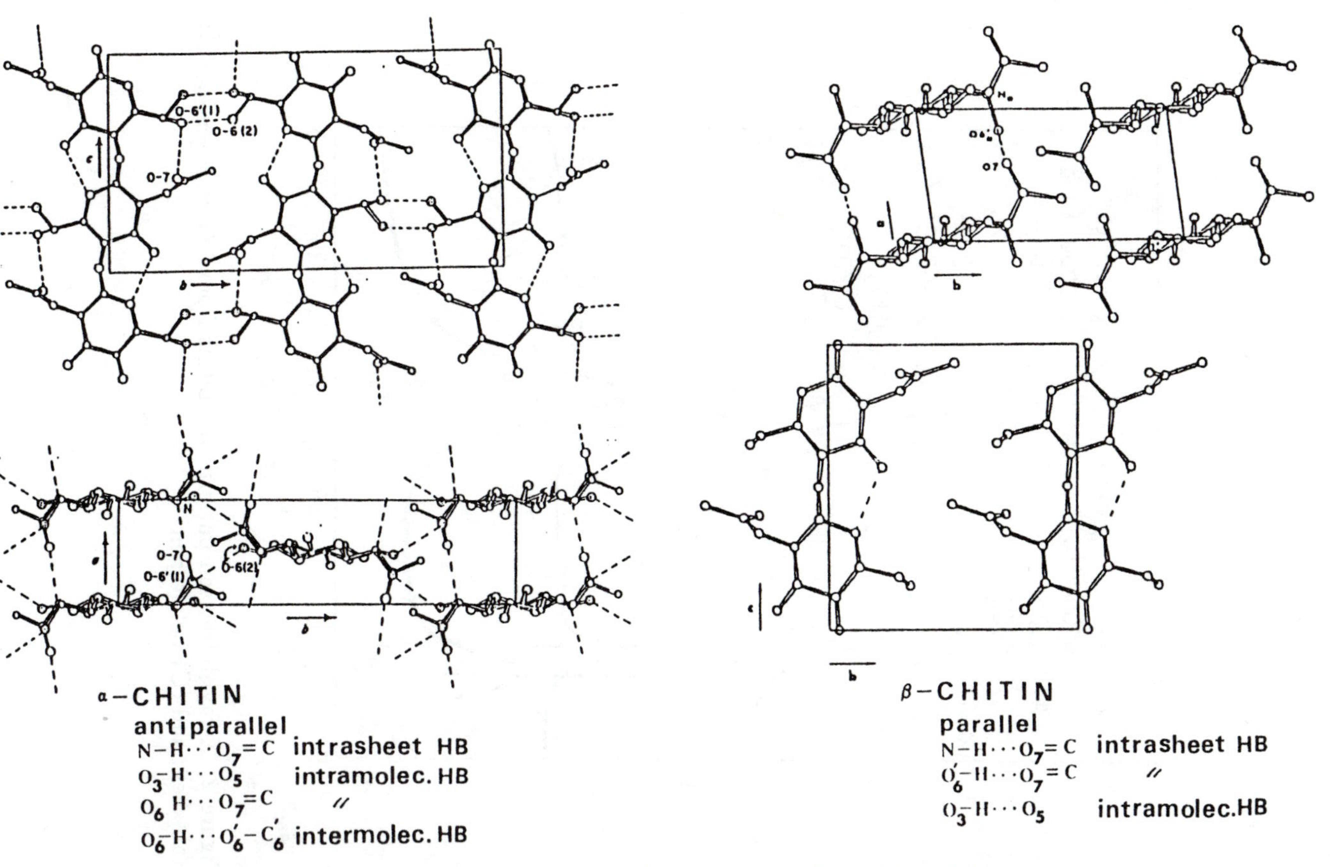

Figure 3. Crystal and molecular structures of α- and β-chitins.

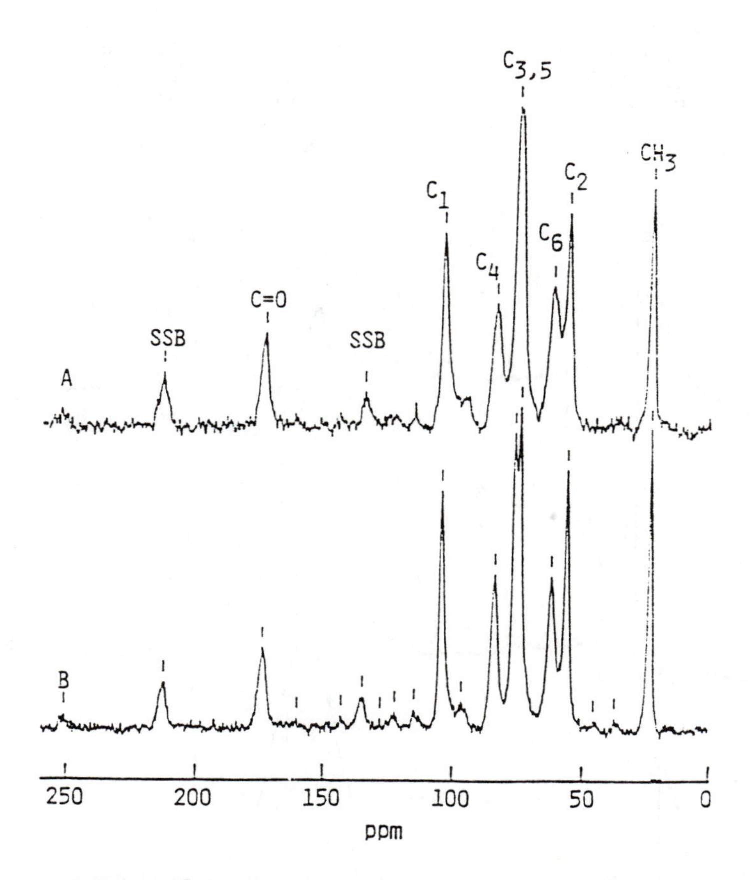

Figure 4. CP/MAS ^{13}C NMR spectra of chitin fibers from Loligo pen: (A) untreated chitin fiber; (B) chitin fiber annealed with superheated steam at 170°C for 2 hr.

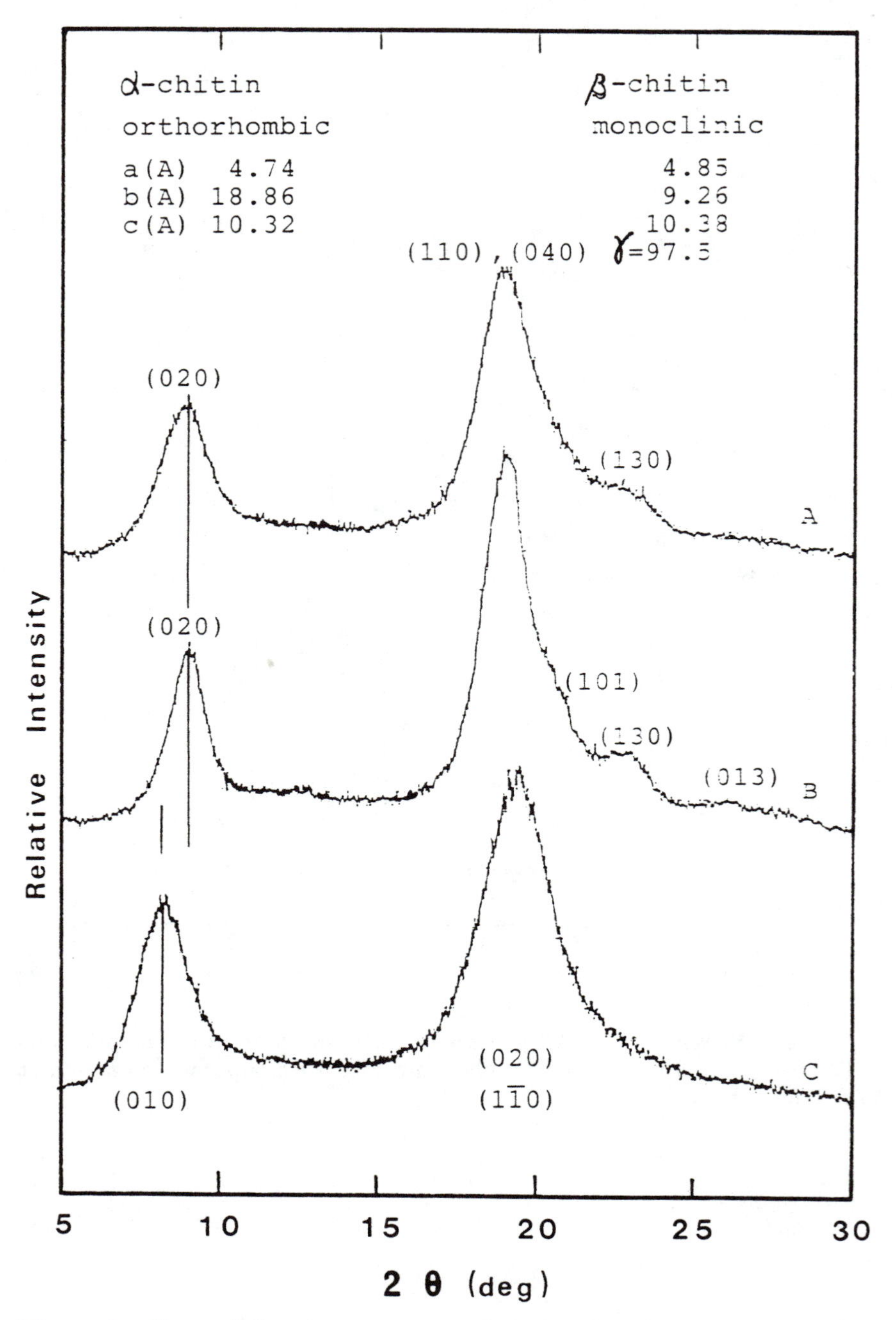

Figure 5. X-ray diffraction patterns of chitin fiber from Loligo pen (A) before and (B) after annealing, and (C) native Loligo pen.

5B shows the pattern of the fiber annealed by superheated steam. This is generally sharper than that of the untreated fiber. These results support the transformation of β-chitin to α-chitin by dissolution.

Physical Properties. Mechanical and physical properties of the handmade or machine-made chitin papers are listed in Table 3. The handmade β-chitin paper shows a high bursting factor of 7.4 and a breaking length of 6.9 Km, compared with those of the α-chitin paper, 1.0 and 3.0 Km, respectively. Nevertheless, the dynamic Young's modulus of 5.1 GPa in the α-chitin paper is about one-half that of the β-chitin paper (8.9–11.3 GPa). Rigidity (or stiffness) which is calculated by E′I or E′I/W [E′, Young's modulus; I, cross-sectional amount of inertia = $b \times d^3/12$; b, width; d, thickness; W, basis weight (g/cm^2)], is quite different from each other. The Loligo pen paper is obviously softer than that from crab shell. Rigidity of the pen paper is dependent on the basis weight, which ranges from 49 in the thinnest paper of 11 g/m^2 to 738 in the thickest of 44 g/m^2. In the crab shell paper of 85 g/m^2, rigidity is 5620, the same order as 7600 in the commercial nonwoven sheet, "Beschitin," developed as an artificial skin, or 31 g/m^2. The marked difference in rigidity could be based on the crystal structure, as mentioned above. Namely, all the hydroxyl groups of the α-chitin molecule are involved in intermolecular, intramolecular, or intrasheet hydrogen bonding. It seems reasonable to assume that α-chitin is more rigid than β-chitin. Furthermore, the permeability of moisture, which ranged from 6100 to 7800 $g/m^2 \bullet$ 24hr • atm, depended upon the thickness of the β-chitin paper sheet. So did the water regain. Especially the water regain for the β-chitin paper is more than three times that for the crab shell paper. This is related to the higher hydrophilicity of β-chitin in which one mole of water is tightly bound to each N-acetyl-D-glucosamine residue (as observed by elemental analysis). This would be an important property if β-chitin paper would be considered a candidate for artificial skin. The last column of Table 3 lists colony formation which is used as a qualitative antibacterial index. Negative colony formation means that *E. coli* does not pass through the paper or sheet. Both the β-chitin and α-chitin papers show antibacterial characteristics. By contrast, the nonwoven sheets from crab shell or bovine collagen that have been developed commercially as artificial skin have no antibiosis and are permeable to moisture. Furthermore, the β-chitin as well as α-chitin papers have a strong affinity for such blood proteins as fibrinogen, albumin, and γ-globulin. The susceptibility toward lysozyme seems to be an important index of biodegradability retention for medical chitin uses. β-chitin Loligo pen dispersed in water using an ultra-disperser for 20 min showed higher susceptibility toward lysozyme than that of α-chitin ground to under 40 mesh size or toward whole Loligo pen. The enhanced susceptibility seems to be a consequence of the loose crystalline structure and high hydrophilicity of the Loligo pen chitin. Thus, the β-chitin paper would be a superior temporary artificial skin.

SEM Photographs. Pictures of the chitin paper surfaces are shown in Figures 6 and 7. The SEM photographs clearly show the difference in mor-

Table 3. Physical Properties of Chitin Sheets Compared with Others

Samples	Basis Weight (g/m^2)	Young Modulus (GPa)	Rigidity E'I	Stiffness E'I/W	Breaking Length (Km)	Bursting Factor	Tearing Factor	Permeability Moisture ($g/m^2 \cdot 24hr$)	Water Regain (%)	Colony Formation (−,+)
Loligo Pen	11	9.3	49	4	6.6	5.3	39	7800	219	−
(hand-made)	22	11.3	271	12	6.9	6.9	37	6600	281	−
	44	8.9	738	17	6.3	7.4	40	6100	305	−
Loligo Pen	10				−	2.9	12			
(machine-made)	16				2.8	3.9	15			
	24				4.7	6.4	15			
Crab Shell	85	5.1	5620	66	3.0	1.0	−	5200	69	−
Nonwoven Sheet										
from crab shell	31	4.3	7600	243	4.1	2.1	96	free	189	+
from collagen	27	3.6	421	16	4.9	3.2		free	118	+
Bacterial Cellulose	11	50.2	195	17	7.9	5.5		4600	240	−
Pulp Sheet (KP)	50	8.0	2720	36	7.4	5.9	139	4400	100	+
News Paper MD	47	8.6	3150	67	4.6	1.5	42	6900	210	+
CD	47	1.4	523	11	2.0	1.6	67			

Figure 6. SEM photograph of α-chitin sheet from crab shell.

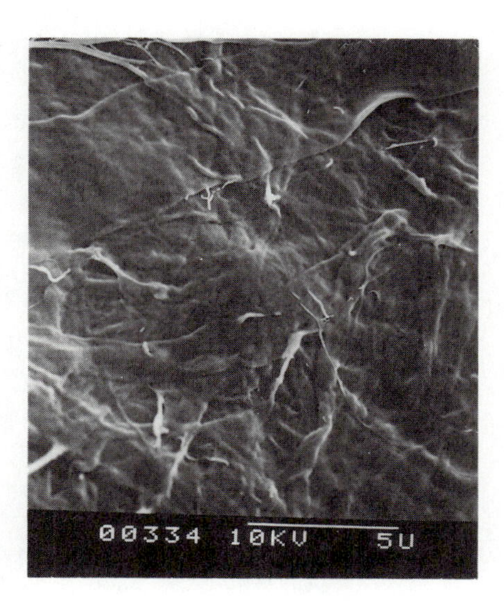

Figure 7. SEM photograph of β-chitin sheet from Loligo pen.

phology between α- and β-chitin papers. The α-chitin paper has a well developed and very pronounced microfibrillar structure of about 100 nm width (Figure 6). By constrast, the β-chitin paper shows a scaly lamellar structure (Figure 7). The morphological difference may be responsible for the different physical properties discussed above.

Literature Cited

1. Porter, W. R. U.S. Patent 3 590 126, 1971.
2. Capozza, R. C. German Patent 2 505 305, 1975.
3. Whistler, R. J.; Kosik, M. *Arch. Biochem. Biophys.* 1971, **142**, 106.
4. Balassa, L. L. U.S. Patent 3 911 116, 1975.
5. Hsu, S. C.; Lockwood, J. L. *Appl. Microbiol.* 1975, **29**, 422.
6. TAPPI T404m-47, TAPPI T403m-47.
7. TAPPI T451m-44.
8. JIS P8117, 1952; JIS Z0208, 1976.
9. Gardner, K. H.; Blackwell, J. *Biopolymers* 1975, **14**, 1581.
10. Minke, M.; Blackwell, J. *J. Mol. Biol.* 1978, **120**, 167.

RECEIVED February 10, 1992

Chapter 4

Effect of Coating Solvent on Chiral Recognition by Cellulose Triacetate

T. Sei[1], H. Matsui[2], T. Shibata[2], and S. Abe[2]

[1]Plastic Business Development and Promotion, Daicel Chemical Industries, Ltd., Osaka 541, Japan
[2]Research Center, Daicel Chemical Industries, Ltd., Himeji, Hyogo 671–12, Japan

We have recently found that cellulose esters coated on chromatographic supports exhibit excellent capabilities of optical resolution as column packing materials for high performance liquid chromatography. In the study on the mechanism of chiral recognition, solid-state CP/MAS ^{13}C NMR analysis has been carried out to investigate the structure of cellulose triacetate (CTA) coated on support. It was revealed that the solvent from which CTA was solidified had an effect on the structure of the polymer, presumably on its conformational regularity. The hysteretic effect of the coating or precipitating solvent on the chiral recognition ability of the polymer was rationalized by the aforementioned structural difference caused by the solvent.

We have reported (1,2) that cellulose triacetate (CTA) coated on macroporous silica gel showed excellent capabilities of chiral recognition as packing material for high performance liquid chromatographic resolution of various enantiomers.

We have performed the study on the mechanism of chiral recognition to develop a more efficient stationary phase.

Recently, we found (3) that the solvent with which CTA has been coated on a support had a considerable effect on the structure of CTA; that of phenol as a solvent was especially outstanding (Figure 1). On the other hand, there were some enantiomers which showed lower selectivity on CTA prepared with phenol as solvent. These results showed that the CTA enhanced a substrate selectivity.

Solid-state CP/MAS ^{13}C NMR analysis has been carried out to investigate the structure of CTA coated on a support. We now wish to discuss the origins of the hysteretic effect of a coating solvent.

0097–6156/92/0489–0053$06.00/0

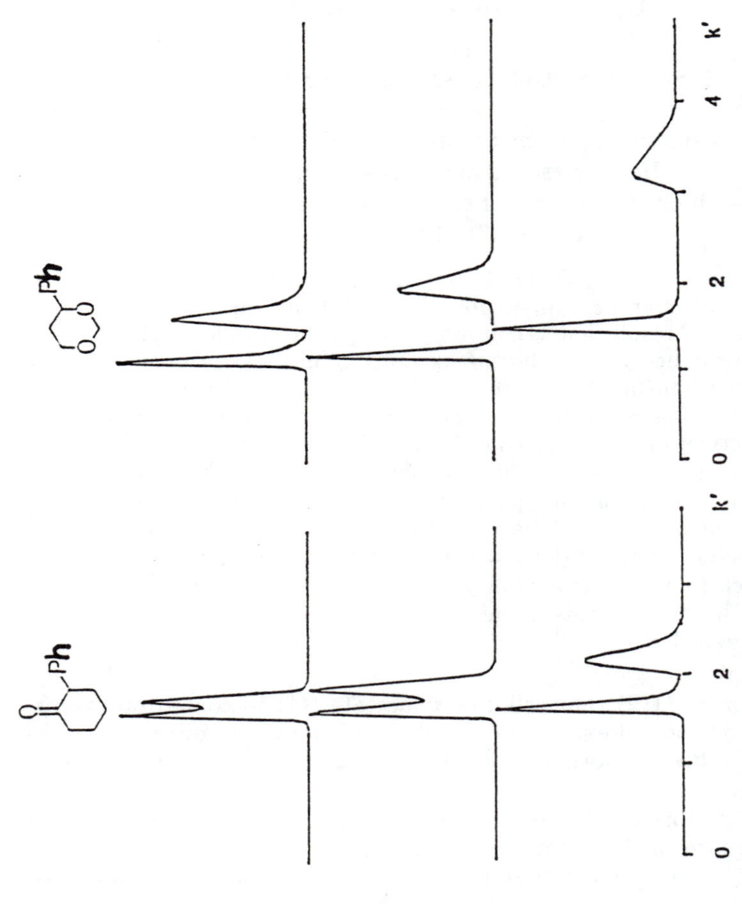

Figure 1. Chromatograms of 2-phenylcyclohexanone (left) and 6-phenyl-1,3-dioxane (right) obtained on cellulose triacetate (CTA) coated on silica gel with dichloromethane only (upper), with a 9 : 5 mixture of dichloromethane and trifluoroacetic acid (middle), and a 8 : 1 mixture of dichloromethane and phenol (lower).

Experimental

Polymer Samples. CTA of DS (degree of substitution) equal to 3.0 and $\overline{Mw}$ (weight-averaged molecular weight) equal to 9960 was used as a starting material. The $\overline{Mw}$ was determined by GPC in chloroform with polystyrene as the standard.

Preparation of CA-1, CA-2, and CA-3 (neat CTA). CTA was dissolved in dichloromethane, a 9:5 mixture of dichloromethane and trifluoroacetic acid, and a 8:1 mixture of dichloromethane and phenol. Under reduced pressure the solvents were removed at 20°C for an hour in the preparation of CA-1 and CA-2, and at 80°C for two hours in CA-3.

Preparation of CA-1D, CA-2D, and CA-3D (CTA coated on silica gel). One portion of CTA samples was dissolved in the same solvents as those of CA-1, CA-2, and CA-3, respectively, and mixed with four portions of silica gel treated with diphenylsilanizing agent. Solvent removal under vacuum gave a powdery substance.

CP/MAS ^{13}C NMR Spectroscopy. CP/MAS ^{13}C NMR experiments were performed at room temperature by a JEOL JNM-GX 270 spectrometer operating at 6.3 Tesla. The matched field strengths ν_{1C} and ν_{1H} of 46.3 kHz were applied to ^{13}C and ^{1}H for 2.0 msec. Recycling delay after the acquisition of FID was 5 seconds. Magic-angle spinning was carried out at a rate of 3.0 to 3.2 kHz. The chemical shifts were measured with respect to that of methylene carbons of adamantan, which was taken as 29.5 ppm downfield from tetramethylsilane.

^{13}C Spin-Lattice Relaxation Time (T_1). T_1 was measured with a slightly modified version of the method developed by Torchia (4). For details of the procedure, see Horii's papers (5).

Results and Discussion

The 67.8 MHz CP/MAS ^{13}C NMR spectra of CTA samples are shown in Figure 2. Chemical shifts of the observed resonances are summarized in Table I. The resonance lines of CTA are assigned on the basis of the previous works in solution (6,7). It is noted that the chemical shifts for carbons do not have large differences between neat CTAs and CTAs coated on silica gel. Therefore, we assumed that the effect of coating solvent on the structure of CTA did not change significantly with coating on silica gel. On the other hand, we noticed that the resonance lines of the C6 carbon for CA-3 and CA-3D shifted downfield more than those for other samples.

The most remarkable feature of the spectrograms is that the line widths of carbons (C1 through C6) in the anhydroglucose unit became narrower in numerical order, especially those of C1 carbons.

The half-widths of C1 signals of CA-1, CA-2, and CA-3 are 323, 273, and 157 Hz, respectively. Those for CA-1D, CA-2D, and CA-3D are 315, 243, and 218 Hz, respectively, showing the same tendency as in CA-1, CA-2, and CA-3.

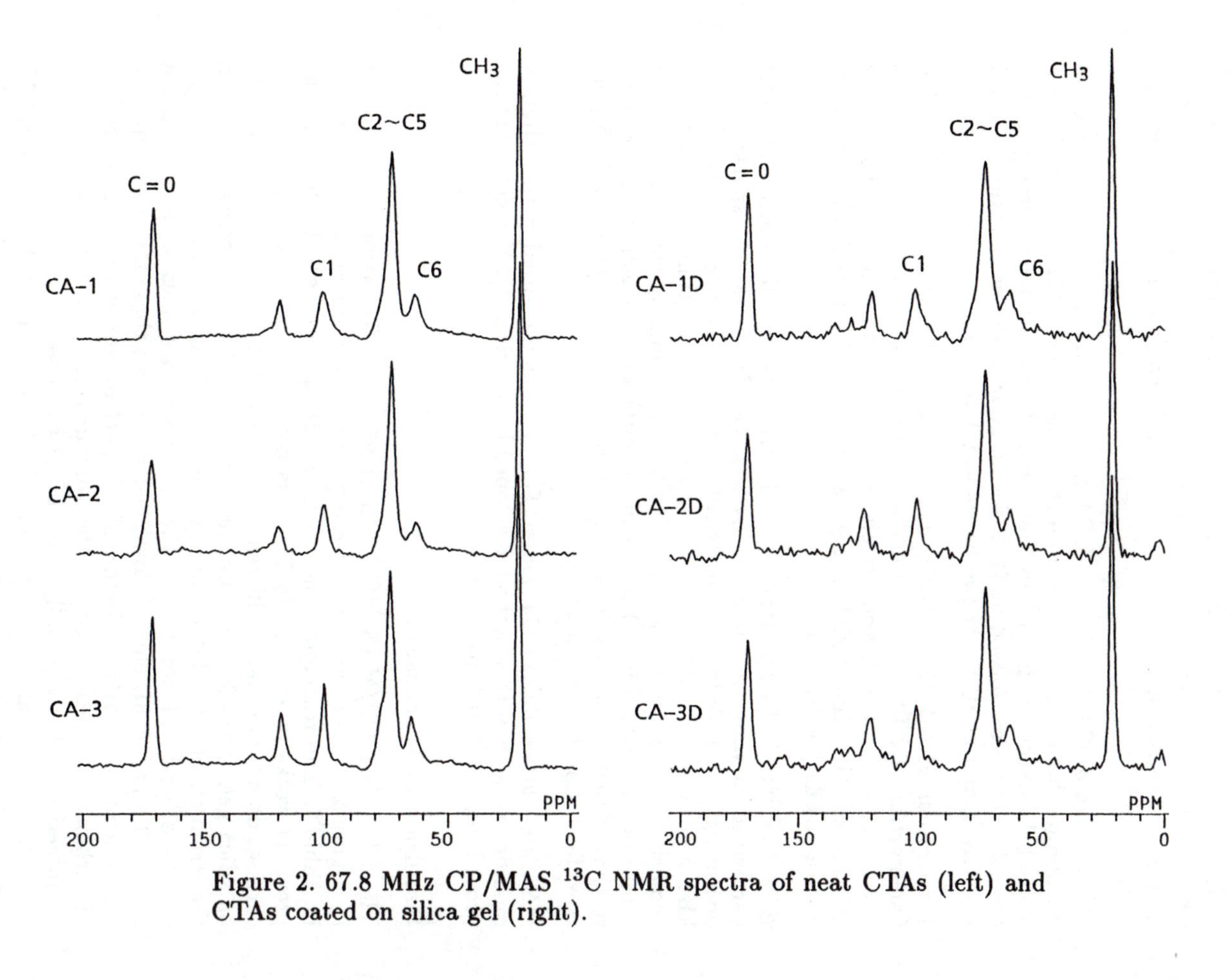

Figure 2. 67.8 MHz CP/MAS ^{13}C NMR spectra of neat CTAs (left) and CTAs coated on silica gel (right).

Table I ^{13}C Chemical Shifts of CTA Samples

Samples	Solvent	^{13}C Chemical Shifts (ppm)				
		C1	C2~C5	C6	C=0	CH_3
CA-1	CH_2Cl_2	101.6	73.2	63.6	170.9	21.1
CA-ID		101.5	72.8	62.8	170.3	20.8
CA-2	CH_2Cl_2/TFA = 9/5	101.1	73.2	63.0	171.8	20.7
CA-2D		100.9	73.0	62.6	170.4	20.5
CA-3	CH_2Cl_2/PhOH = 8/1	101.1	73.8	65.2	171.3	21.1
CA-3D		101.6	73.1	63.8	170.9	20.8

It is generally considered that the line width of a NMR signal depends upon regularity of molecular conformation and molecular mobility. In order to reveal which of the above is the dominant factor, we measured spin-lattice relaxation time (T_1).

There was no significant difference among T_1 values of the major components of CTA. The minor components with lower T_1 values than those of the major ones appeared in CA-1D and CA-3D. Table II shows T_1 values of the C1 carbon and the proportion of each component. We assumed from this result that the narrowing of line width of carbon signals was not the outcome of enhanced molecular mobility, but of the higher regularity of molecular conformation. The narrowing of the C1 signal leads us to the assumption that the line width for C1 depends on regularity of the conformation about the $\beta - 1,4$ glycosidic linkage. And this higher uniformity of the conformation can explain the observed higher selectivity of CTA. In addition, we have found another fact that supports our proposal about the effect of phenol on the structure of CTA. Figure 3 shows chromatograms of 2-phenylcyclohexanone obtained on CTA coated on silica gel. It proved that the magnitude of the effect that phenol gave to a completely acetylated cellulose was much larger than that given to an incompletely acetylated one. It can be rationalized that the higher structural regularity attained with the phenolic solvent is not so much beneficial to the chemically disordered CTA; in other words, CTA of a lower DS.

We speculated that the peculiar property of CTA prepared with phenol can be attributed to the interaction between CTA and phenol. Figure 4 shows ^1H NMR spectra of CTA solution in chloroform containing various amounts of phenol. In Table III, the difference of chemical shifts of CTA protons in solution caused by the addition of various substrates was summarized. It is noteworthy that H1(H6), H4 and H5 protons tend to shift upfield, while H2 and H3 protons shift downfield. Another remarkable feature is that the protons of the acetoxy group attached to the C6 position shifted upfield considerably; on the other hand, the chemical shifts of the acetoxy protons attached to C2 and C3 changed little. Though 3-methyl phenol, benzyl alcohol and benzene showed similar tendencies to phenol, the change of chemical shifts was not so remarkable. These results suggest that phenol interacts with CTA in some particular manner. It is presumed that this peculiar interaction causes the higher conformational regularity of CTA molecules. Detailed studies on the interaction between phenol and CTA are in progress.

Until now, we have discussed the interaction between phenol and CTA. However, phenol should be washed away from CTA when it is actually used in a chromatography column. Figure 5 shows CP/MAS ^{13}C NMR spectra of CTA after removal of phenol by washing with methanol. The half-width of the C1 signal was 214 Hz, and the T_1 value for the C1 atom was 28.0 seconds. It was concluded from these results that the CTA still retained a highly ordered structure even after complete removal of phenol.

Table II ^{13}C Spin–Lattice Relaxation Times (T_{1C}) of CTA Samples

Samples	Solvent	T_{1C} (sec) for C1	Proportion on of components(%)
CA – I	CH_2Cl_2	31.9	~100
CA –3	CH_2Cl_2 / PhOH	28.0	~100
CA – 1D	CH_2Cl_2	30.7 1.3	84 16
CA – 3D	CH_2Cl_2 / PhOH	33.8 1.3	87 13

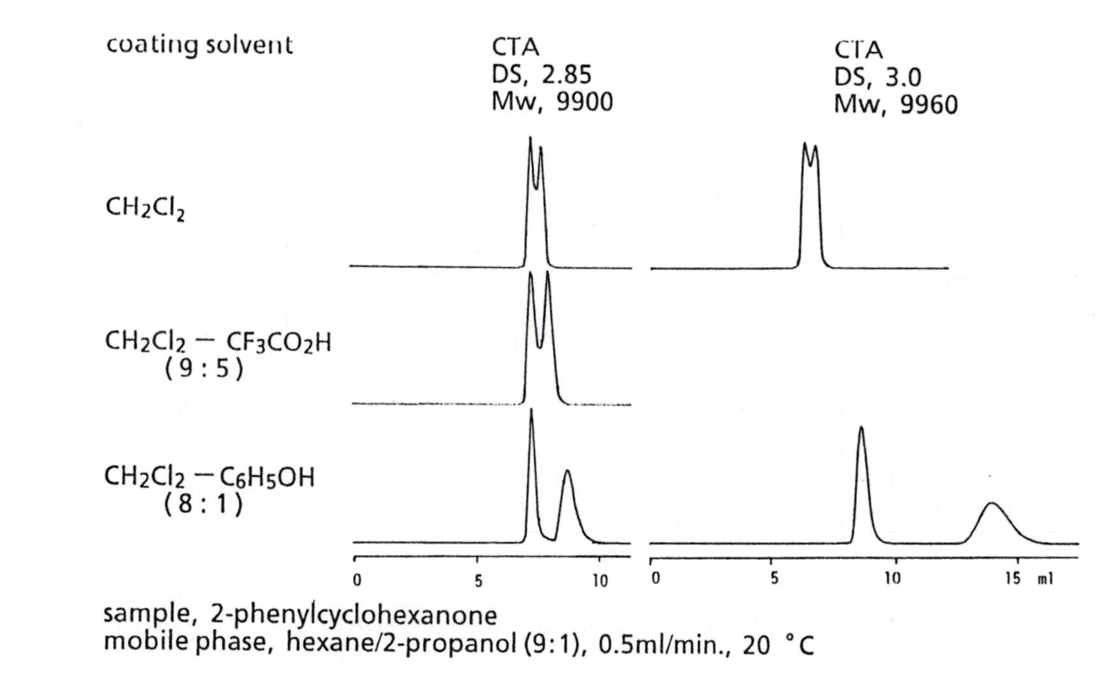

Figure 3. Chromatograms of 2-phenylcyclohexanone obtained on CTA of DS (degree of substitution) = 2.85 (left) and of DS = 3.0 (right).

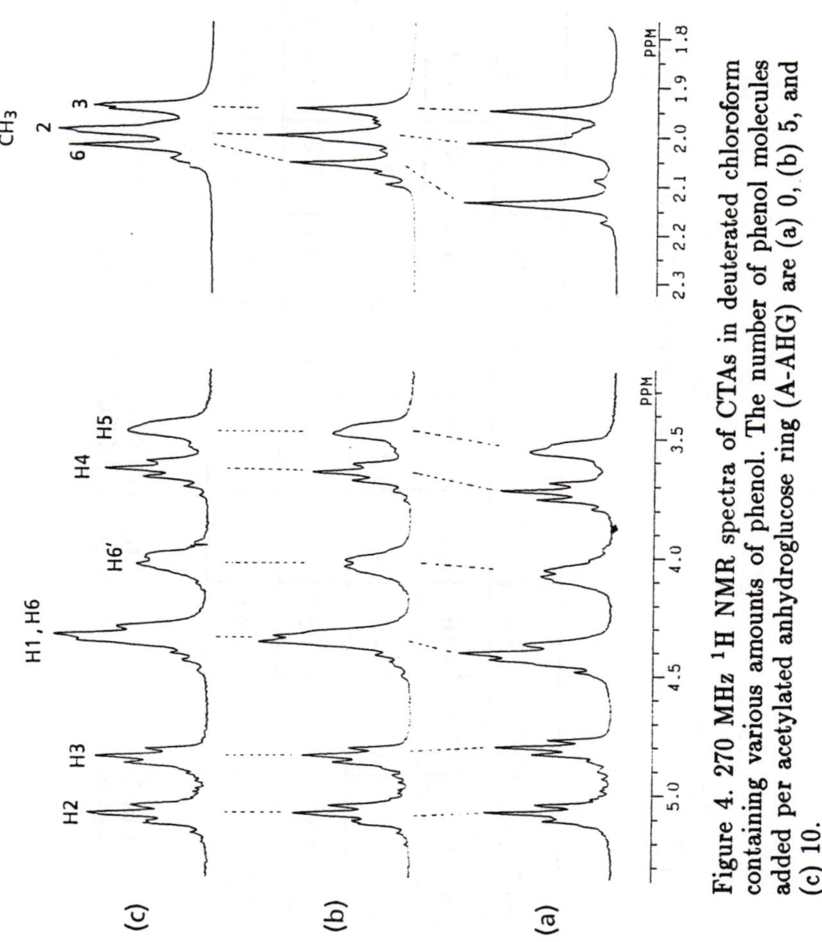

Figure 4. 270 MHz ^{1}H NMR spectra of CTAs in deuterated chloroform containing various amounts of phenol. The number of phenol molecules added per acetylated anhydroglucose ring (A-AHG) are (a) 0, (b) 5, and (c) 10.

Table III Difference of Chemical Shifts of CTA Protons (Hz)

substrate \ proton	H1(H6)	H2	H3	H4	H5	H6′	acetyl methyl 2	3	6
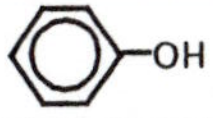 phenol (C6H5–OH)	+ 21.4	− 16.6	− 5.9	+ 24.4	+ 24.4	+ 10.8	+ 5.8	+ 2.0	+ 31.3
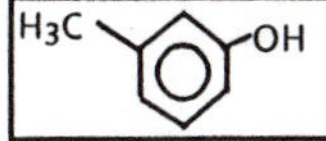 m-cresol (H3C–C6H4–OH)	+ 13.7	− 3.9	− 4.0	+ 14.7	+ 17.6	+ 5.9	+ 2.9	− 1.0	+ 23.4
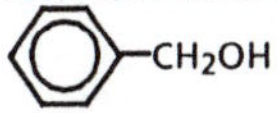 benzyl alcohol (C6H5–CH2OH)	−	+ 2.0	+ 3.9	+ 12.7	+ 12.7	+ 9.8	+ 8.8	+ 7.8	+ 18.6
 benzene (C6H6)	+ 4.9	− 5.8	− 4.9	+ 7.8	+ 9.7	+ 3.0	+ 5.8	+ 3.0	+ 13.7

+ upfield substrate / A–AHG = 10 / 1 (molar ratio)
− downfield CTA 50mg / $CDCl_3$ 1 ml

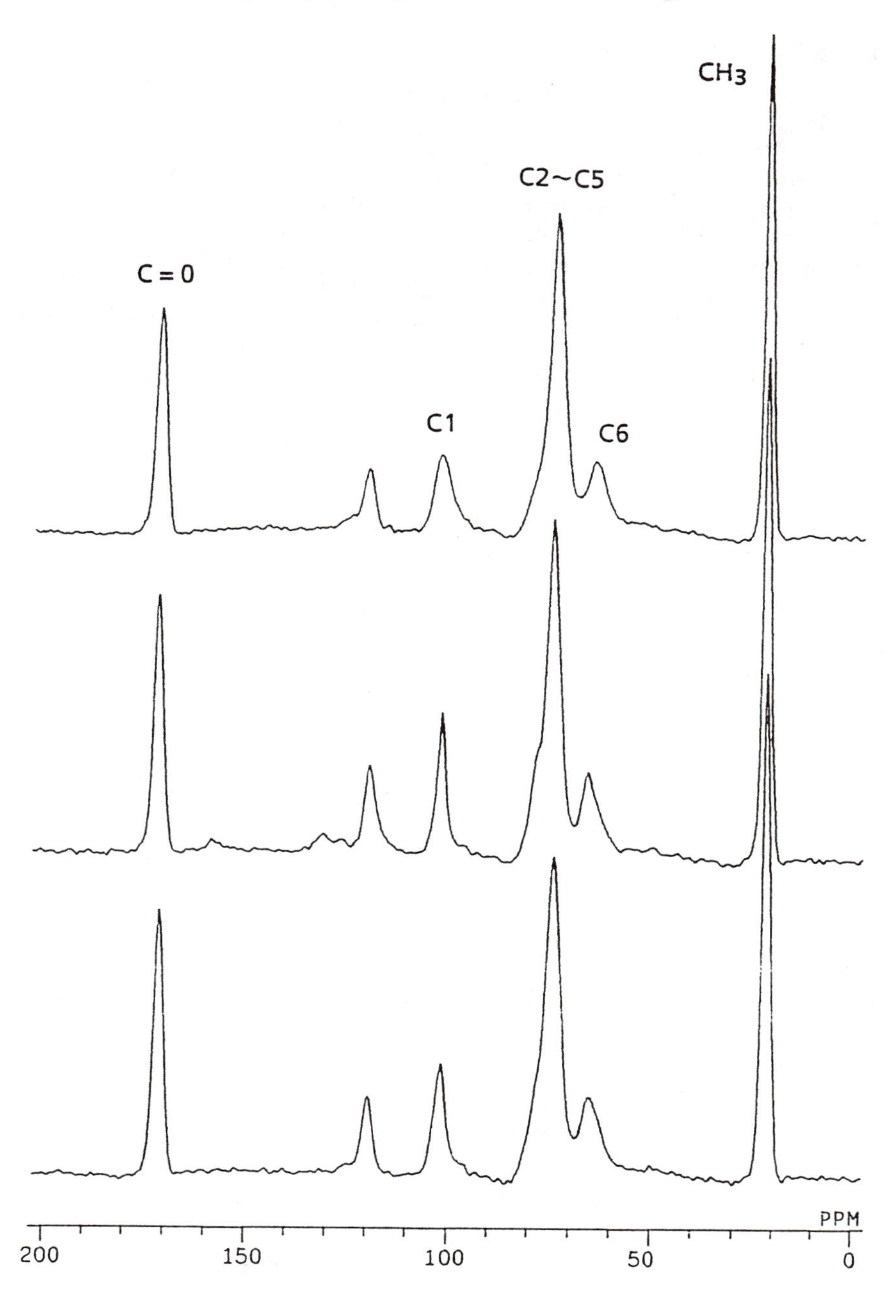

Figure 5. CP/MAS ^{13}C NMR spectra of CTAs, CA-1 (upper), CA-3 (middle), and CA-3 washed with methanol (lower).

LITERATURE CITED

1. Shibata, T.; Okamoto, I.; Ishii, K. *J. Liq. Chromatogr.* 1986, **9**, 313.
2. Ichida, A.; Shibata, T.; Okamoto, I.; Yuki, Y.; Namikoshi, H.; Toga, Y. *Chromatographia* 1984, **19**, 280.
3. Shibata, T.; Sei, T.; Nishimura, H. *Chromatographia* 1987, **24**, 552.
4. Torchia, D. A. *J. Magn. Reson.* 1978, **30**, 613.
5. Horii, F.; Hirai, A.; Kitamaru, R. *J. Carbohydr. Chem.* 1984, **3**, 641.
6. Miyamoto, T.; Sato, Y.; Shibata, T.; Inagaki, H.; Tanahashi, M. *J. Polym. Sci., Polym. Chem. Ed.* 1984, **22**, 2363.
7. Sei, T.; Ishitani, K.; Suzuki, R.; Ikematsu, K. *Polym. J.* 1985, **17**, 1065.

RECEIVED December 16, 1991

Chapter 5

Mechanical Properties of Chemically Treated Wood

A. Björkman and Helena Lassota

Technical University of Denmark, Lyngby DK–2800, Denmark

Molecular configurations of cellulose, hemicelluloses and lignin molecules are well documented. All three polymers seem to constitute separate entities in wood, but the ultrastructure is not known; neither is known how the polymers act together to impart the properties which determine the industrial utilization of wood and its fibers. In an attempt to gain more knowledge, species of solid wood are treated with chemicals to produce swelling, substitution, etc., with measurements of mechanical properties on the same slender sample before and after treatment. Methods to measure axial stiffness, creep and axial compression strength are developed. The treatments give distinct changes, which may partly be related to individual polymers. Stiffness, creep and strength change "independently." Changes are sometimes slow and associated with a transient increase of creep.

Wood is a unique composite, optimized by nature over a very long time span to form trees and forests. The way wood properties are created constitutes possibly the greatest enigma in wood science. The paper gives an outline of extensive investigations on solid wood in an additional effort to attain a better comprehension of the conjoint performance of cellulose, hemicelluloses ("hemi") and lignin in the wood fiber wall, and also to understand better the molecular framework—the ultrastructure—which endows wood with its properties and utility as material and fibre source, etc.

It is postulated that cellulose, "hemi," and lignin have tantamount functions in forming the mechanical/rheological properties of the fiber wall; all three are necessary constituents. Thus—as a notion—lignin should not be regarded as a padding in the wall structure.

Molecular structure and chemical properties are nowadays almost completely unravelled for cellulose, "hemi," and lignin, but the same cannot be

0097–6156/92/0489–0065$06.00/0

said about the ultrastructure of the components in plants. Cellulose forms fibrils which constitute separate regions in the fiber wall, while considerable uncertainty prevails about detailed distribution and molecular configuration and orientation of "hemi" and lignin.

Research Program Notions

The *primary basic* principle behind the new way to study the ultrastructure is an indirect approach. Measurements are made of changes in physical properties of solid wood during or after treatments of various kinds, which influence the properties of the wood composite, *e.g.*, solvent action, introduction of a substituent, or careful removal of some lignin and/or "hemi." It is endeavored to find treatments which have a selective influence on one wood component.

A general problem when dealing with wood is its natural inhomogeneity, which is normally tackled by measurements on many specimens and statistical elaboration of the experimental results. The *second basic* principle is to circumvent the problem by measuring the alteration of properties on one and the same wood specimen. An attempt to base the investigation on many samples for statistical reasons would have been impossible time- and costwise and the alternative is doubtful anyway.

The study has centered around three selected measurements of mechanical properties: axial stiffness modulus (E_d); creep index (CI) on bending; and axial compression strength (ACS).

The measurements intend to give distinctive types of information about wood properties, but only the first two are nondestructive. Stiffness is determined by a vibrational method, where no creep may take place. By limiting load and time the creep method is expected to give only minimal damage on the wood structure. The strength measurement by compression is a complex matter and interpretation may be inferior to the other methods, but no alternative has been found.

Selection and development of treatment methods has been a major effort in this study. An extensive literature search has been made but it cannot be rendered in this communication. The paper presents briefly a few results to illustrate the potentialities of the new approach.

Wood Specimens

Of the species studied so far, only spruce, birch, and aspen are included in this paper. The fundamental problem is to select the specimen dimensions which allow both uniform treatment on impregnation and reasonably good measurements, using the mechanical tests. This lengthy "optimization" procedure will not be reviewed in this presentation. Only axial specimens allow reliable measurements, being amenable to physical calculations according to theory and interpretation.

The preferred dimensions of the "ideal" specimen have been 7 × 9 × 152 mm (a "microbeam"), with the radial direction being parallel to the 9 mm side of the cross-section and the axial ring pattern perfectly parallel to the longitudinal direction. This specimen may have ray cells at the 9 x 152 mm surface, but *not* either early- or latewood cells only, which is important since the bending of the specimen is made perpendicularly to this surface and outermost cells dominate the strength.

Extractives are removed from the specimen, which is dried carefully over P_2O_5 (60°C). This "bone-dry" specimen is the starting material for all treatments. Cycling of a spruce wood specimen from this state to a water uptake at RH 50 (23°C) did not cause much difference in weight and stiffness. It may be inferred that the use of bone-dry wood as the reference is acceptable.

Physical Measurements

Since the developed methods are innovations, the methods are presented in some detail.

Stiffness (E_d). The method applied is well known in principle. The arrangement used in our investigations is illustrated in Figure 1. The specimen dimensions are partly determined by theoretical requirements. The specimen rests with its 9 x 152 mm surface on silicon rubber bars, located at the nodes of the first normal mode of natural vibration of a uniform free beam. Both ends are provided with small metal staples, pressed into the wood. One end is under the influence of a magnetic field, transduced from a 2-channel analyzer (Brüel and Kjær), producing simultaneous oscillations within a range of frequencies. A resonance vibration is actuated by the frequency proper.

At the other end the amplitudes of the vibrations are registered by the analyzer for all frequencies within the range, whereby the resonance frequency is seen on the instrument screen and identified by moving a cursor line to the frequency tip (Figure 2). (More than one tip may indicate some irregularity. Attempts to study radially cut specimens from wood have not been successful.)

When the frequency tip is not distinct, more than one pair of staples may be used. The dynamic vibration modulus is calculated by the formula (SI units):

$$E_d = \frac{1}{12.68} \bullet \nu^2 \bullet \frac{\text{weight} \bullet l^3}{I} GN \bullet m^{-2}$$

where ν is the resonance frequency, and $I = b \bullet h^3/12$, l, b and h being shown in Figure 1. A correction for staples weight is necessary and has been determined empirically. The measurement is simple. E_d is not much different from E_b, the modulus for bending (cf. below), which may be used as a control. If the wood specimen contains a volatile liquid, it may be placed in a small tight plastic box to keep the weight constant by avoidance of evaporation or uptake of water from air.

The possible physical effect as such of a liquid on E_d is an intricate matter, which has been examined. The weight of an inert liquid will change the frequency ν but not E_d, if the presence of micro-bubbles of air is eliminated, but a larger mass of liquid (*e.g.*, in lumen) may oscillate a little

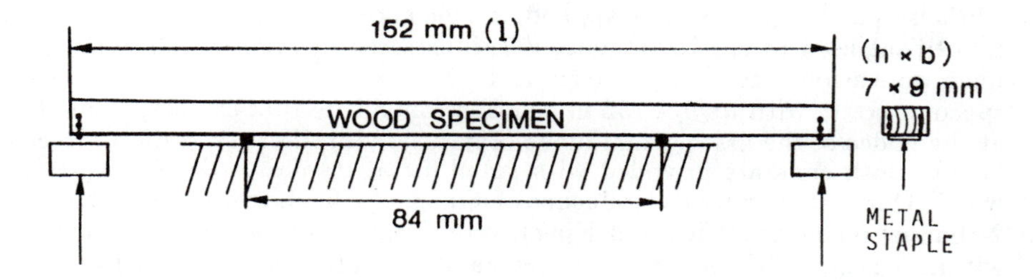

Figure 1. Measurement of stiffness as vibrational modulus (E_d). The transducers for generating oscillations and measuring amplitudes are indicated at specimen ends.

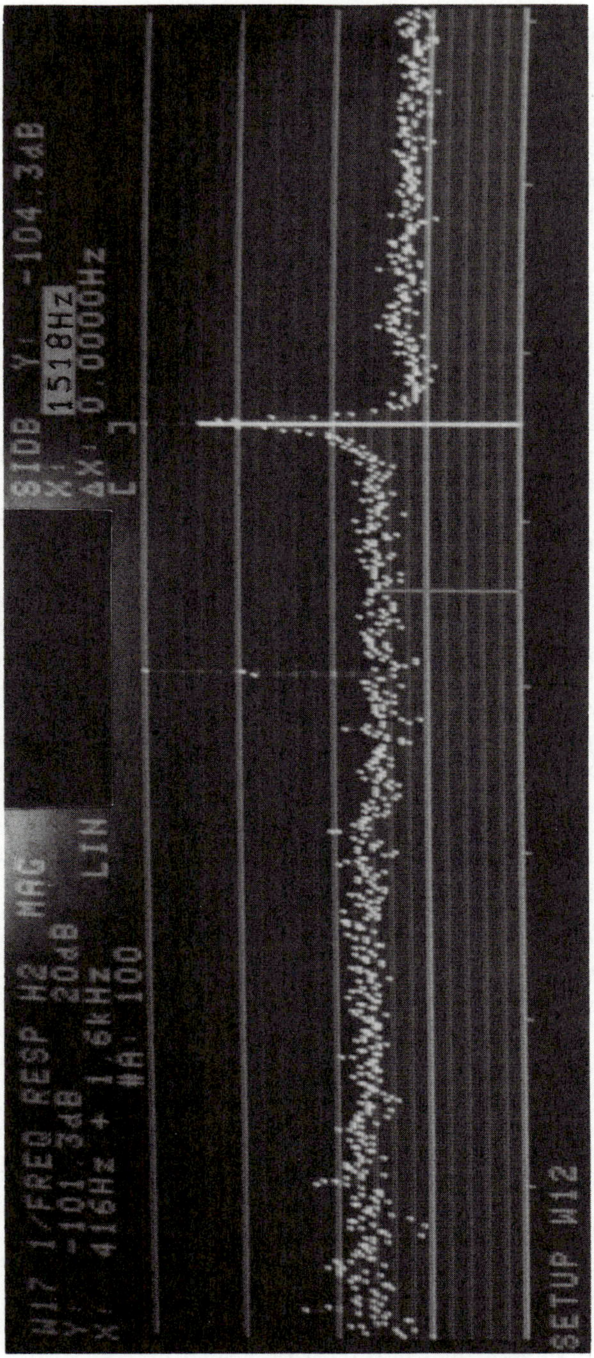

Figure 2. Measured frequencies from oscillating wood specimen (Figure 1), indicating resonance at 1518 Hz.

because of its compressibility. The indications are that such an effect is quite small, if it may be observed at all.

Though stiffness can be measured easily and precisely, the interpretation of changes in the modulus E_d may be difficult. Specimens made from a piece of wood usually deviate in E_d, but if the E_d-value after treatment is normalized by division with the original figure, reproducible values are obtained in parallel experiments.

Creep Index (CI). Though creep is well known as a phenomenon, it may so far only be measured on an empirical basis. With polymers, the viscoelastic changes of molecule conformation are part of the process, but for an inhomogeneous material like wood, macromorphological changes may also occur, sometimes designated as slippage. Still, variations in creep behavior may reveal changes in mutual polymer interactions after various treatments.

The progress of creep is at times described by a mathematical model, usually an exponential function. The method chosen in this investigation is a graphical computation, from which is contrived a measure named *creep index* (CI), as explained below.

Creep is normally measured by homogeneous tension on elongation of an axial specimen. This would hardly be possible with the "microbeam" without some destruction of its ends. An obvious option is bending of the "microbeam" as illustrated by Figure 3. The ends are inserted in holders, one having a groove and resting on a fixed edge perpendicular to the wood specimen, the other having a central groove along the specimen and resting on a steel ball (on a steel surface). The distance between edge and ball is the specimen length (152 mm).

The load applied is less than 10% of the breaking load. The instantaneous depression is determined by the bending modulus E_b, but a precise measurement of E_b is very difficult, since creep starts immediately on bending. In order to limit the time required for creep, a sensitive position meter is used, which feeds the information to a datalog/computer, which records the positions at predetermined times. After deloading, these values are displayed on a screen. The depression is registered every 1/20 of a second during 2.00 sec and then every 12 sec up to a suitable limit, usually 10 min.

Treatment of the creep data to obtain CI as a normalized (dimensionless) figure requires some manipulation, since a reference value for E_b is needed. Assuming the instantaneous bending depression to be $\Delta u_o = (u_o)_{\text{loaded}} - (u_o)_{\text{unloaded}}$, a value for the bending modulus may be calculated:

$$E_b = \frac{1}{48} \bullet \frac{P \bullet l^3}{\Delta u_o I} \qquad I = \frac{1}{12} b \bullet h^3$$

where P is the load. However, the precise registration of $(u_o)_{\text{loaded}}$ is not possible. Instead, by the assumptions that E_b is close to E_d and that the relation $E_b = k \bullet E_d$ holds, a precise auxiliary quantity, Δu_d, may be calculated from E_d, using the same formula.

A depression value u_i at the time t_i may then be transformed into a normalized depression φ_i as:

$$\varphi_i = \frac{u_i - (u_o)_{\text{unloaded}}}{\Delta u_d}$$

In diagrams it is easier to insert the quantity $(\varphi_i - 1)$.

It has been found in this work that $(\varphi_i - 1)$ *vs.* $\log(t_i - t_o) = \log(\Delta t_i)$ is a linear relation. This is why we may define CI as the slope of this line, *i.e.*,

$$CI = \frac{\Delta\varphi}{\Delta \log(\Delta t_i)} \bullet 10^3$$

a dimensionless number in the range 10–300. As indicated above, the linear curve may display (sooner or later) a break, indicating a change in creep mechanism (cf. Figure 4), but a linear progress has been observed up to 16 hrs. The measurement is empirical, and the results must be judged with care. It should be emphasized that the conventional creep test, measured by axial tension, is much different by being homogeneous, while the creep on bending is a mixture of tension and compression.

The loading arrangement is placed in a box, which may be closed airtight, *e.g.*, for work in water-free air (over P_2O_5), or with obnoxious solvents, etc. (Figure 5).

Axial Compression Strength (ACS). Every strength test is destructive and marks the end of the specimen life. In order to get something of an "absolute" quantity, it would be proper to measure toughness, using a rapid fracture method like the hammer swing. This possibility has been tested on sections of the "microbeam," but an inordinate spread of values made the test impractical. Among several slow strength test methods studied, which all implicate creep and some kind of slow deformation, axial compression was selected.

In the standard method developed, the "microbeam" is cut into several 14 mm pieces (precisely with right angle ends). The compression is made in an Instron instrument, placing the small specimen in a cage (Figure 6). The cage plates have rough surfaces to avoid gliding, and the bottom plate movement is 5 mm/min^{-1}. The compression normally occurs at the top end of the specimen. When the specimen buckles or is split axially, the strength value usually is below the normal range. Due to individual properties of the piece of wood, the maximum load recorded in a diagram is registered after 10–25 sec.

A main difficulty with a destructive test is to estimate the value of a specimen before treatment. Use of "twin specimens" cut in parallel from a piece of wood as reference was found to be doubtful. Since the variation of ACS within one specimen appears to be less than between two specimens, the best thing to do is to use a separate specimen cut in halves, measuring the reference value on one half and the strength after treatment on the other. When this was not done, an average figure for a species is the only reference (this figure given below in parentheses).

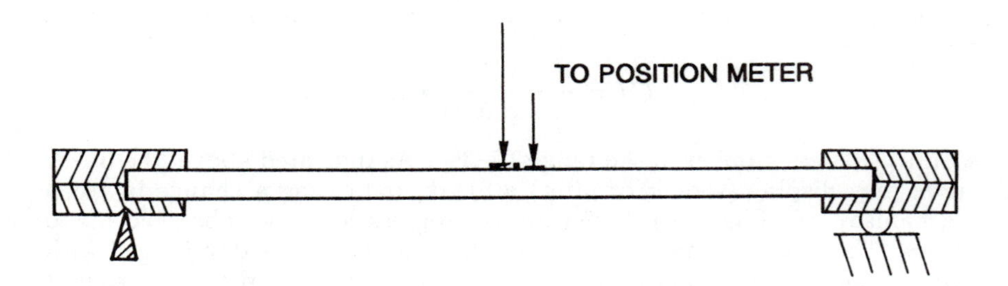

Figure 3. Creep measurement. Bending load applied on the middle of specimen (same as in Figure 1).

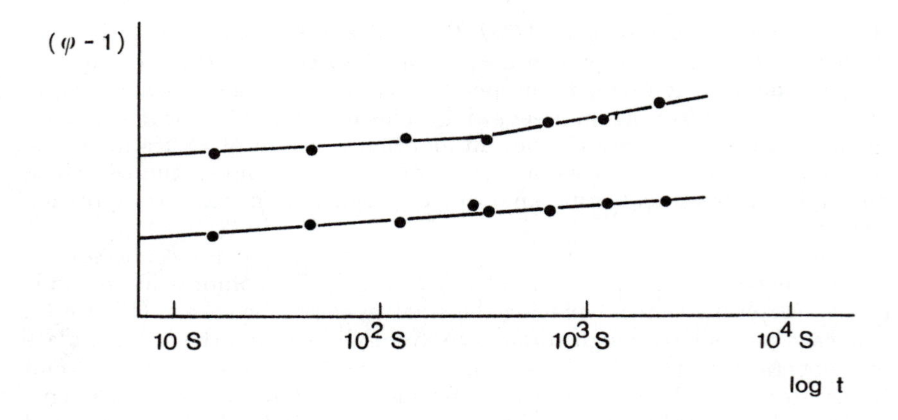

Figure 4. Typical curves from creep measurements (Figure 3).

Figure 5. Loading arrangement for creep measurement in box. Front window removed. Note position meter above specimen.

Figure 6. Compression cage with axial wood specimen (Instron).

Specimen Treatments, General

It is important that the impregnation of wood is always as complete as possible. When measurement of E_d is made on liquid-filled wood, no gas should be left in cavities. In standard procedure the wood specimen is placed in a vessel, which is evacuated thoroughly, and the liquid is admitted before the vacuum is broken. (The procedure is not used on treatment with liquid NH_3). Thereafter, the vessel with specimen in liquid is placed in an autoclave and pressurized at 3 MPa for 2 hrs or more.

A specimen may either be measured bone-dry (over P_2O_5) or filled with the liquid, which changes its polymer properties. After treatment with a liquid, with or without chemicals, different schemes are applied to bring the wood always back to a bone-dry state (over P_2O_5). Extraction is normally done with acetone. Chemicals may be removed by placing the specimen in running water for several days, possibly—for acetic acid—followed by keeping it over moist alkali pellets.

Results

In this introduction to the new methodology, some examples of experimental results are given to illustrate its potentialities, but systematic conclusions and statements must await an elaboration of the entire experimental material. For this reason, rather few experimental details are included. Generally, reference to the action of solvents/reagents on isolated polymers is not included. Also, there is no room to discuss intricate chemistry. The measured values for specimen weight and volume and for E_d after treatment are normalized by division with figures for the original specimen, but it is not done for CI and ACS because of lower accuracy.

As the E_d varies between specimens of spruce, corresponding changes are found in CI, as shown in Figure 7. However, generally in the treatments the values of E_d and CI, as well as ACS, have been found to increase/decrease independently. This observation is illustrated by the alkaline treatment of spruce at pH approximately 12 and 14 (Table I). The uptake of liquid by the wood specimens exhibit monotonous parallel increases (higher for pH 14), while E_d falls immediately and stabilizes at a somewhat lower value. A dramatic change appears in CI, which has an early peak value and then falls to moderate levels. Thus, during the primary transient state the wood composite is less resistant to creep.

Aspen was treated with succinic anhydride, dissolved in dimethylsulfoxide (DMSO), with an organic catalyst. As seen in Table II, the degree of substitution seems to be proportional to density. The minor change in (normalized) E_d is about the same in both samples. The results illustrate the reproducibility, using normalized figures. CI also remained unchanged, but a notable increase in ACS was observed, also with other wood species. Thus, the introduction of carboxyl groups increases intramolecular bonding but does not affect viscoelasticity. The latter observation was also made on acetylation with acetic anhydride, all species giving weight increases of the order 10 to 20%, while the dimensions as well as E_d, CI and ACS remained

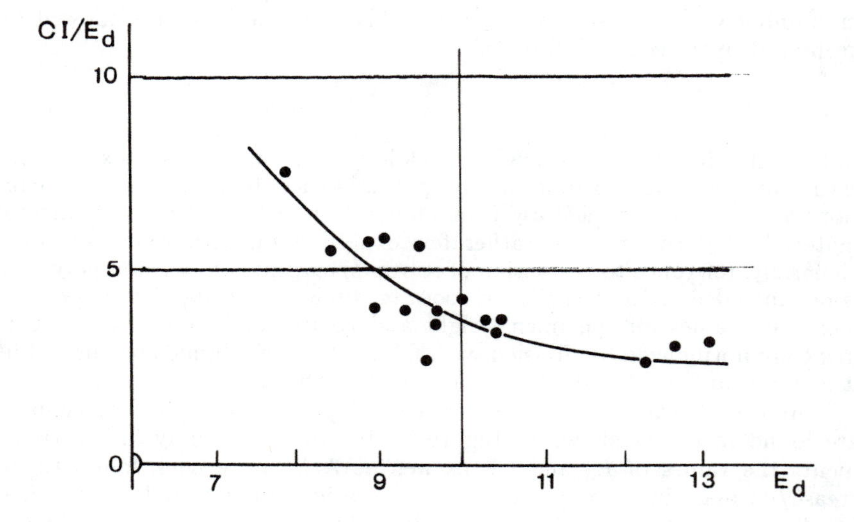

Figure 7. The ratio CI/E_d as a function of E_d for dry spruce wood.

Table I. Changes of aspen wood properties on succinylation

Sample	1	2
1. Density, g cm^{-3}	0.619	0.608
2. Weight after treatment (normalized)	1.14	1.11
Ratio 2./1.	1.84	1.83
E_d, GN m^{-2}, before treatment	15.0	12.4
After treatment (normalized)	0.96	0.98

Table II. Effect on spruce wood of alkaline treatment

Time	Weight		E_d		CI	
	pH 12	pH 14	pH 12	pH 14	pH 12	pH 14
0	4.85	3.68 g	18.8	13.2 GN•m^{-2}	29	28
1 hour	1.62	2.25	0.71	0.56	182	266
1 day	1.69	2.27	0.65	0.54	75	165
1 week	1.91	2.59	0.70	0.56	52	(213)
2 weeks	2.04	2.78	0.68	0.57	45	100
3 weeks	2.07	2.80	0.67	0.53	86	167
9 weeks	2.27	3.05	0.67	0.53	66	24

Note: Weight and E_d after treatment are normalized.

largely unchanged. Acetyl groups evidently do not influence strength. The conclusions of course are only valid for dry wood.

The next example is the strong action of liquid NH_3 on wood, performed at −40°C with careful avoidance of absorption of moisture. The main part of the change in wood properties occurred during the first 24 hrs. After the treatment ammonia was removed by evacuation and keeping the wood specimens in vacuum for several days over 90% H_2SO_4 before drying over P_2O_6. The effect on spruce and birch wood of three days in liquid ammonia is shown in Table III. Only a small amount of wood substance is extracted, while the volume shrinkage is considerable.

In contrast to all other treatments, a change in the axial dimension was noted, viz. a reduction of 3 mm. In spite of the considerable increase in density (for birch from 0.564 to 0.863 g cm^{-3}, i.e., 53%) the measured E_d decreased moderately. It is even more remarkable that the CI also changed moderately. The change in ACS is more difficult to examine, since the starting value is not known. However, with correction for the density change, the compression strength appears to be reduced. Thus, along with the volume reduction the mechanical properties measured seem to have

Table III. Transformation of wood on three days treatment in liquid NH_3

	Spruce	Birch
Dissolved wood, %	0.69	1.75
Resulting volume (norm.)	0.76	0.61
E_d, before	15.3	14.9
E_d, after (norm.)	0.89	0.95
CI, before	18	50
CI, after	46	43
ACS, before	(650)	(675)
ACS, after	758	1003

deteriorated. Several conformational changes of the cell wall components may have contributed. The transformation of Cellulose I to another form may be a principal reason.

The ammonia-treated wood may be compared with the wood resulting from the action of alkali as discussed above (Table I), after thorough removal of the alkali with water. As shown in Table IV, the weight loss was slightly higher than with ammonia, but the E_d increased, contrary to the notion with NH_3, while the moderate change in CI is similar. The increase in density was less than 10%. Thus, a higher ACS was definitely obtained by alkali treatment. Again, the independent changes of E_d, CI and ACS are demonstrated.

Table IV. Changes of spruce wood on treatment with alkali

pH	12	14
Weight*	0.97	0.95
Volume*	0.94	0.88
E_d*	1.02	1.08
CI, orig.	29	28
After treatment	45	43
ACS, orig.	(650)	
After treatment	827	824

*Normalized.

Another experiment within the same regime was made on birch with the strongly swelling agent N,N-dimethylacetamide, containing 8% LiCl, with the results given in Table V. As usual, the weight increased monotonously, but otherwise it looks like little or nothing happened during the second week, the main changes occurring during the first 24 hrs. However, after six weeks a further decrease in E_d is noted and CI is augmented markedly, not exhibiting a maximum like under the action of alkali. Since

Table V. Solvent action on birch wood of dimethyl acetamide with 8% LiCl

Time	0	5 hrs	24 hrs	1 wk	2 wks	6wks	DMAc/LiCl Removed (after 6 mos.)
Weight*	4.913	2.03	2.08	2.28	2.29	2.41	0.99
Volume*	8.34	1.05	1.14	1.22	1.23	1.23	0.68
Density*	0.589	—	—	—	—	—	0.849
E_d	15.8	0.96	0.77	0.70	0.70	0.67	1.30
CI	20	291		194	183	257	50
ACS	(675)	—	—	—	—	—	1315

*Normalized.

the solvent swells cellulose, a main reason could be that also the cellulose fibrils were affected in the end, though more weakly than with NH_3.

After being in the solvent six months, the normalized volume of the specimen had decreased to 1.14, but the wood had become quite soft, so E_d could no longer be measured. The chemicals were removed by an extended washing with running water and the specimen was dried. As seen in the last column of Table V, little or no material had been removed from the specimen, while its density had increased from 0.589 to 0.849 (44%). A slight reduction of the specimen length, smaller than at ammonia treatment, was noted. E_d had increased considerably and a large increase was evident for ACS, also after making a correction for the density change. Notwithstanding these large increases, the wood had become weaker with regard to CI. These remarkable effects are most interesting, but very difficult to interpret on a molecular level. A possible remainder of solvent in the wood could possibly explain the higher CI, but nothing else.

More or less drastic alterations of the wood structure of the kind reported above may be produced with water, but only if high temperatures are employed. It should be added that a small amount of water may have a favorable strength effect. When wood with a normal water content is dried completely, the CI gets somewhat larger.

Finally, attention will be directed to the effect of lignin solvents. Table VI illustrates the difference between the solvent actions on spruce of DMSO and acetylmethyldioxane (AMD). DMSO has an immediate effect on E_d and further changes of measured properties are very slow, though at last the wood gets rather soft, further E_d measurements being unfeasible, and the ACS value gets very low (compared to the average of 650). The likely explanation is that DMSO dissolves both "hemi" and lignin, why 60% of the wood structure is swelled by the solvent. (In spite of the strong effect, the wood specimen regains its original properties, as measured, after removal of DMSO). AMD, on the other hand, is a lignin solvent, which may swell "hemi" but to a much lower extent. The decrease of E_d is therefore slower and more limited. The ACS, however, is reduced notably, which can be understood, since lignin is considered decisive for the compression strength of wood.

Table VI. Action of solvents on spruce wood

Solvent	E_d before GN • m^{-2}	E_d, impreg., norm. 1 hr	1 day	15 days	53 days	ACS, final kg • cm^{-2}
Dimethyl-	11.01	0.613	0.606	0.595	—	120
sulfoxide	11.87	0.619	0.604	0.603	—	125
Acetylmethyl-	9.86	0.914	0.921	0.801	0.762	235
dioxane	12.87	0.832	0.828	0.790	0.748	240

A more lignin-selective solvent than AMD is the mixture dichloroethane/-ethanol in the volume ratio 2:1. The two solvents in the pure form each have little or no effect on wood properties. Thus, if the mixture does act on wood it is likely to be on lignin only. A typical result with spruce is given in Table VII. Again, a monotonous increase in weight is noted, while the specimen volume displays an increase at first but then seems to stay at this level. E_d is reduced, but in the end most of the original stiffness seems to be regained. CI is increased and stays at the higher level. With other wood species the CI is increased considerably at first, but turns back slowly towards the original level. ACS undergoes a considerable reduction, which—as for AMD—may be attributed to the lignin contribution to compression strength of wood.

Table VII. Solvent action on spruce wood of $ClCH_2CH_2Cl/EtOH$ 2:1

Time	0	1 day	1 wk	2 wks	4 wks	8 wks
Weight*	3.86	1.37	1.77	1.90	2.23	2.57
Volume*	9.46	1.07	1.09	1.09	1.10	1.07
E_d*	13.2	0.76	?	0.91	0.89	0.97
CI	23	50	48	48	57	49
ACS	(650)	—	—	—	—	307

*Normalized.

Concluding Remarks

As already indicated, this short presentation of a new approach does not lend itself to systematic or final conclusions. The experimental methods have for the most part been found to be reproducible and consistent. New effects have been disclosed—some being rather unexpected—like the "independence" of changes in stiffness, creep and compression strength on treatments, the deterioration of all measured mechanical properties by the action of liquid NH_3 on wood, the transient increases in creep, and the slowness of some changes, lasting weeks or months.

However, it was conceived from the beginning that observations of mechanical properties are not sufficient to reveal the behavior of the polymers in the wood structure. Therefore, a Nordic collaboration project has been established and several of the wood specimens prepared have been sent to L. Salmén at STFI (Stockholm, Sweden) for a study of rheological properties and to J. J. Lindberg at Helsinki University (Finland) for solid state NMR and FTIR studies. The results of this successful cooperation will be reported in separate publications.

Apart from the ambition of this investigation to contribute to the disclosure of wood ultrastructure and the roles of cellulose, hemicelluloses, and lignin, the new methods may contribute to a practical goal: wood characterization. Whether we deal with wood as such or with wood as a raw material for pulps, our meager understanding of the variability of wood, all species, is an obstacle for industry. New methods of characterization may offer a partial remedy to the problems.

Acknowledgments

The authors thank the Danish Technical Research Council, the Nordic Fund for Technology and Industrial Development, and Junckers Industries A/S for financial support.

RECEIVED February 10, 1992

Chapter 6

Wood Fiber Reinforced Composites

C. Klason[1], J. Kubát[1], and P. Gatenholm[2]

[1]Department of Polymeric Materials and [2]Department of Polymer Technology, Chalmers University of Technology, Gothenburg S–412 96, Sweden

The reinforcement of polypropylene (PP) using wood cellulose has been studied. The strength and stiffness of the composites are improved by a prehydrolytic treatment of the cellulose as well as by the use of maleic anhydride-modified PP (MAPP) as a coupling agent for promoting interfacial adhesion. The embrittlement of the cellulosic component brought about by the hydrolysis facilitates fine dispersion of fibers in the shear field of the compounding extruder. The E-modulus of composites based on prehydrolyzed fibers is significantly improved as compared with untreated fibers and exceeds values calculated theoretically with the Halpin-Tsai equation. Strength values increase with increased fiber content, which is a remarkable characteristic for such composites. Significant improvement of the modulus and strength in composites with hydrolyzed fibers indicates that such fibers are disintegrated into microfibrils characterized by very high modulus and strength values. Selecting suitable processing conditions was found to be crucial for the chemical reaction between the MAPP coupling agent and fiber surface, which produces improved adhesion.

Natural fibers such as wood, cellulose and jute are renewable materials with very attractive mechanical properties. For instance, cellulose fibers with moduli up to 40 GPa can be separated from wood by a chemical pulping process. A growing awareness of environmental problems and the importance of energy conservation have made such renewable reinforcing materials of great importance. The use of wood fibers as fillers and reinforcements in thermoplastics has been documented in several reports (1, 3-5). However, the limited heat stability of cellulose unfortunately reduces

0097–6156/92/0489–0082$06.00/0

its useful range in a number of possible matrix materials. In common polymers, such as polyethylene (PE), polystyrene (PS), polyvinyl chloride (PVC), and polypropylene (PP), wood fibers yield relatively low stiffening effects (1,3,6). The discrepancy between the reinforcing effect obtained in composites and the reinforcement potential of wood fiber, particularly its constituents, cellulose fibers and fibrils, has intrigued scientists.

One explanation for the rather poor mechanical properties of wood fiber-polymer composites is a lack of interfacial adhesion between components. Several attempts have been made to improve interfacial adhesion in wood fiber-polymer composites. The addition of a coupling agent such as maleic anhydride-modified polypropylene (MAPP) has been reported to improve bonding between cellulose and polypropylene matrix (1). Reports on an improvement in tensile strength for PP-based composites owing to the addition of a MAPP coupling agent (7) indicate an increase in strength from 18 MPa for the unfilled matrix to 30 MPa for a composite containing 60% fiber. The corresponding modulus increase was relatively small, from 1.2 to 2.1 GPa. Infrared spectroscopic studies of extracted fibers have shown that the OH groups of cellulose were esterified by the anhydride groups of MAPP (7). A coupling agent improving the mechanical parameters of PMMA-based composites is polymethylene phenylene isocyanate (5), which was also found to be effective with wood flour/PP materials (8). In the latter case, the tensile strength increased from 31 to 38 MPa when 40% wood fiber was added to the matrix. The simultaneous modulus change was only from 1 to 2.3 GPa, mainly due to a low value of the aspect ratio of the filler. Kokta and coworkers reported on the use of grafted wood fibers as reinforcement in several polymers (9, 10). Grafting appears to provide a fair degree of adhesion between the fibers and the matrix. This has also been demonstrated for PMMA-based composites (11).

Another problem in using wood and cellulose fibers together with a polymer is the occurrence of agglomeration due to insufficient dispersion. This obviously contributes to poor mechanical properties, for example because agglomeration decreases the effective fiber aspect ratio in composites. Hydrolytic treatment of the cellulose fibers prior to compounding resulted in a significant increase in the modulus of the injection-moulded samples (1). This appears to be caused by the embrittlement of the fibers brought about by hydrolysis. In the shear field of the processing machinery the fibers are broken down into smaller fragments, which include an unidentified fraction of submicroscopic microfibrils known to possess unusually high modulus values (1,2). The presence of such entities in the composites was indicated by a comparison of the mechanical parameters of the samples with the results of theoretical calculations using the Halpin-Tsai equation.

This paper is a brief account of a series of experiments performed to investigate the reinforcement of injection-moulded PP with prehydrolyzed cellulose fibers (chemithermochemical pulp, CTMP, and dissolving grade cellulose). The results presented below are part of a systematic study of the properties of composites based on prehydrolyzed cellulose fibers and a thermoplastic matrix (1,11,12). Such materials do not attract a great deal of interest among researchers in this field, despite their obvious technical potential. The experiments also included the use of MAPP as a coupling agent intended to improve adhesion between the different components of the composite. The beneficial effect of this coupling agent was noticeable, provided that the mixing of the components in the compounding extruder was carried out at a temperature exceeding the activating point of MAPP. The presence of chemical bonds between MAPP and cellulose was established by IR and ESCA analysis, which explains the improvement of adhesion between different phases, as reflected in increased tensile strength levels at higher fiber loadings. Unhydrolyzed cellulose behaved in a similar manner.

Experimental

Materials. The matrix polymer used was an injection moulding grade of PP (copolymer type, GYM 621, ICI, density 905 kg/m^3, melt index $MI_{230/16}$ 13, broad MW distribution with M_n 6500 and M_w 83600, as determined by GPC, supplied as powder). The coupling agent was a maleic anhydride-modified polypropylene (Hercoprime G from Hercules, T_m 155°C, density 900 kg/m^3, powder). M_n and M_w were 5000 and 39000, respectively, and the acid value was 59, which corresponded to 6 weight percent of maleic anhydride.

The pulps used were dissolving pulp (Ultra, Billerud, spruce, DP 764 (239), R18 94.2%, R10 89.8%, ethanol extract 0.2%, ash 0.04%), denoted as α-pulp in the text, and chemithermomechanical pulp, CTMP (15% pine, 85% spruce). DP values in parentheses relate to hydrolyzed samples.

The hydrolysis was carried out using 3% oxalic acid in water solution at 0.2 MPa pressure and a temperature of 120°C for 1 hour. After hydrolysis, the pulp was washed to pH 7 and dried at 70°C for 24 h in a circulating air oven. Both the untreated and hydrolyzed cellulose fibers were disintegrated in a rotating knife mill (Rapid GK 20, screen 4 mm). Disintegration in the plastics granulator fitted with a large screen was necessary in order to obtain a relatively free-flowing mixture with the PP powder, which could be fed into the compounding machine without difficulty. The fiber length after this mild milling procedure was approximately 1 mm for the unhydrolyzed fibers and 0.5 mm for the hydrolyzed fibers. The disintegrated fibers were dried at 100°C for 24 hours in vacuum before compounding. The hydrolyzed pulps are designated as HC pulps in the text.

Compounding and Mechanical Testing. The components were homogenized in a Buss-Kneader compounding extruder (Model PR 46, screw diameter 46 mm, L/D 11, melt temperature 180°C). To achieve a good wetting of the coupling agent (MAPP) onto the fibers, a precompound of the pulp (95%) with MAPP (5%) was prepared at 180°C. The extruder was run with the die taken off, the resulting blend coming out as a fine powder (HC-pulps) or as grains (unhydrolyzed pulps). This precompound was then mixed with PP in the extruder to obtain the desired fiber loading. Possible moisture uptake by the fibers during feeding of the extruder was eliminated in the degassing zone of the machine (vacuum). The extruded strands were granulated before injection moulding (Arburg 221E/17R, melt temperature 180°C). The test bars (DIN 53455) with a cross-section of 10 × 3.5 mm and an effective length of 75 mm were conditioned at 23°C and 50% RH for 24 h before testing. The tangent modulus, E, the stress at yield, σ_y, and the corresponding elongation, ε_y, were determined from the stress-strain curves (Instron Model 1193, extensometer G51-15MA, deformation rate $4.5 \times 10^{-3} s^{-1}$). The Charpy impact strength, IS, was measured on unnotched samples (Frank, Model 565 K, DIN 53455). The mechanical parameters were evaluated from data on at least 10 test bars for each composition. The arithmetic mean value and the standard deviation were evaluated, the latter of which increases with increased fiber loading. The vertical error bars in the figures mark the standard deviation for the unfilled materials at the highest fiber content.

Filling levels are given as weight percentages.

ESCA. The surfaces of the cellulose fibers were studied with an ESCA AEI ES200 Spectrometer with an Al(K_α) X-ray source at 14 kV and 20 mA. The energy scale was chosen so that the binding energy of C ls in aliphatic hydrocarbons was 285 eV. A Tektronix 4051 graphics terminal was used for background subtraction, peak separation and peak area measurements. The C ls spectrum was resolved under the assumption of a Gaussian distribution. Two parameters, the full width at half-maximum and the peak position, were varied until correspondence with the observed spectrum was obtained.

IR Spectroscopy. Transmission spectra were obtained with the KBr technique, using a Nicolet DX-10 FTIR spectrophotometer at a resolution of 2 cm^{-1} with the coaddition of 32 scans.

Scanning Electron Microscopy (SEM). The tensile fracture surfaces of the composite samples were studied with a Jeol JSM-U3 scanning electron microscope operated at 25 keV.

Results

Mechanical Properties. The results of the measurements of the mechanical characteristics of the injection moulded PP-based composites containing untreated and prehydrolyzed cellulose fibers are reproduced in Figures 1-4. Figure 1 shows the variation of the tensile modulus with the fiber content (α-pulp and CTMP). The untreated fibers appear to have a fairly low reinforcing effect on the PP matrix. This applies to both the lignin-rich CTMP grade and to the nearly pure cellulose which constituted the α-pulp. On modification by hydrolysis, the two fiber types appear to behave differently when incorporated into the matrix material. Although a significant increase in the modulus value is noted for the α-fibre upon hydrolysis, there is only limited improvement with the lignin-rich CTMP fibers unless they are precompounded with the MAPP coupling agent (cf. Experimental). MAPP elevates the modulus relatively little when used together with the α-pulp.

Injection moulding of the tensile test bars containing more than 40% α-pulp was limited by the high viscosity of the filled melts. For untreated

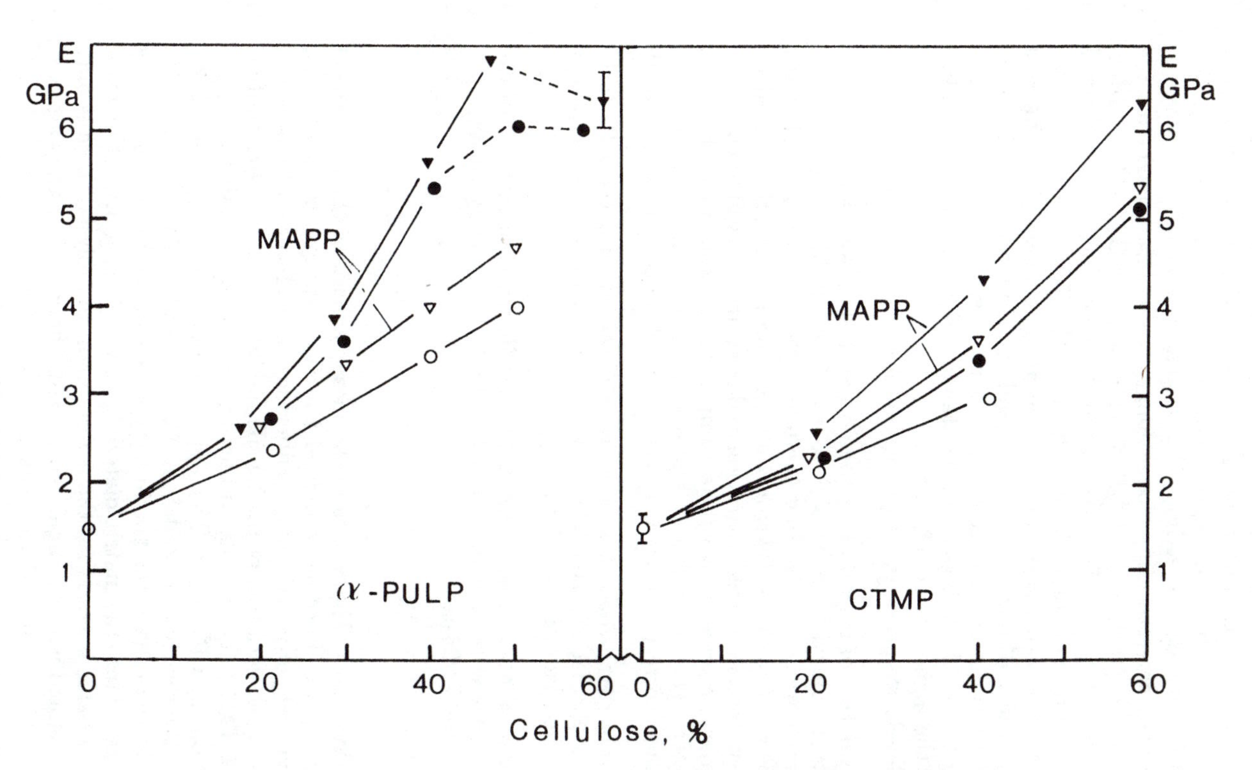

Figure 1. E-modulus vs. fiber content in PP for dissolving pulp fibers (left) and CTMP fibers (right). Filled symbols refer to hydrolyzed fibers; open symbols are unhydrolyzed fibers. MAPP denotes fibers precompounded with MAPP (coupling agent).

fibers it was necessary to use the maximum injection pressure of the machine at only 40% loading. This limit was reached at about 60% by hydrolyzing the fibers. A previous paper (1) reported on the discoloration occurring because of pyrolysis. Discoloration is found for both CTMP and cellulose fibers, and the degree of discoloration increases with filling content. Cellulose fibers yield a light yellow color and CTMP gives a gray color to the injection-molded test bars.

An important feature of the moulded samples is the degree of dispersion of the fibers in the matrix and the overall homogeneity of the composite structure. The prehydrolytic treatment produced a highly significant improvement in this respect. With regard to the compounding technique, precompounding of the fibers with the MAPP coupling agent prior to the addition of the PP matrix material gave a higher degree of dispersion and better homogeneity than one-step compounding of all three components.

With unhydrolyzed α-pulp a significant improvement of the composite modulus (stiffness) was recorded after precompounding with MAPP, indicating a better adhesion between the fibers and PP. Untreated α-fibers were difficult to disperse in the unpolar matrix. The relatively low polarity of CTMP appears to explain the observation that they were somewhat more easily dispersed in the matrix.

Figure 2 shows the influence of the fiber content on the tensile stress at yield, σ_y, for the two pulps. Also included is the effect of the addition of MAPP. As can be seen, the coupling agent produces a remarkable improvement in σ_y with increasing fiber loading, especially for the composites containing α-cellulose. In our earlier work, we studied the effect of the amount of added MAPP on mechanical properties (13), and an addition of 5% was found to give optimum strengthening. This effect was of about the same magnitude for both untreated and prehydrolyzed fibers. With regard to the variation of σ_y with the fiber content, hydrolyzed CTMP without the coupling agent produced results inferior to those obtained with untreated pulp. The marked improvement in modulus resulting from the hydrolytic treatment as such (no MAPP added), as shown in Figure 1, is thus not duplicated by the behavior of the corresponding σ_y data.

The elongation at yield, ε_y, and the impact strength of the α-pulp and CTMP-pulp, are reproduced in Figures 3 and 4, respectively. As can be expected for filled polymers in general, both these properties are reduced as the filler content increases. This is true particularly for the impact strength. Hydrolysis does not appear to have any influence on the behavior of these parameters. On the other hand, a certain beneficial effect of the coupling agent is recorded. This confirms earlier data relating to an improvement of ductility due to MAPP (1).

As evident from Figures 3 and 4, unhydrolyzed fibers produce composites with very low impact strength levels. This applies to both fiber types, even at relatively low fiber contents (20%). The reason appears to be the uneven dispersion of the fibers and their fragments in the matrix. The impact strength is significantly improved by hydrolysis and MAPP addition, the α pulp performing slightly better than CTMP in this respect.

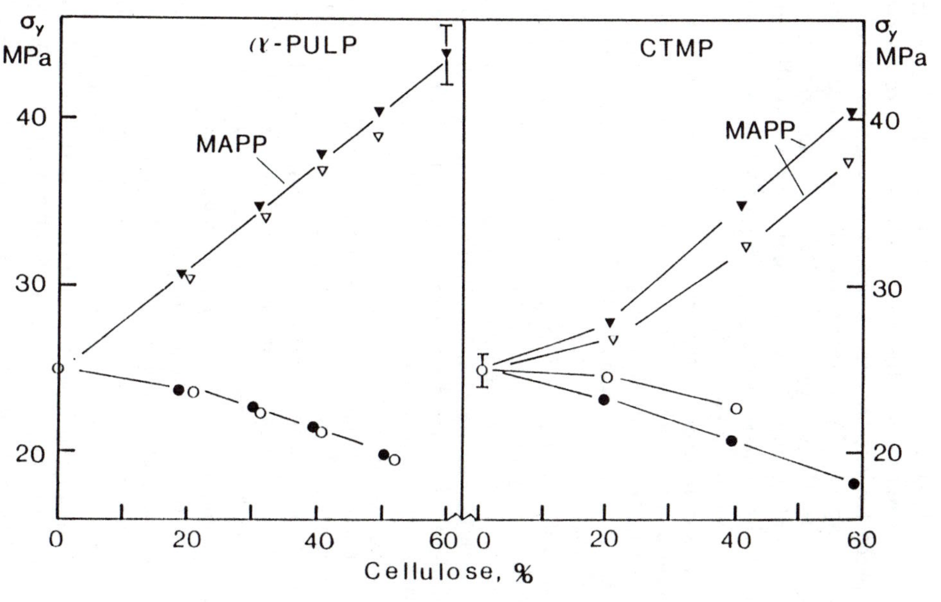

Figure 2. Tensile strength vs. fiber loading in PP. Filled symbols refer to hydrolyzed pulps and open symbols refer to unhydrolyzed pulps. MAPP marks coupling agent-treated fibers.

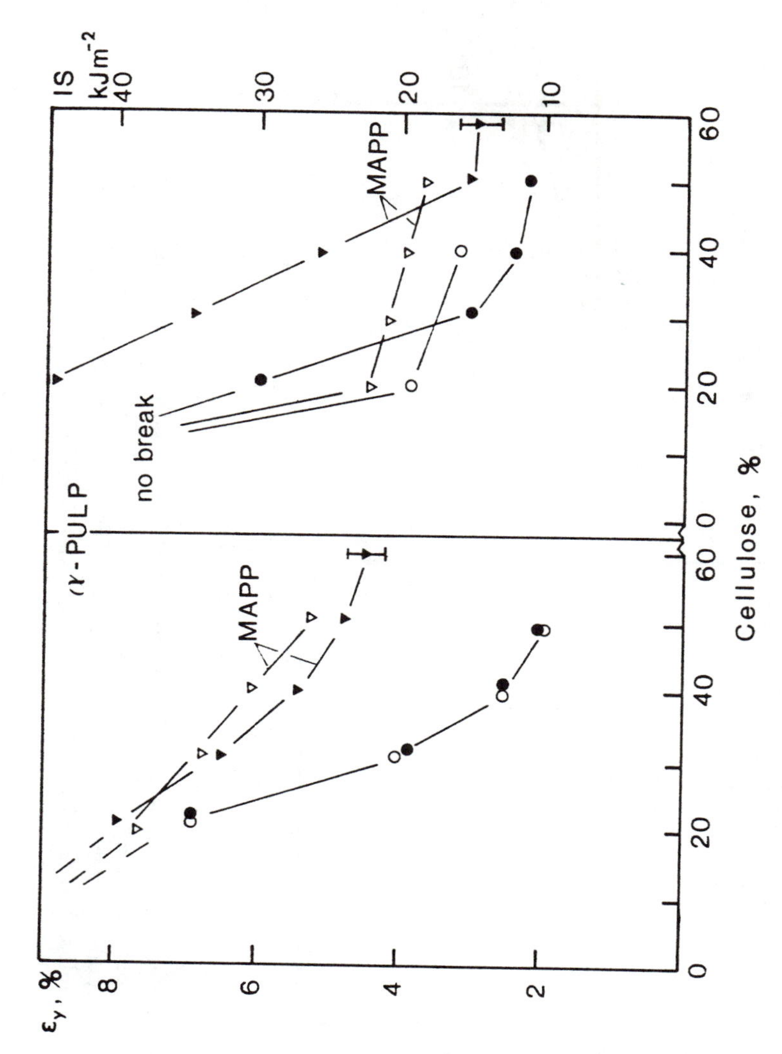

Figure 3. Variation of elongation at yield (left) and impact strength (unnotched, Charpy) (right) with the cellulose fiber content (dissolving pulp) in PP. Open symbols – unhydrolyzed pulp; filled symbols – hydrolyzed pulp. MAPP – fibers treated with coupling agent. Unfilled PP: 10% elongation at yield, and no break indication in impact strength.

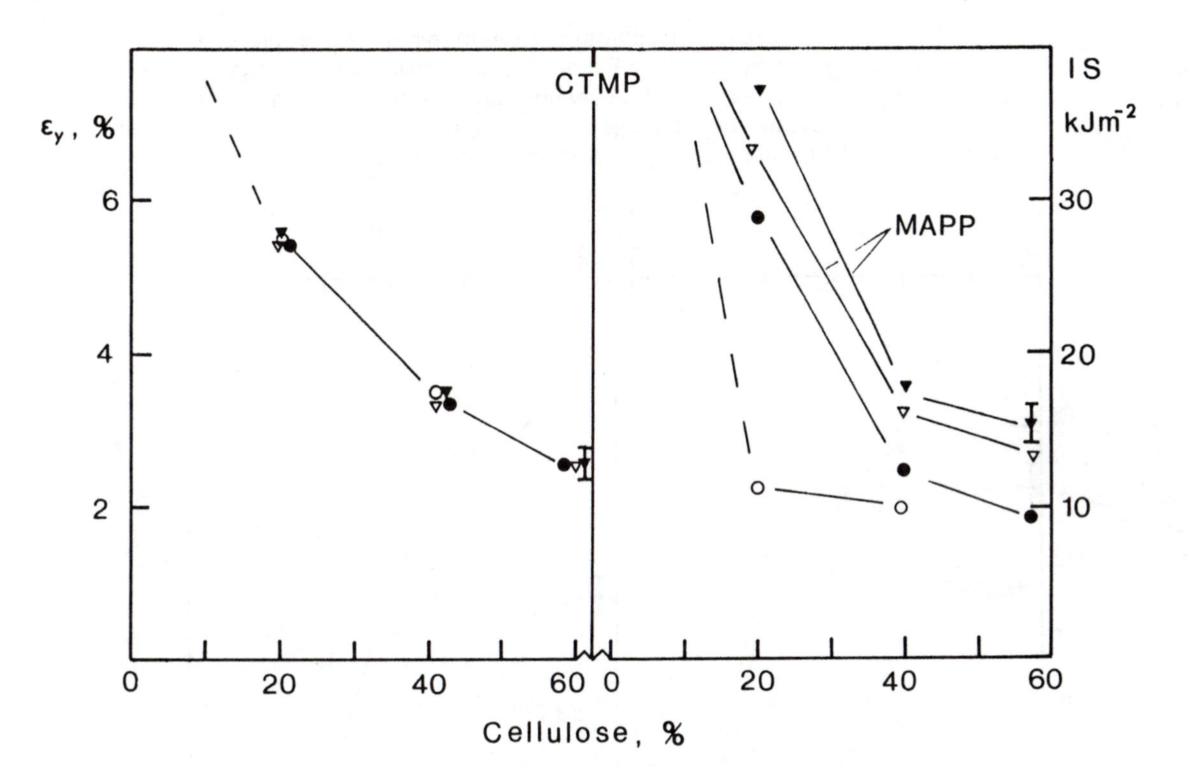

Figure 4. Variation of elongation at yield (left) and impact strength (unnotched, Charpy) (right) with the loading of CTMP fibers in PP. Open symbols refer to unhydrolyzed pulps; filled symbols refer to hydrolyzed pulps. MAPP – fibers treated with coupling agent.

The changes in strength at rupture and the corresponding elongation with the fiber content and MAPP addition were largely similar to those of σ_y and ε_y. The increase in ductility produced by MAPP appears to be caused by improved wettability of the fibers by the matrix material. This is confirmed by an investigation of the fracture surfaces (at room temperature) of samples with and without MAPP (unhydrolyzed fibers, 20%), using SEM. Figure 5 provides evidence of the different degrees of adhesion in composites without a coupling agent (Figure 5a) and with the addition of MAPP (Figure 5b). Samples containing CTMP behaved similarly, as illustrated in Figure 6, which shows SEM pictures of fractured samples with 40% CTMP (untreated, hydrolyzed and hydrolyzed with MAPP). A fair degree of wetting is found only for the MAPP-treated material.

Coupling Reactions. The FTIR spectra of the MAPP coupling agent in the as-received form and after heat treatment (180°C) are shown in Figure 7. The peak at 1717 cm^{-1} is characteristic of the dimeric form of a dicarboxylic acid, indicating the presence of two carboxyl groups in the as-received substance. When heated above 160°C, the two carboxyl groups are converted into the more reactive anhydride ring, and were then able to react with cellulose during compounding at 180°C. Details of the reaction mechanism have recently been reported (14). The results obtained with FTIR explain our observations that the compounding temperature has a crucial effect in adhesion promotion. The mechanical properties were not improved when the precompounding step was carried out at temperatures below 180°C.

Following the pre-compounding step, the cellulose fiber surfaces were further characterized using the ESCA method. Figure 8 shows the high resolution ESCA spectra of untreated α-cellulose. The shapes of the carbon (1s) and oxygen (1s) peaks and the ratio of area under them are related to the composition of the material. As indicated in the lower part of Figure 9, the spectra of cellulose pulp are changed after precompounding with MAPP. The ratio between the oxygen and carbon peaks is reduced and the shape of the carbon peak is changed. While C–O bonds were dominant in the untreated sample, C–C functionalities are preponderant after MAPP treatment, thus verifying that the coupling agent has reacted with the cellulose surface. This explains the excellent contact between the fibers and the matrix in MAPP-treated composites, as seen directly in the SEM micrographs (Figures 5 and 6) and as reflected in the improvement in mechanical properties of the composites.

Modelling of the E-modulus. The modulus is one of the properties of a composite containing a fiber phase which can be calculated relatively easily using the Halpin-Tsai model or its simplified form proposed by Lewis-Nielsen (15):

$$\frac{E_c}{E_1} = \frac{1 + ABv_2}{1 - Bpv_2}$$

Here, E_c and E_1 denote the elastic moduli of the composite and the matrix, respectively, and v_2 is the volume fraction of the filler. The factor

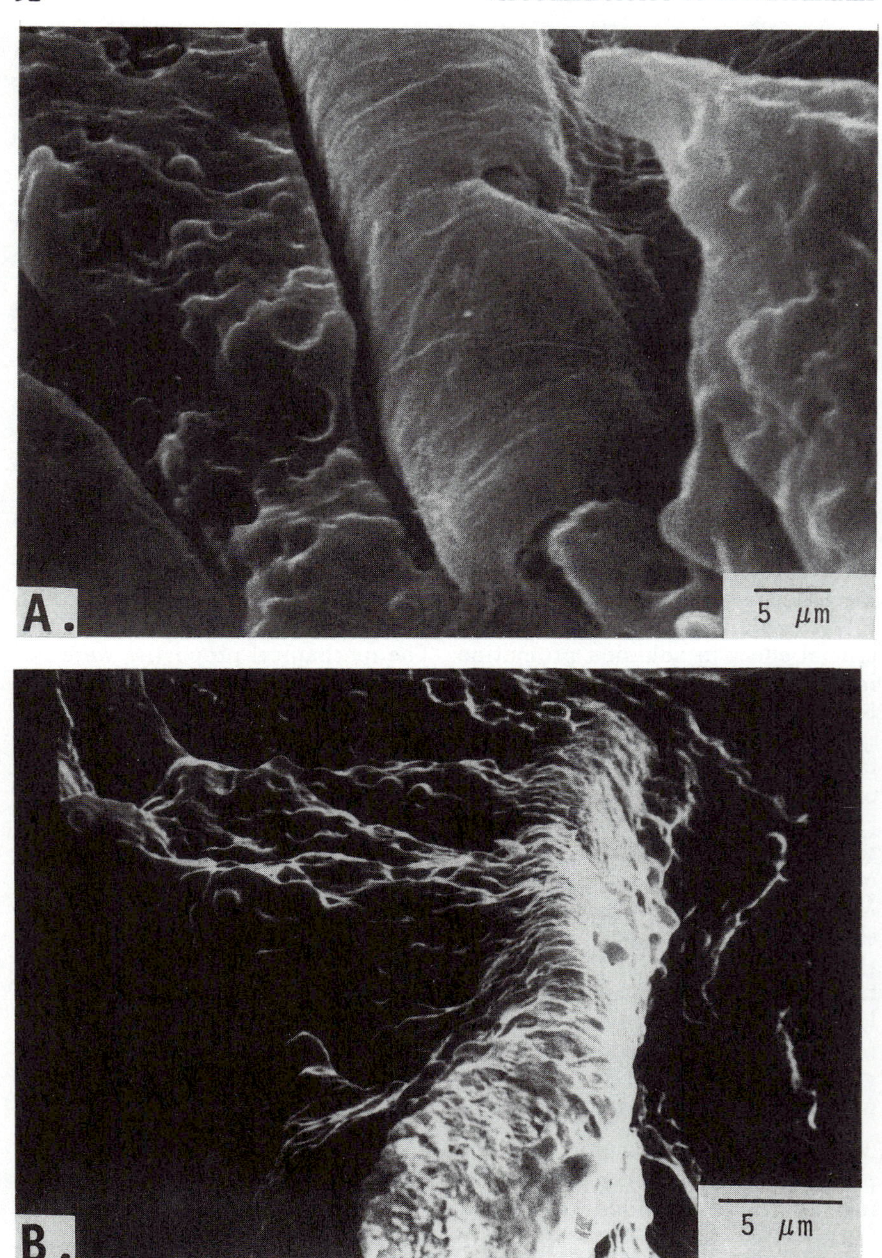

Figure 5. SEM micrograph of fracture surface of dissolving pulp fibers in PP. Fiber loading 20%. A – untreated fibers (magnification 2000); B –MAPP-treated fibers (magnification 3000).

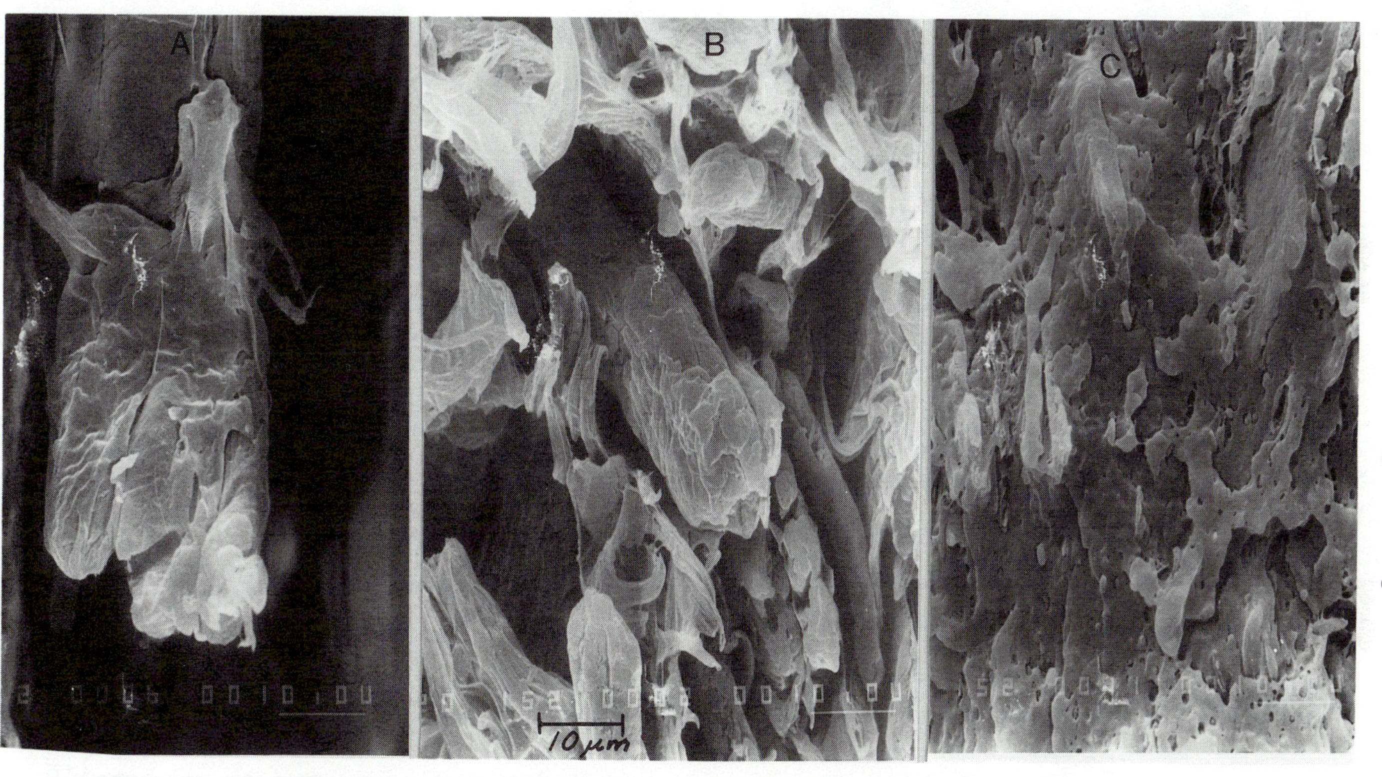

Figure 6. SEM micrograph of fracture surface of CTMP fibers in PP at 40% loading level. A – untreated fibers; B – untreated, hydrolyzed fibers; C – MAPP-treated, hydrolyzed fibers. Magnification is 1500 X-70.

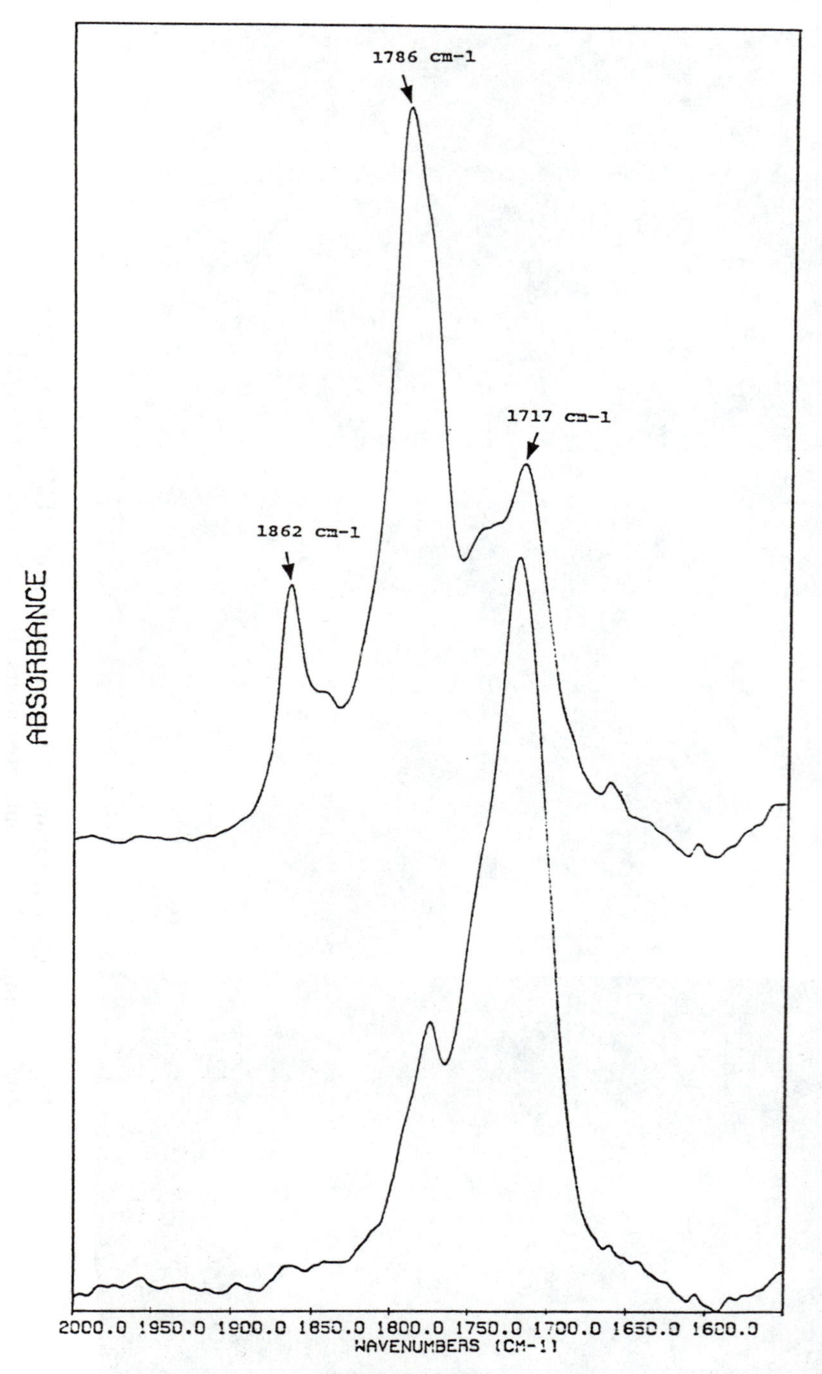

Figure 7. FTIR spectra of the MAPP coupling agent as received (lower spectrum) and after heat treatment at 180°C (upper spectrum).

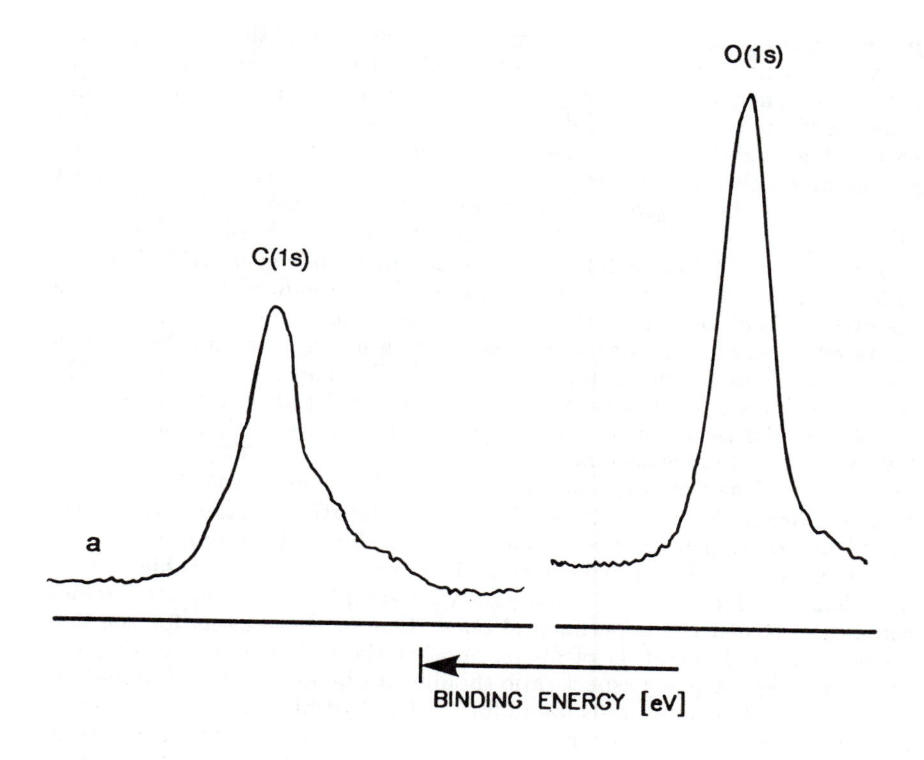

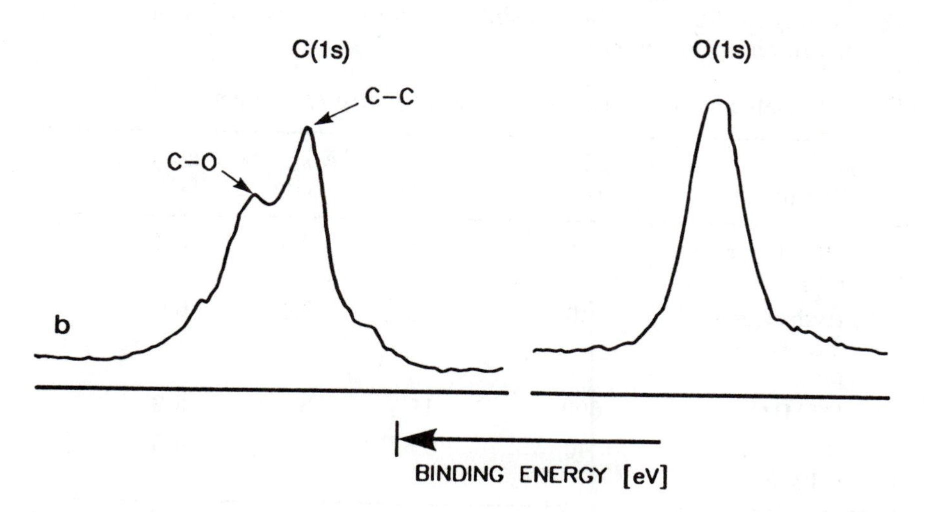

Figure 8. High resolution ESCA spectrum of the (a) untreated (upper part) and (b) MAPP-treated α-pulp (lower part).

p takes into account the maximum packing fraction of the filler, v_m, and is expressed as $p = 1 + (1 - v_m)v_2/v_m^2$. For close hexagonal packing, v_m is 0.74 for spheres and 0.9 for parallel hexagonal packing of fibers. For the cellulose fibers used here it is difficult to estimate a correct value of v_m due to the deformability of the fibers during injection moulding. The factor p is assumed therefore to be equal to unity: $A = 2$ (L/D) for fibers with aspect ratio L/D (length/diameter). Finally, the parameter B is given by $B = (E_2/E_1 - 1)/(E_2/E_1 + A)$. E_1 is 1.5 GPa (see Figure 1), and the modulus of the cellulose fibers, E_2, is assumed to be 40 GPa (16). In order to determine the L/D ratio, parts of the moulded test bars were dissolved in hot o-dichlorobenzene. The solutions were filtered and the fibers retained on the filter were washed with hot xylene and dried. The size of the fibers was determined microscopically. The average fiber length, which was independent of the fiber loading, was 100 (40) μm for the α-pulp fibers and 600 (40) μm for the CTMP fibers (values within parentheses for HC-pulps); the corresponding aspect ratios were 6.5 (3.7) and 15 (2.5), respectively. The fiber degradation for both natural fibers and HC-fibers is thus significant. For the hydrolyzed fibers, this length reduction is desirable, the ideal situation being the conversion of all fibers into microfibrils.

The calculated and experimental E values are given in Table 1 for a fiber loading of 40%. For the unhydrolyzed samples, the calculated values somewhat exceed the experimental ones. One of the reasons for the discrepancy is agglomeration of fibers, another the deformability during the moulding step. A lower aspect ratio should thus be used in the Halpin-Tsai calculations. The opposite is found for the two hydrolyzed pulps, i.e., there is a large discrepancy between calculated and experimental data, the latter being substantially higher. This is indirect evidence that at least a part of the hydrolyzed fibers has been disintegrated into microfibrils which cannot be seen in an optical microscope.

Table 1. Calculated E-values in comparison with experimental data

Sample	Fiber Length μm	Fiber L/D	E_{theory} GPa	$E_{exp.}$ GPa
Dissolving pulp	100	6.5	4.3	3.4
Hydrolyzed dissolving pulp	40	3.5	3.1	6.0
CTMP	600	15	6.6	3.2
Hydrolyzed CTMP	40	2.5	2.6	4.5

Final Remarks

The chief result of this report is the demonstration that a hydrolytic treatment of the cellulose fiber and the resulting embrittlement produce a significant improvement in the mechanical characteristics of composites containing such fibers as reinforcement. The results reported in this study for a PP matrix confirm earlier data for other polymers (1, 11). The modulus levels obtained with such composites fall considerably above those found with carefully compounded, untreated fibers. Another improvement compared with the latter fiber type is a significantly higher degree of homogeneity. Although the main part of the prehydrolyzed fibers has been broken down into irregular fragments, there are signs of a certain degree of degradation of the cellulosic component to high-modulus microfibrils. Halpin-Tsai data appear to support such a notion.

The mechanical breakdown of cellulose fibers is an important factor determining the mechanical parameters of the composite material. Prehydrolyzed and untreated fibers appear to behave differently in this respect. While the fiber length reduction has a detrimental effect on the mechanical parameters of composites containing untreated fibers, an improvement is noted in the case of fibers which have been embrittled before being incorporated into the matrix. Apparently the degree of homogeneity and dispersion plays an important role in this connection. Our work has clearly demonstrated that untreated fibers are noticeably difficult to disperse in a thermoplastic matrix, despite a certain degree of mechanical length reduction. Because of an unusually high embrittlement encountered with prehydrolyzed fibers, the dispersion is rendered much less difficult, the ultimate result being a significant improvement of mechanical parameters. In addition, the apparent production of high modulus microfibrils seems to contribute to the improved property profile. The easier processing of materials based on embrittled fibers, where fiber contents up to about 60% can be handled without difficulty, is also of practical importance. The corresponding figure for untreated fibers is less than 40%.

Compared with the modulus, the strength of a composite is crucially dependent upon the adhesion between the phases, high strength implying a high degree of wetting of the fibers by the matrix material. The results presented above demonstrate that a pretreatment of the fibers with MAPP is a highly suitable means of fulfilling requirements of this type. A direct chemical bonding between the coupling agent and the cellulose surface was found for both the CTMP fibres and the α-pulp. This result is in agreement with previous data relating to wood flour/PP composites modified with MAPP (7).

In summary, we find that prehydrolyzed cellulose modified by suitable coupling agents shows several attractive features when used as a reinforcing phase in composites based on thermoplastic matrices. Among the advantages of such fibers are a highly improved processability, higher than expected modulus values, a strength which increases with fiber content, and, finally, an outstanding homogeneity and surface finish.

Acknowledgments

The authors wish to express their thanks to the National Swedish Board for Technical Development for financial support for this project. Thanks also go to Mr. J. Felix and Mr. A. Mathiasson for experimental assistance.

Literature Cited

1. Klason, C., *et al. J. Polymeric Mater.* 1984, **10**, 159; and 1987, **11**, 229.
2. Battista, O. A. In *Microcrystal Polymer Science;* McGraw-Hill: New York, 1975; p 1.
3. Xanthos, M. *Plastics Rubber Proc. Appl.* 1983, **3**, 223.
4. Michell, A. J.; Vaughan, J. E.; Willis, D. *J. Polymer Sci. Symp.* 1976, **55**, 143.
5. Maldas, D.; Kokta, B. B.; Daneault, C. *Intern. J. Polymeric Mater.* 1989, **12**, 297.
6. Raj, R. G.; Kokta, B. V. In *Wood Processing and Utilization;* Kennedy, J. F.; Phillips, G. O.; Williams, P. A., Eds.; Ellis Horwood Ltd.: Chichester, 1989; p 251.
7. Kishi, H.; Yoshioka, M.; Yamanoi, A.; Shiraishi, N. *Mokuzai Gakkaishi* 1988, **34**, 133.
8. Kokta, B. V.; Raj, R. G.; Daneault, C. *Polymer-Plastics Tech. Eng.* 1989, **28**, 247.
9. Beshay, A. D.; Kokta, B. V.; Daneault, C. *Polymer Composites* 1985, **6**, 261.
10. Kokta, B. V.; Daneault, C. *Polymer Composites* 1986, **7**, 337.
11. Klason, C.; Kubát, J. In *Polymer Composites;* Sedlácek, B., Ed.; Walter de Gruyter and Co.: Berlin, 1986; p 153.
12. Klason, C.; Kubát, J.; Gatenholm, P. In *Cellulosics Utilization;* Inagaki, H.; Phillips, G. O., Eds.; Elsevier Appl. Sci.: London, 1989; p 87.
13. Dalväg, H.; Klason, C.; Strömvall, H.-E. *Intern. J. Polymeric Mater.* 1985, **11**, 9.
14. Felix, J.; Gatenholm, P. *J. Appl. Polym. Sci.* 1991, **42**, 609.
15. Lewis, T. B.; Nielsen, L. E. *J. Appl. Polym. Sci.* 1970, **14**, 1449.
16. Nissan, A. H.; Hunger, G. K.; Sternstein, S. S. *Encyc. Polym. Sci. Techn., Vol. 3;* Bikales, N. M., Ed.; Interscience: New York, 1965; pp 133-135.

RECEIVED February 10, 1992

Chapter 7

Thermomechanical Properties of Polyethylene–Wood Fiber Composites

R. G. Raj and B. V. Kokta

Centre de Recherche en Pâtes et Papiers, Université du Québec à Trois-Rivières, C.P. 500, Trois-Rivières, Québec G9A 5H7, Canada

The effects of different chemical fiber treatments (CTMP aspen) on mechanical properties of HDPE-wood fiber composites were studied. Composites of HDPE filled with pretreated wood fibers (silane, polyisocyanate, epolene) produced higher tensile strength and modulus compared to untreated wood fiber composites. Elongation and Izod-impact strength decreased with the increase in filler concentration and were less influenced by fiber treatment. DSC was used to characterize the matrix-filler interaction. Some of the factors which influence the interaction at the filler-matrix interface were examined.

The increasing demand coupled with the rising cost of common thermoplastics have encouraged the use of low-cost fillers, as extenders or reinforcements, in plastics. Different inorganic fillers such as glass fiber, mica, talc, clay, etc., are more commonly used in thermoplastic and thermosetting resins to achieve economy as well as to enhance the physical and mechanical properties of the final product (1, 2). In recent years, cellulosic fillers have attracted greater attention because they are relatively cheap and biodegradable. So far the use of wood fibers in thermoplastic polymers is limited, mainly because of (i) poor compatibility between hydrophobic polymer and the hydrophilic filler; (ii) thermal degradation of wood fibers at higher processing temperature; and (iii) the increased moisture absorption of wood fibers.

A necessary condition for the effective performance of wood fibers in thermoplastics seems to be the nature of adhesion between the fiber and polymer matrix. Many attempts have been made to improve the polymer-filler adhesion by surface modification of cellulose fibers (3-7) or the polymer matrix (8-10). The surface characteristic of the reinforcing fiber is an important factor because of its role in the effective transfer of stress from

0097–6156/92/0489–0099$06.00/0

the matrix to fiber. Chemically, the hydroxyl-rich surface of lignocellulosic material is advantageous because it provides the potential for reaction with different coupling agents, e.g., titanate, isocyanate, and maleic anhydride (11-14).

The object of this study is to improve the polymer-filler interaction, by modification of the filler surface, to effect an improvement in the strength of the finished product. In the present work, the wood fiber surface is pretreated with different coupling agents such as polyisocyanate, maleated propylene wax and silanes, and then compounded with the polymer. The effects of filler-matrix interface and the variation in filler concentration on mechanical properties of the composites are examined. DSC is used to characterize the matrix-filler interaction. Some of the factors which influence the interaction at the filler-matrix interface are also examined.

Experimental

High density polyethylene (HDPE 2906) was supplied by DuPont, Canada. The reported properties of HDPE are: melt index, 0.75 dg/min; density, 0.960 g/cc. Chemithermomechanical pulp (CTMP) of aspen was prepared in a Sunds defibrator (15). The average fiber aspect ratio (L/D) was 11.9. Table I shows the list of coupling agents that were used for the pretreatment of wood fiber.

Table I. Coupling agents

Coupling agents
Polymethylenepolyphenyl isocyanate (PMPPIC, Polysciences)
Maleated propylene wax (Epolene E-43, Eastman Kodak)
Silane coupling agents (Union Carbide):
γ – Methacryloxypropyltrimethoxysilane (silane A-174)
γ – Aminopropyltriethoxysilane (silane A-1100)

Pretreatment of Wood Fiber. CTMP aspen fiber was dried at 60°C in an oven for 24 hours. The oven-dried fibers were treated with different coupling agents in a two-roll mill (C. W. Brabender laboratory group, mill No. 065). In a typical coating procedure, the mixing of ingredients, wood fiber (30 g), isocyanate (3%), and polymer HDPE 2906 (5%), were done at room temperature. The above mixture was gradually added to a preheated two-roll mill at 160°C and mixed well for about 10 min. to achieve a better dispersion of coupling agent on the fiber surface. The pretreated fibers were cooled to room temperature and then ground to mesh 20. The procedure for silane pretreatment of wood fiber was described in an earlier publication (16).

Preparation of Composites. Compounding of HDPE and pretreated wood fiber (0-40% by weight) was done at 150-160°C in a two-roll mill. After mixing for 5 minutes, the mixture was ground to mesh 20 and then compression molded in a Carver laboratory press. The molding pressure was

3.2 MPa. After heating the mold for 15 min. at 150°C, the samples were slowly cooled to room temperature with the pressure maintained during the process.

Mechanical Tests. The samples were conditioned for 14 hours at 23°C and 50% RH prior to testing. Instron model 4201 was used to study the tensile properties of the composites. The full-scale load was 500 N and the cross-head speed was 10 mm/min. The test results were automatically calculated by a HP86B computing system using the Instron 2412005 General Tensile Test Program. An average of six specimens were tested in each case. The coefficient of variations of the reported properties were: tensile strength, 3.4-6.7%; elongation, 2.8-7.1%; tensile modulus, 2.1-5.8%; and Izod-impact strength, 3.9-8.3%.

Results and Discussion

Surface treatment is of prime importance in the use of wood fiber as a filler/reinforcing agent for thermoplastics. Good bonding of the matrix to the fiber is essential in order to utilize the full strength of the fiber in a composite. Different fiber surface treatments were carried out to improve the bonding at the fiber-matrix interface.

PMPPIC or Epolene Treated Fiber. The effect of fiber treatment on mechanical properties, tensile strength, elongation and tensile modulus of the composites were studied at different concentrations of wood fiber (Figures 1-3). The composites containing untreated wood fibers showed a steady decrease in tensile strength with an increase in filler concentration (Figure 1). The PMPPIC or epolene-treated fibers produced higher tensile strength, at 10 and 20% filler concentrations, than that of untreated fiber composites. However, at higher filler concentrations (30 and 40%), a rapid drop in tensile strength was observed in all the cases. This deterioration in tensile strength can be attributed to a higher filler agglomeration, caused by fiber-to-fiber interaction, resulting in a poor dispersion of fiber in the polymer matrix and a weak filler-matrix interface.

Figure 2 shows that elongation generally decreased with the increasing amount of filler. The treatment of wood fiber with PMPPIC or epolene did not have much effect on elongation of the composites. The decrease in elongation is related to the increase in the stiffness of the matrix as a result of the addition of fillers (12, 14). Tensile modulus increased steadily with the filler concentration as can be seen from Figure 3. Compared to unfilled HDPE (1.1 GPa), a significant rise in modulus was observed at 40% filler concentration in HDPE filled with epolene treated fibers (1.9 GPa). Different fiber surface treatments were carried out to improve the bonding at the fiber-matrix interface.

Silane-Treated Fiber. Tensile properties of HDPE filled with untreated and silane-treated wood fibers are presented in Figures 4-6. Silane treatment of fibers has a positive effect on tensile strength of the composites, as shown in Figure 4. The results show that tensile strength increased steadily with

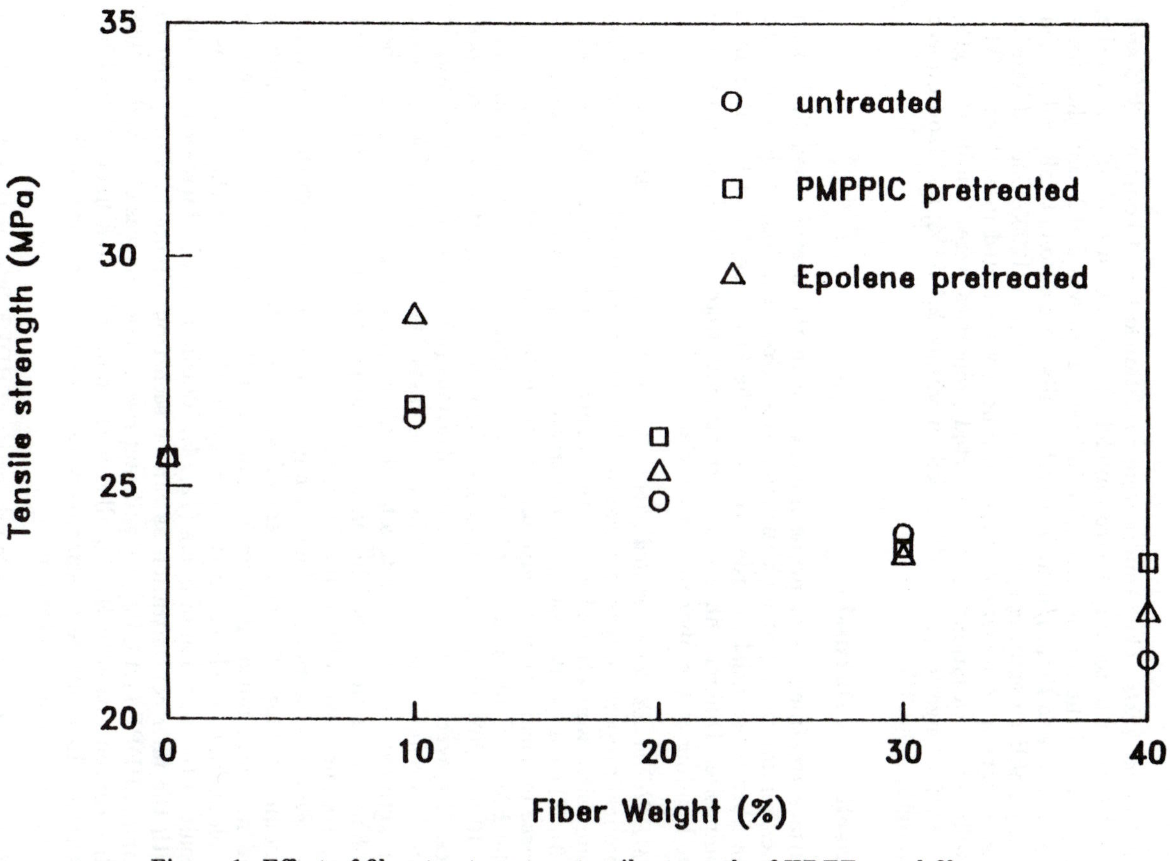

Figure 1. Effect of fiber treatment on tensile strength of HDEE-wood fiber composites.

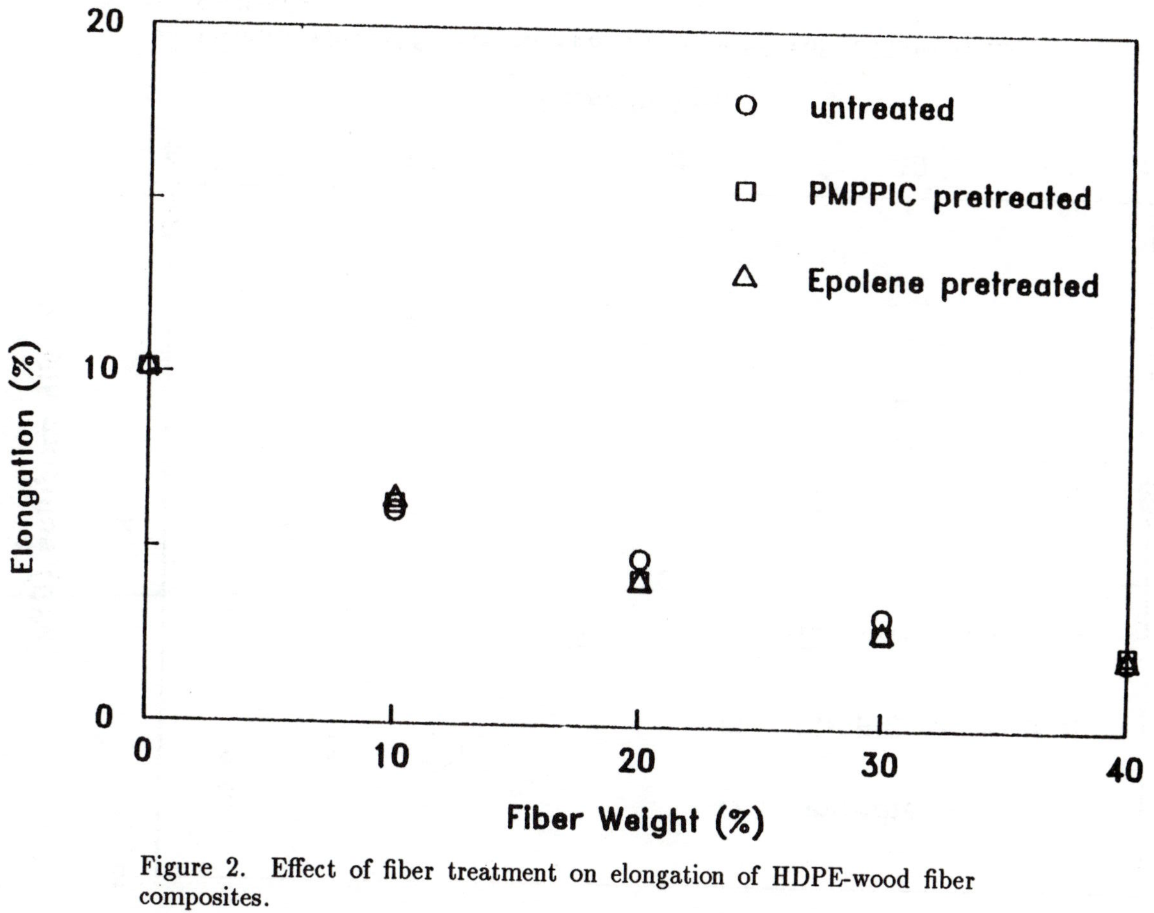

Figure 2. Effect of fiber treatment on elongation of HDPE-wood fiber composites.

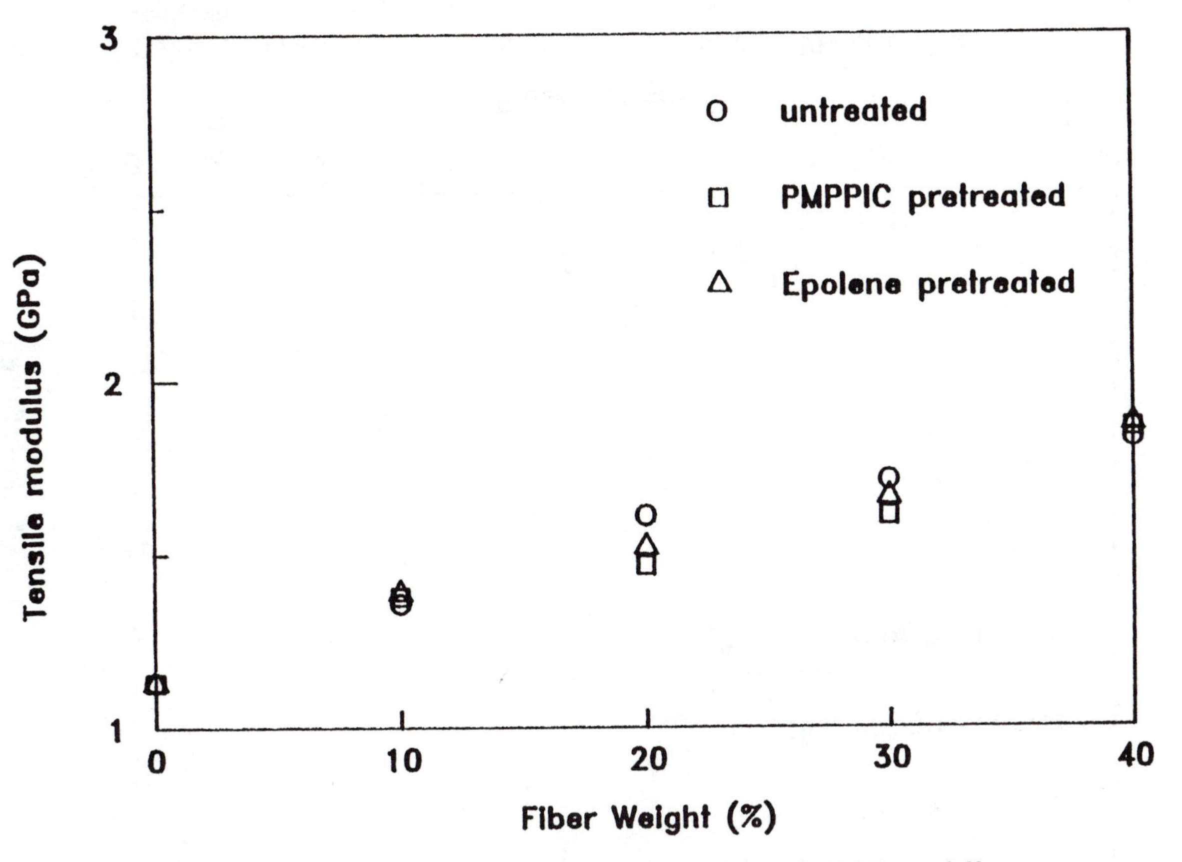

Figure 3. Effect of fiber treatment on tensile modulus of HDPE-wood fiber composites.

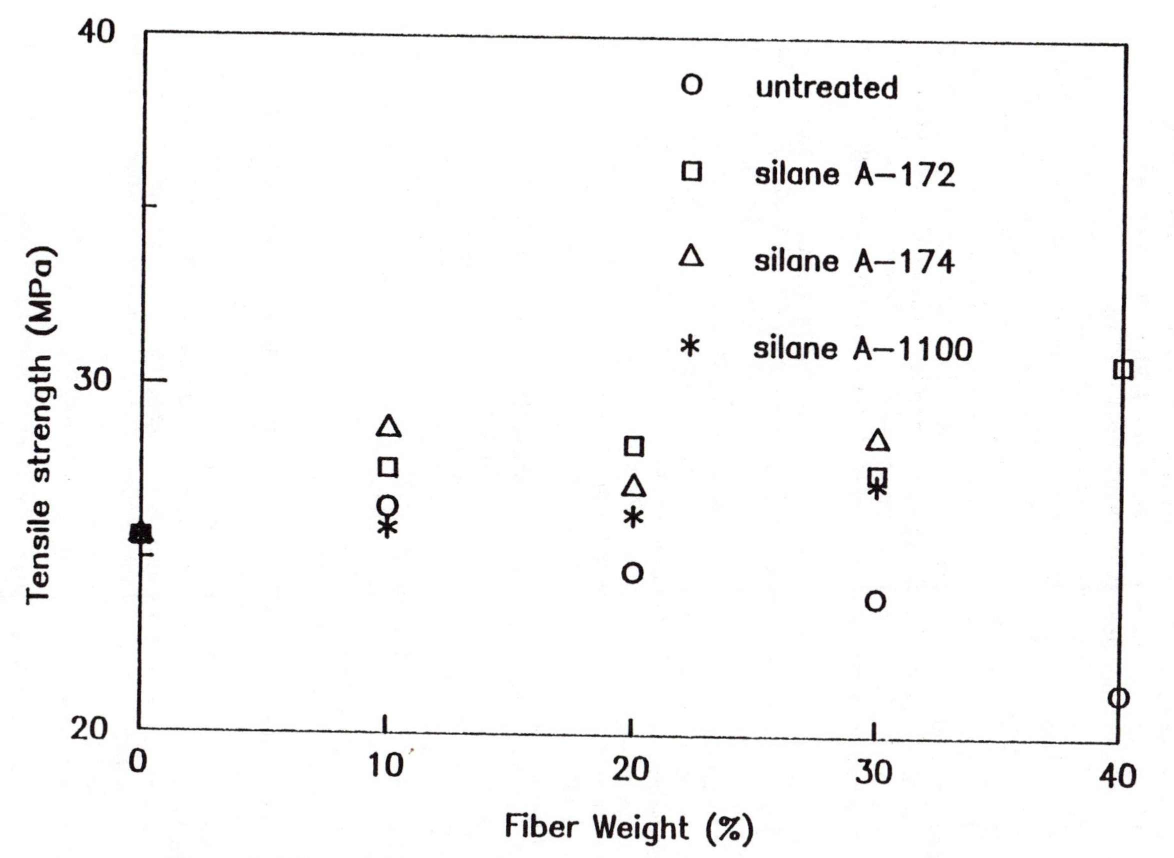

Figure 4. Effect of different silane treatments on tensile strength of HDPE-wood fiber composites.

the filler concentration in silane-treated fibers, which clearly indicates that the silane treatment influences the bonding at the interface. At 40% fiber level, tensile strength of the silane A-172 treated wood fiber composites reached 30.9 MPa (coefficient of variation 5.1%), compared to 21.3 MPa of untreated fiber composites at the same filler loading.

A possible coupling mechanism at the filler-matrix interface can be explained as follows: During fiber pretreatment, the reaction of the coupling agent with the hydroxyl groups of wood fiber produces the hydrolysis product silanol, $-Si(OH)_3$. A coupling agent silanol can be linked to the fiber by the formation of either covalent (siloxane bond) or hydrogen bonds with OH groups of cellulose. Subsequent reaction of the functional organic group of the coupling agent with the polymer completes the establishment of the molecular bridge between the polymer and fiber. Although the above coupling mechanism seems rather simplified, the actual picture is much more complex.

The results also indicate that silane A-174 (methacryl functional group) performed better as a coupling agent than silane A-172 or A-1100. This can be due to the stronger interaction of methacryl groups with the filler. In an earlier study on silane coupling agents, Ishida observed that silanes tend to be ordered in the interphase and the degree of organization depends largely on the organofunctionality of the silane. The orientation and organization of the silane affects the reinforcement mechanism (17). The performance of the silane coupling agent in reinforced composite may depend as much on the chemistry of the organofunctional silane.

Figure 5 shows that the increase in filler concentration has a negative effect on the elongation of the composites. The elongation decreased sharply with an increase in filler content in the polymer. The silane treatment had little influence on elongation of the composites. The effect of fiber treatment on tensile modulus of the composites is presented in Figure 6. The modulus increased from 1.1 GPa (unfilled HDPE) to 2.1 GPa (coefficient of variation 3.8%), at 40% filler loading, in silane A-172 treated fiber composites. The results also show that the fibers pretreated with silane A-172 seem to be more effective in improving the modulus.

Fiber Dispersion. The effect of fiber treatment on dispersion of fiber in the polymer matrix was investigated. Optical micrographs of the fractured surfaces of the samples are shown in Figures 7 and 8. As can be seen from Figure 7, the untreated fiber composite shows a large number of fiber aggregates. Associated with the fiber aggregates is a poor dispersion of fiber in the polymer matrix. While in silane A-172 treated wood fiber composite the number and size of fiber aggregates are greatly reduced. In this case, a better dispersion of fiber in the matrix was observed (Figure 8).

Comparison of Different Fiber Treatments. A comparison of different fiber treatments is shown in Figures 9-11. The silane A-172 treatment had a positive effect on both tensile strength (Figure 9) and modulus (Figure 10) at 30% filler concentration. Clearly, the tensile strength is more influenced by the degree of adhesion at the fiber-matrix interface which is seen from

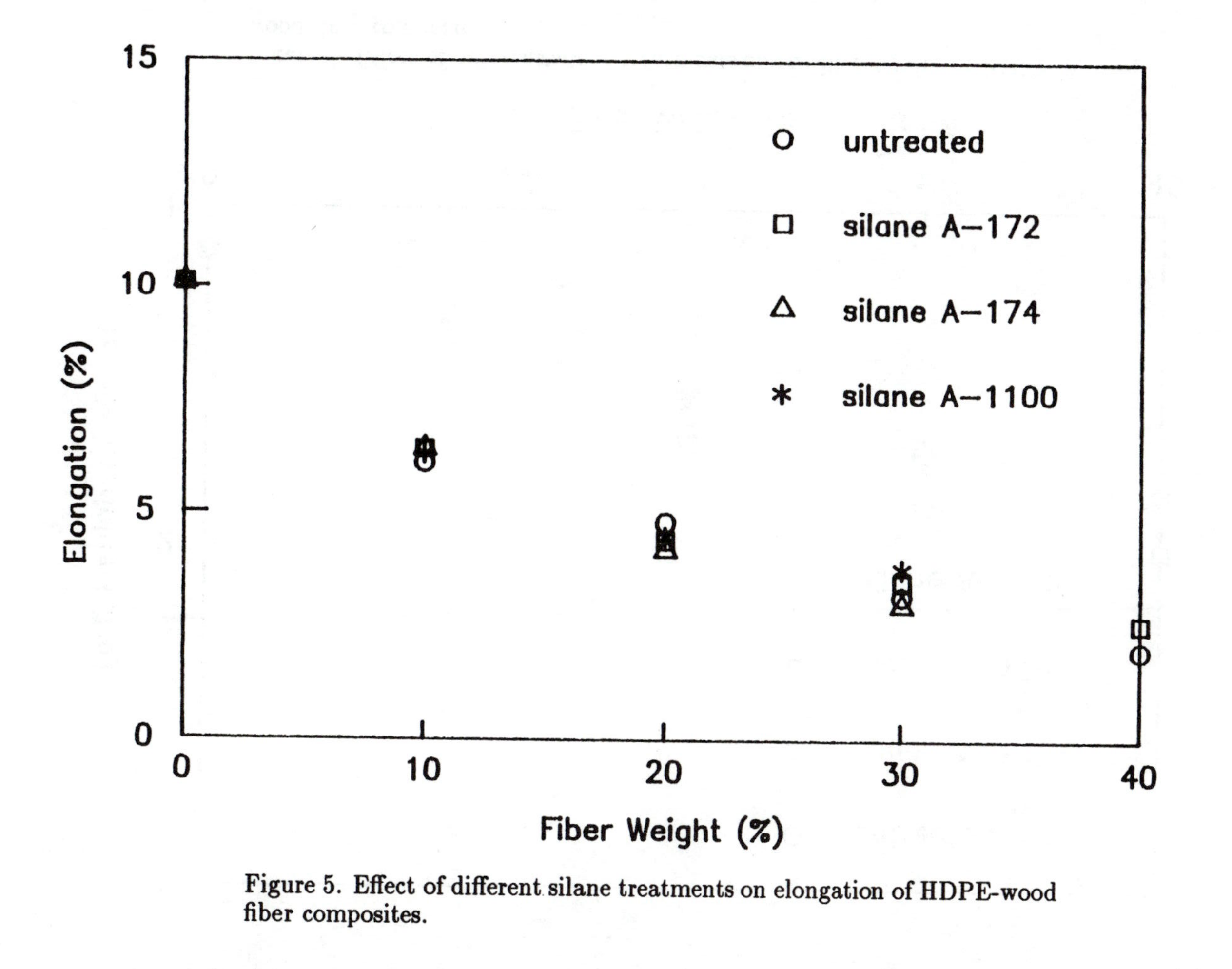

Figure 5. Effect of different silane treatments on elongation of HDPE-wood fiber composites.

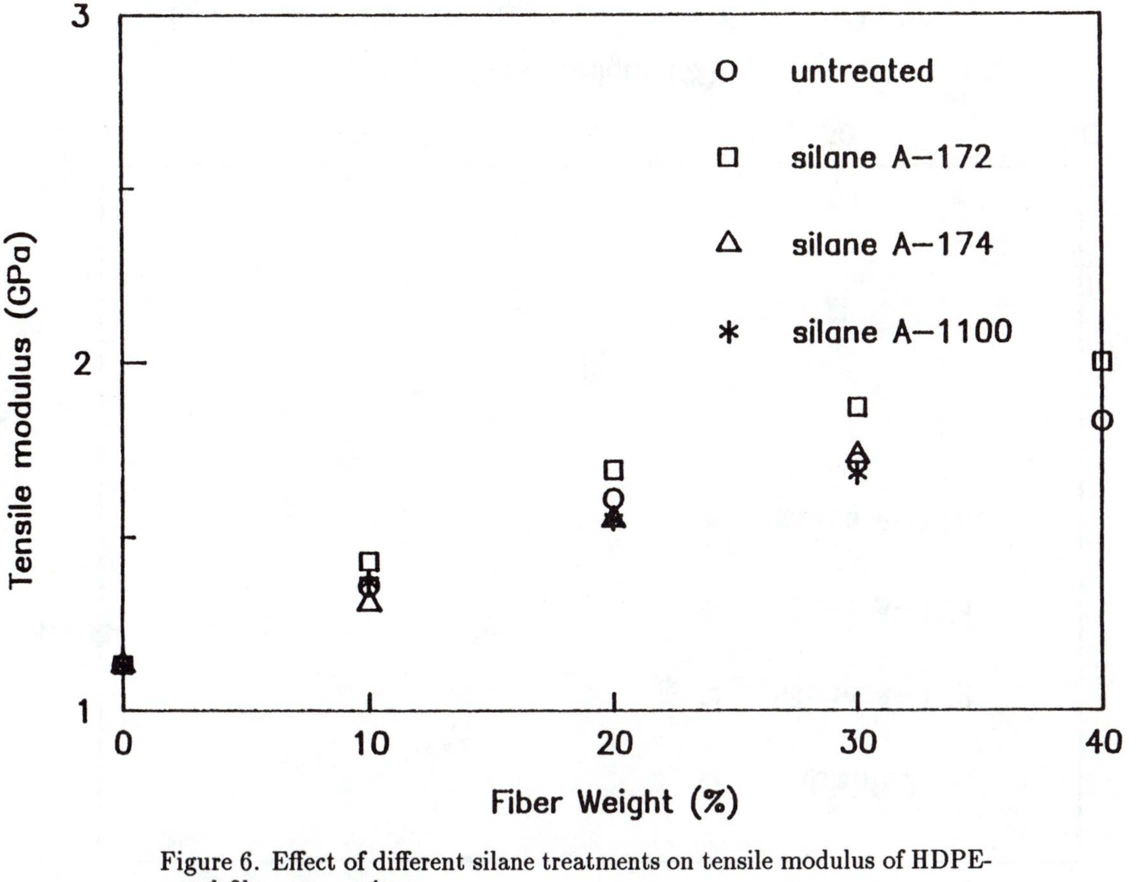

Figure 6. Effect of different silane treatments on tensile modulus of HDPE–wood fiber composites.

Figure 7. Optical micrograph of HDPE-untreated wood fiber composite (30% fiber) 250X.

Figure 8. Optical micrograph of HDPE-silane A-172 treated wood fiber composite (30% fiber) 250X.

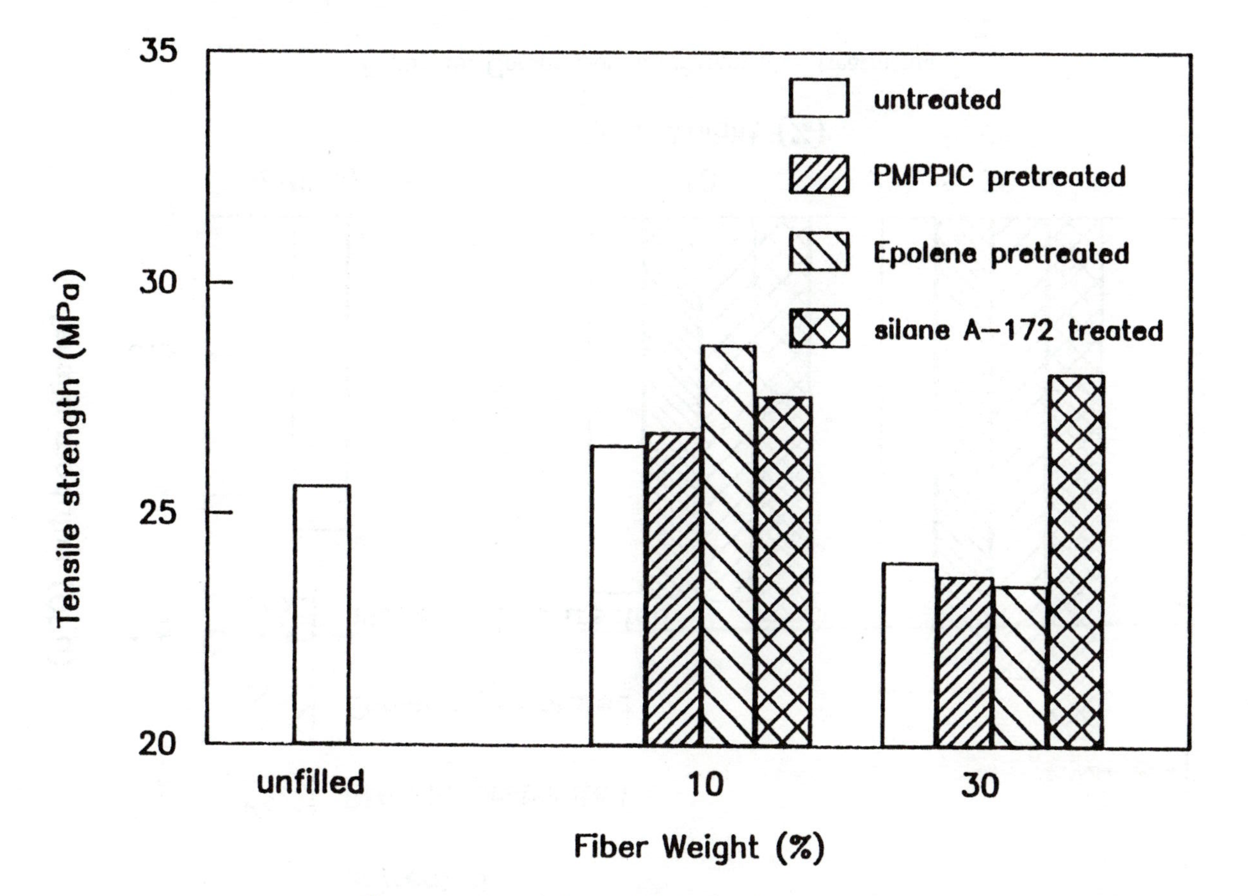

Figure 9. Comparison of different fiber treatments.

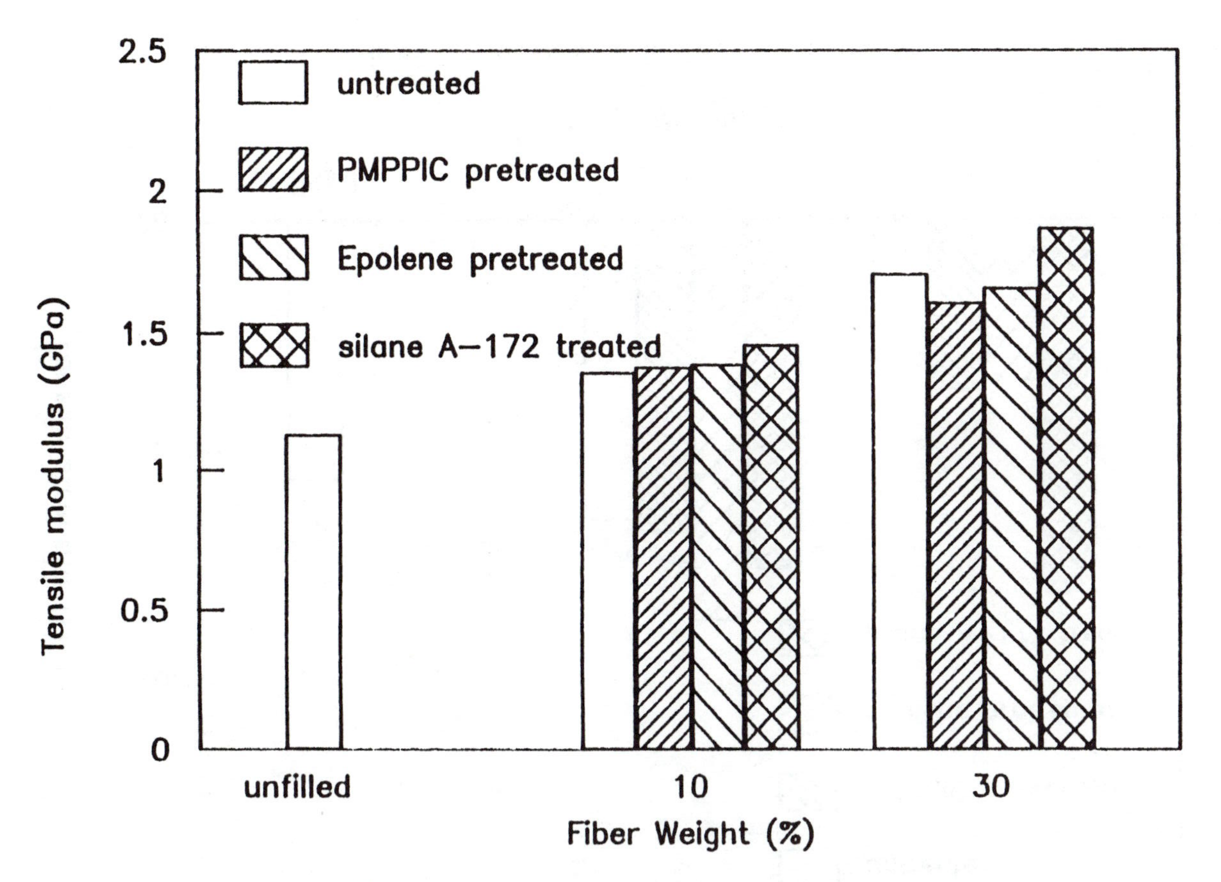

Figure 10. Comparison of different fiber treatments.

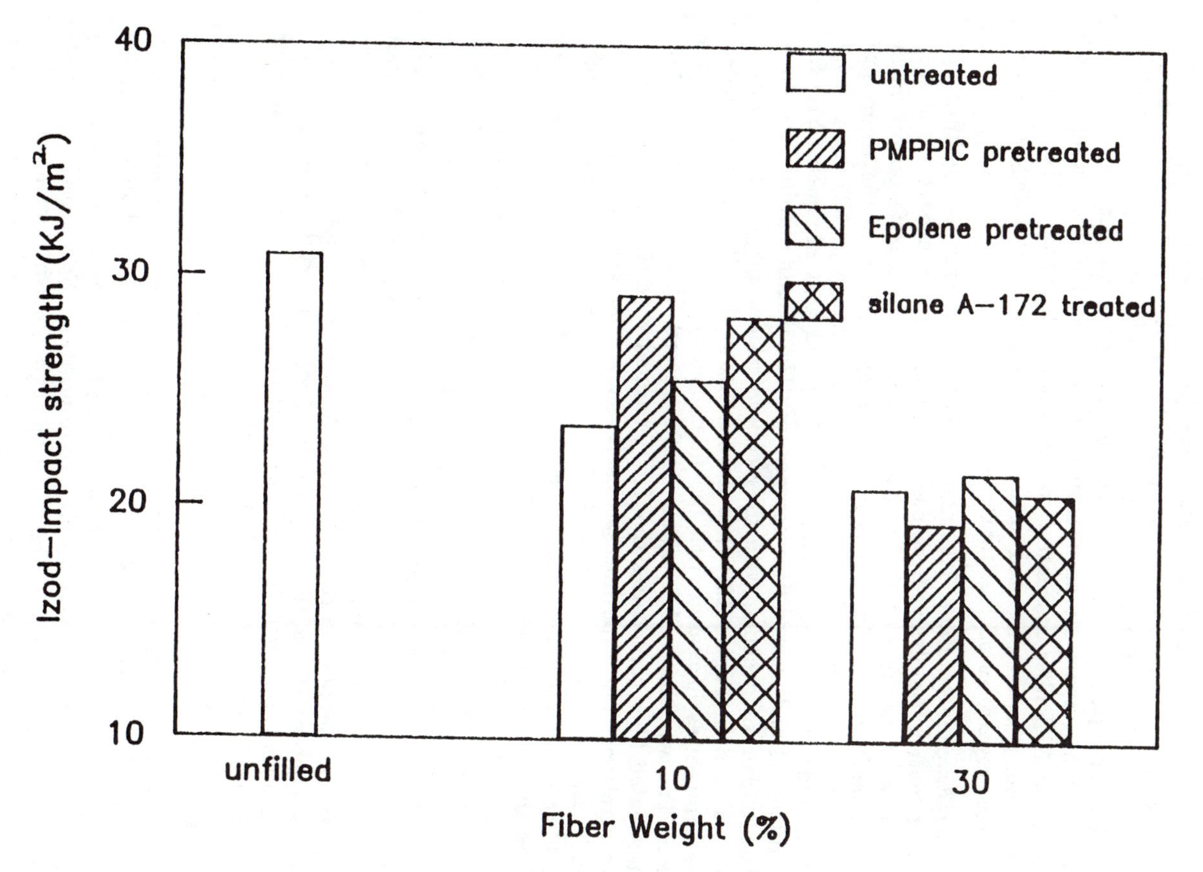

Figure 11. Comparison of different fiber treatments.

the increase in tensile strength (28.1 MPa, coefficient of variation 4.5%) of silane A-172 treated fibers compared to untreated fiber (23.8 MPa) composites. Silane A-172 treated fibers also exhibited a higher tensile modulus, though it is less pronounced compared to tensile strength. Izod-impact strength generally decreased with the increase in filler loading. At 10% filler concentration, epolene and silane-treated wood fiber composites produced slightly higher impact strength compared to untreated fiber composites. However, at 30% filler concentration the impact strengths of the composites were much lower compared to the unfilled polymer, and they remained unchanged regardless of fiber treatment.

Polymer-Filler Interaction. The interface between the reinforcing fiber and polymer matrix has always been considered as a crucial aspect of polymer composites. A number of studies have described the interfacial region in terms of shifts in the glass transition temperature (Tg) of the composite (18, 19). In this study, differential scanning calorimetry (DSC) was utilized to determine the melting temperature (Tm) and heat of fusion (ΔH) for the untreated and treated fiber composites. The results obtained for three different fiber treatments, with filler concentrations ranging from 0-30%, are presented in Figure 12. The data indicate that Tm of the treated fiber composites increased with the filler concentration. The maximum rise in Tm was observed in silane A-172 treated fiber composites, while in untreated fiber composites the Tm decreased with the increase in filler concentration. The experimental results show that the change in Tm is mainly affected by the intensity of the interaction between the polymer and filler at the interface. This is consistent with the higher tensile strength values obtained in silane-treated fiber composites.

The effect of fiber treatment on ΔH is presented in Figure 13. The results indicate that relative to untreated fiber composites, the silane- and PMPPIC-treated fiber composites showed higher ΔH values. At the same time, ΔH generally decreased with the increase in filler concentration. The results suggest that the variation in ΔH is influenced by the fiber surface treatment, the relative strength of the interfacial interaction, and the concentration of the filler in the polymer matrix. However, other factors, such as stress concentrations, the morphology of polymer (crystallization), physical and surface properties of the filler must be taken into consideration for the effective characterization of filler-matrix interface.

Conclusions

The addition of wood fiber in HDPE produced a significant increase in modulus while improvement in tensile strength can be achieved with a suitable pretreatment of wood fiber. Elongation and Izod impact strength generally decreased with the increase in filler concentration. Silane coupling agents performed well in improving the adhesion at the fiber-matrix interface. Optical micrographs showed a better dispersion of fiber in the matrix when silane-treated fibers were used. The characterization of matrix-filler interaction by DSC showed that fiber surface treatment and filler concentration can strongly affect the interaction at the interface.

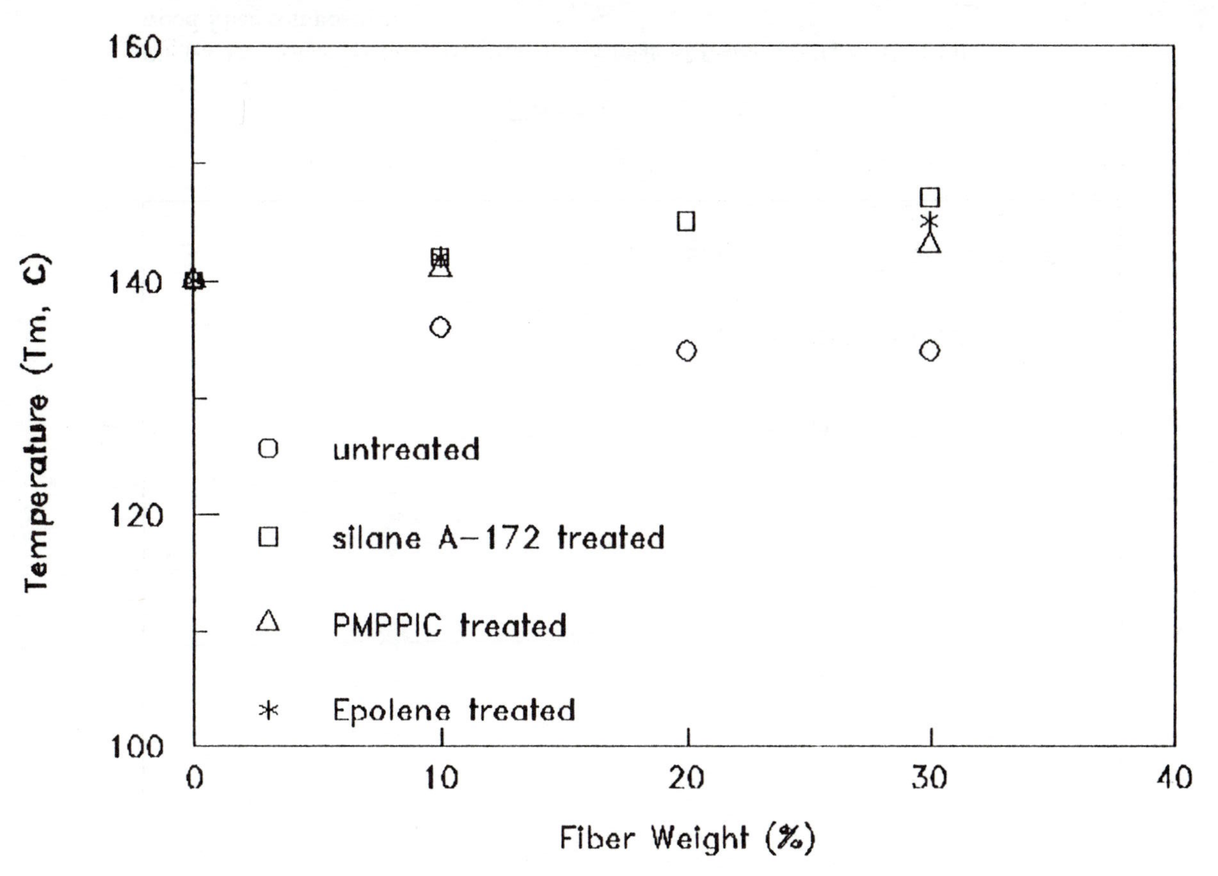

Figure 12. Effect of fiber treatments on melting temperature (Tm) of HDPE-wood fiber composites.

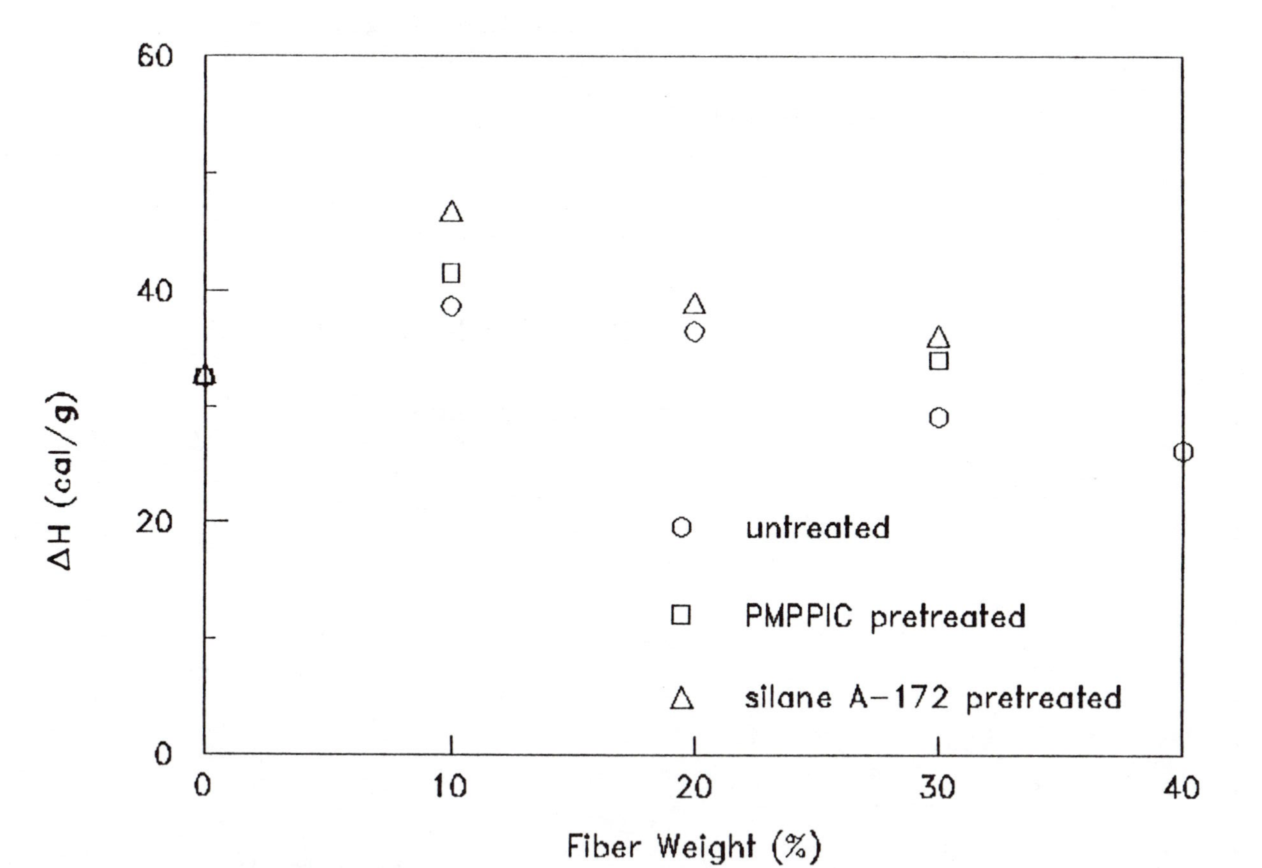

Figure 13. Effect of fiber treatments on heat of fusion (ΔH) of HDPE0-wood fiber composites.

Literature Cited

1. Stedfeld, R. *Mater. Eng.* 1981, **5**, 93-97.
2. Theberge, J. E.; Hohn, K. *Polym. Plast. Technol. Eng.* 1981, **16**, 41-52.
3. Chia, L. H. L.; Kom, K. H. *J. Macromol. Sci. Chem.* 1981, **16**, **4**, 803-809.
4. Kokta, B. V.; Chen, R.; Daneault, C.; Valade, J. L. *Polym. Composites* 1983, **4**, 229-232.
5. Goettler, L. A. U.S. Patent 4 376 144, 1983.
6. Gautts, R. S. P.; Campbell, M. D. *Composite* 1979, **10**, 228.
7. Nakajima, Y. Japanese Patent Kokai 127 632, 1981.
8. Coran, A. Y.; Patel, R. U.S. Patent 4 323 625, 1982.
9. Iwakura, K.; Suto, S.; Fujimura, T. *Kobunshi Ronbun* 1978, **35**, **9**, 595-597.
10. Natow, M.; Wassilewa, S.; Georgijew, W. *Plast. Kautschuk* 1982, **29**, **5**, 227-231.
11. Monte, S. J.; Sugerman, G. *Polym. Plast. Technol. Eng.* 1981, **17**, 95-101.
12. Raj, R. G.; Kokta, B. V.; Daneault, C. *Polymer Composites* 1988, **9**, **6**, 404-411.
13. Morrell, S. H. *Plastics and Rubber Processing and Applications* 1981, **1**, 179-184.
14. Dalvag, H.; Klason, C.; Strömvall, H.-E. *Intern. J. Polymeric Mater.* 1985, **11**, 9-38.
15. Beshay, A. D.; Kokta, B. V.; Daneault, C. *Polymer Composites* 1985, **6**, 261-271.
16. Raj, R. G.; Kokta, B. V.; Daneault, C. *J. Appl. Polym. Sci.* 1989, **37**, 1089-1103.
17. Ishida, H. *Polymer Composites* 1984, **5**, 101-123.
18. Droste, D. H.; DiBenedetto, A. T. *J. Appl. Polym. Sci.* 1969, **13**, 2149-2153.
19. Fritschy, G.; Papirer, E. *J. Appl. Polym. Sci.* 1980, **25**, 1867-1874.

RECEIVED February 10, 1992

Chapter 8

Thermoplasticization of Wood

Esterification

David N.-S. Hon and Lan-ming Xing

Wood Chemistry Laboratory, Department of Forest Resources, Clemson University, Clemson, SC 29634–1003

Esterification of wood was demonstrated by using maleic and succinic anhydrides reacted between 60 and 200°C. At low temperature, only monoester was formed. At temperatures beyond 140°C, a large amount of diester was also formed. Esterified woods exhibited viscoelastic properties. DMTA studies revealed the presence of α, β, γ and δ transitions between -150 and 200°C. Decarboxylation of esterified woods took place between 200 and 300°C, and this was confirmed by DSC and TGA techniques. Wood esterified with succinic anhydride displayed better flow properties than that esterified with maleic anhydride. The higher the diester content, the lower the flow properties.

Interest in the chemical modification of wood and other lignocellulosic materials has been rekindled in the past fifteen years. In an era of uncertainties about oil supplies and prices and rising concern over environmental issues, wood and other lignocellulosic materials have become increasingly attractive to a variety of materials-based industries. Accordingly, academic and industrial sectors are actively involved in this area of research in order to find ways to convert wood and lignocellulosic materials into value-added engineering products (1,2). By the recognition of wood waste as a valuable natural resource, it is hoped that the use of this waste material can at least, in part, alleviate the shortage of raw materials as well as the burden on landfills.

Since wood contains cellulose, hemicelluloses, and lignin, it possesses primary and secondary hydroxyl groups which readily undergo classical etherification and esterification. Depending on the degree of substitution, etherified and esterified woods exhibit thermoplasticity properties which can be molded or extruded into various products (3). In the etherification

0097–6156/92/0489–0118$06.00/0

reaction, benzylated, cyanoethylated, and hydroxyalkylated products have been made (4-6). In the esterification reaction, acetylated and acylated products have also been made (3,7,8). Commercial applications of these products have been successful, particularly in the Japanese market.

Esterification of wood by using monocarboxylic acid has been reported by Shiraishi (7). The use of dicarbocylic acid has been limited due to its low reactivity. Recently, however, a series of esterifications was conducted by Matsuda and his coworkers (9–12). They concluded that dicarboxylic acid anhydrides could be introduced into wood as an intermediate for subsequent modification into new products. Based on Matsuda *et al.*'s work, maleic and succinic anhydrides, in the absence of solvents, were used to prepare esterified wood in this study. The thermal behavior of esterified wood was evaluated by means of a differential scanning calorimeter (DSC) and a thermal gravimetric analyzer (TGA).

Experimental

Materials. Southern yellow pine (*Pinus* spp.) was used as the wood raw material. A ground sample (40 mesh) was successively extracted with ethanol-benzene (1 : 2 v/v), hot distilled water, and was finally air-dried to eliminate extractives. Maleic and succinic anhydrides and anhydrous sodium carbonate used in this study were reagent grade and used without further purification.

Esterification. Two grams of air-dried sawdust of southern yellow pine, 70 g of maleic anhydride, and 0.2 g of anhydrous sodium carbonate were placed into a 500 ml three-necked flask equipped with a mechanical stirrer, reflux condenser, and a thermometer. The top of the condenser was attached to a drying tube containing calcium chloride. The flask was heated in an oil bath to a designated temperature and maintained at that temperature for 3 hrs with stirring. After the reaction, the esterified sawdust was filtered and washed successively with acetone and distilled water. To ensure complete removal of unreacted reagents, the product was further extracted with acetone in an extractor for 6 hrs, and then vacuum dried at 70°C until a constant weight was obtained. A similar procedure was applied to esterification with succinic anhydride. In one case, sawdust was reacted with succinic anhydride in a solid phase, i.e., at 60°C. The melting point for succinic anhydride is 120°C. For simplicity, maleic anhydride-treated wood and succinic anhydride-treated wood are designated as Wood-MA and Wood-SA, respectively.

Characterization. The extent of esterification reaction was characterized by weight gain (%) and acid value (%). Monoesters and diesters were formed during the reaction. They were qualitatively analyzed by using the following equations:

Acid Value:

$$AV(eq/kg) = \frac{A * N}{W}$$

where: $A =$ volume of base (ml)
$N =$ normality of the base
$W =$ weight of sample (g)

Saponified Value:

$$SV(eq/kg) = \frac{(V-P)f}{W} \times \frac{1}{2}$$

where: $V =$ volume of standard base (ml)
$P =$ volume of base used for 0.4N NaOH and 0.42N HCl (ml)
$f =$ normality of standard base
$W =$ weight of sample (g)

Monoester Content (%):

$$ME(\%) = (A_1 - \frac{A_o W}{1000}) \times \frac{100 M_1}{W}$$

where: $A_1 =$ acid value of esterified wood
$A_o =$ acid value of wood
$M_1 =$ molecular weight of the reagent
$W =$ weight gain (%)

Diester Content:

$$DE(\%) = [S_1 - 2A_1 - \frac{(S_o - 2A_o)W}{1000}] \times \frac{50 M_2}{W}$$

where: $S_1 =$ saponified value of esterified wood
$S_o =$ saponified value of wood
$A_1 =$ acid value of esterified wood
$A_o =$ acid value of wood
$W =$ weight gain (%)
$M_1 =$ molecular weight of anhydride
$M_2 = M_1 - 18.01$

FTIR Analysis. A Nicolet 20-DX Fourier Transform Infrared (FTIR) spectrometer was used to analyze changes in chemical structure of esterified woods. All measurements were carried out by using the potassium bromide pellet technique and in a nitrogen atmosphere. Each sample was scanned 20 times.

Thermal Analysis. The esterified woods were thermally analyzed by means of a thermogravimetric analyzer (Perkin-Elmer TGA7) and a differential scanning calorimeter (Perkin-Elmer DSC4). For TGA and DSC studies, heating rate was set at 10°C/min and 20°C/min, respectively. For each measurement, 5 mg samples were used. All measurements were carried out in nitrogen atmosphere.

Viscoelasticity Analysis. Softening temperature and thermal transition properties of the esterified woods were determined by means of a dynamic mechanical thermal analyzer (Polymer Laboratories, PL-DMTA). A dual cantilever technique, operating at a fixed frequency of 3 Hz, ×4 strain, and a constant heating rate of 3°C/min over a temperature range of −150 −250°C, was used. To prepare specimens for DMTA testing, 0.6 g of esterified wood powder was placed into a metal mold and pressed into a pellet with a dimension of 12 × 30 × 1.1 mm^3 on a hot press under 5000 psi at 120°C and 140°C for Wood-MA and Wood-SA, respectively.

Results and Discussion

Esterification. Since wood is a composite of cellulose, hemicelluloses and lignin, it is very difficult to determine the sites of reaction taking place in the individual polymers. Hence, the overall weight gain of reaction products was measured as an indication of degree of esterification. The acid content was also calculated to monitor the extent of reaction. The relationship between weight gain and ester content as a function of temperature for maleic and succinic anhydrides are shown in Figures 1 and 2, respectively. The correlationship between ester content and acid value, and the amount of monoester and diester formed are summarized in Table 1. It is clear that both weight gain and acid content were increased steadily as the reaction temperature increased. Maleic anhydride appeared to achieve higher weight gain and ester content than the succinic anhydride. It is interesting to note that succinic anhydride with a melting point of 120°C was able to react with wood at 60°C in which the reaction proceeded in solid state.

FTIR Analysis. The esterification reaction scheme between wood and maleic or succinic anhydride is shown in Figure 3. It is obvious that Wood-MA and Wood-SA are bearing, respectively, propenoyl and propionyl side chains with a free carboxylic acid group. The FTIR spectra of these products are illustrated in Figure 4. The absorption peak at 1725 cm^{-1} for both acid anhydrides indicated that carboxylic and ester groups were introduced into wood. An additional peak at 1640 cm^{-1} was observed from wood reacted with maleic anhydride. This particular peak was due to the presence of a double bond in the maleic acid. Broad absorption peaks at 3350 cm^{-1} were also detected. This implies that residual hydroxyl groups are still present in wood, and the introduction of carboxylic groups to wood may also absorb a certain amount of moisture.

Thermal Properties. Heat treatments may induce physical or chemical transitions on the esterified wood. These changes may be detected by DSC, TGA and DMTA.

DSC Analysis. The DSC thermograms of wood, Wood-MA and Wood-SA are shown in Figure 5. Endothermal peaks at about 120 – 125°C were detected for all samples due to the vaporization of water. Wood is a hygroscopic material. It always contains a certain amount of water. In these samples, not all the hydroxyl groups in the wood are esterified and the

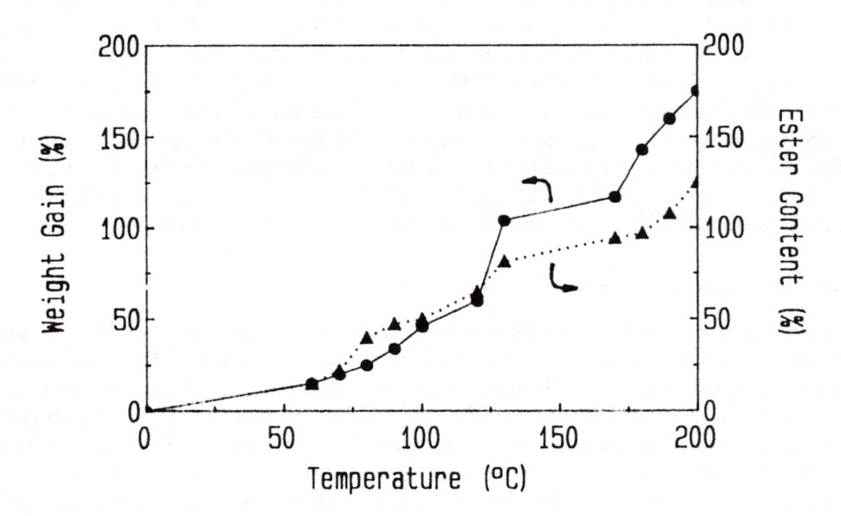

Figure 1. Effect of temperature on weight gain and ester content of wood reacted with maleic anhydride for 3 hrs.

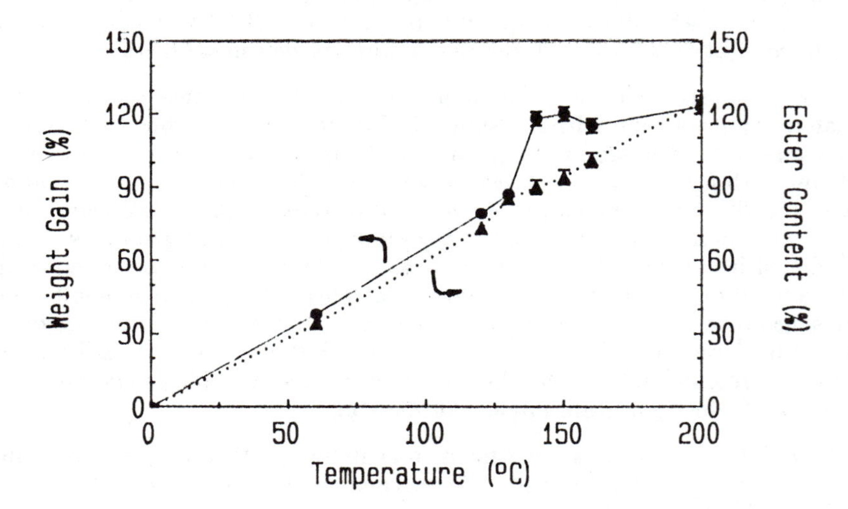

Figure 2. Effect of temperature on weight gain and ester content of wood reacted with succinic anhydride for 3 hrs.

Table 1. Esterification of wood with maleic anhydride (MA) and succinic anhydride (SA)

Sample	Reaction Temperature (°C)	Weight Gain (%)	Acid Value (eq/kg)	Monoester (%)	Diester (%)	Saponification value (eq/kg)	Ester Content (%)
2-MA	60	15.00	0.13	0.48	5.91	1.02	6.30
1-MA	70	20.00	0.95	10.13	----	----	----
10-MA	80	25.00	2.50	29.66	10.02	3.75	39.86
9-MA	90	34.50	2.76	35.41	11.92	4.08	47.33
8-MA	100	46.35	2.50	34.89	15.16	3.99	50.05
7-MA	120	59.65	1.99	30.16	----	----	----
4-MA	130	104.00	1.93	37.60	43.90	4.76	81.50
6-MA	170	117.50	3.00	62.90	31.30	4.93	94.20
3-MA	180	143.50	2.96	69.70	20.72	4.14	90.42
5-MA	190	159.50	3.00	75.33	32.40	4.67	107.73
4-SA	130	82.50	2.77	48.31	40.49	5.56	88.80
1-SA	150	98.10	2.82	54.50	54.71	6.31	109.21
2-SA	160	109.65	2.84	57.73	----	----	----
3-SA	175	112.19	2.15	43.59	----	----	----

Wood with maleic anhydride:

$$\text{WOOD-OH} + \underset{\text{(maleic anhydride: HC=CH, O=C, C=O, O)}}{} \longrightarrow \text{WOOD-O-}\overset{O}{\overset{\|}{C}}\text{-}\overset{H}{\overset{|}{C}}\text{=}\overset{H}{\overset{|}{C}}\text{-}\overset{O}{\overset{\|}{C}}\text{-OH}$$

Wood with succinic anhydride:

$$\text{WOOD-OH} + \underset{\text{(succinic anhydride: }H_2C\text{—}CH_2\text{, O=C, C=O, O)}}{} \longrightarrow \text{WOOD-O-}\overset{O}{\overset{\|}{C}}\text{-}\underset{H}{\overset{H}{C}}\text{—}\underset{H}{\overset{H}{C}}\text{-}\overset{O}{\overset{\|}{C}}\text{-OH}$$

Figure 3. General scheme of esterification.

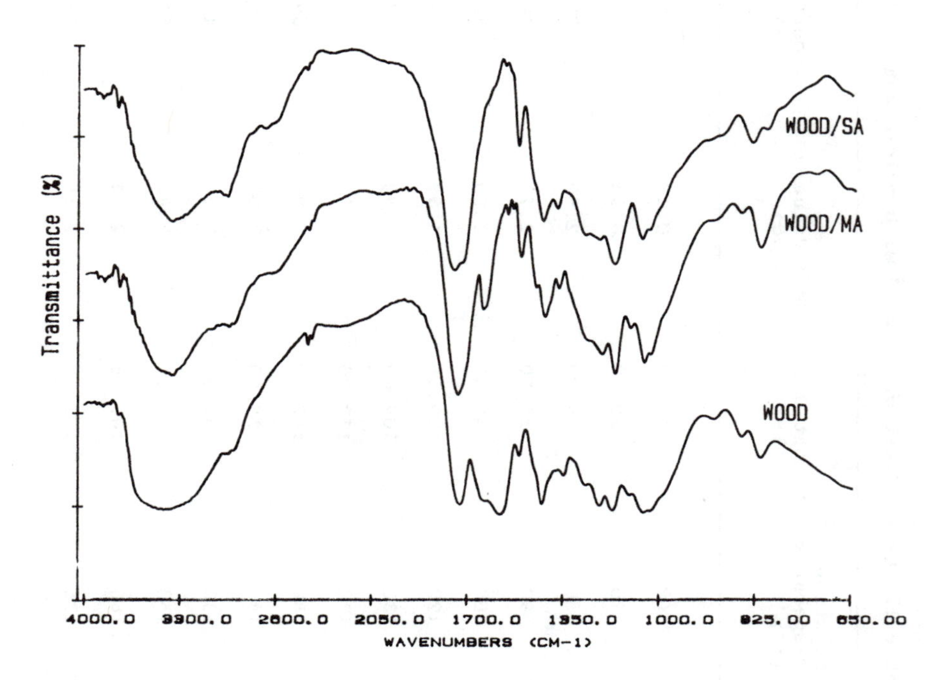

Figure 4. FTIR spectra of esterified wood.

newly introduced carboxylic groups are also capable of absorbing moisture. This phenomenon is also observed from the FTIR spectra in which a broad absorption peak at 3550 cm^{-1} due to hydroxyl group was detected. An exothermal peak at about 315°C was observed from the untreated wood but not from the Wood-MA and Wood-SA. This exothermic reaction may be due to the primary thermal decomposition of hemicelluloses and the cleavage of the glycosidic linkages of cellulose (13) as well as the decomposition of lignin (14). It has been reported that depolymerization of the polysaccharides by transglycosylation at about 300°C provides a mixture of levoglucosan, other monosaccharide derivatives, and a variety of randomly linked oligosaccharides (15). The depolymerization may be accompanied by dehydration of sugar units in cellulose which give unsaturated compounds and a variety of furan derivatives (16). The rearrangement and condensation reactions which occurred during thermal depolymerization contribute to the exothermic reactions. For the untreated wood, the exothermal peak was followed by an endothermal at about 400°C. This endothermal may be attributed to the fission of sugar units which provide a variety of carbonyl compounds, such as acetaldehyde, glyoxal, and acrolein (16). These compounds readily evaporate, resulting in an endothermal peak in the DSC. Similar endothermal peaks were also detected from Wood-MA and Wood-SA. This suggests that esterification of wood does not influence the rate of pyrolysis of wood polymeric components. Nonetheless, endothermal peaks at 235°C and 270°C were detected from Wood-MA and Wood-SA, respectively. This is due to the decarboxylation reaction. Since α,β-unsaturated carboxylic acid decomposes more readily than does saturated acid (17), the decarboxylation of Wood-MA takes place at a lower temperature than that of Wood-SA.

TGA Analysis. Wood and esterified woods lost their weight as a function of increasing temperature. TGA results are shown in Figure 6. Untreated wood started to lose its weight at about 200°C. This is probably due to the generation of non-combustible gases such as CO_2, CO, formic acid and acetic acid (14). Significant loss of weight was observed at 280°C indicating the onset of pyrolysis. It is known that this is the initial temperature for generation of combustible fuel gases (14). Increasing temperature led to continuous loss of weight until it reached 410°C.

Wood-MA and Wood-SA lost weight at a faster rate than the untreated wood. Although esterified woods decreased slightly in weight at about 150°C, they started to lose weight rapidly at 200°C, especially for Wood-MA at temperatures between 200 and 280°C. This acceleration of weight loss indicated that esterified woods decomposed rather rapidly at this temperature zone compared to untreated wood. Decarboxylation played a major role in this weight loss. Another loss of weight was observed from both samples at 290°C due to the onset of pyrolysis, which is very comparable to that of the untreated wood.

Viscoelastic Properties. All polymers exhibit viscoelastic properties (18). A dynamic mechanical thermal analyzer (DMTA) is one of the best in-

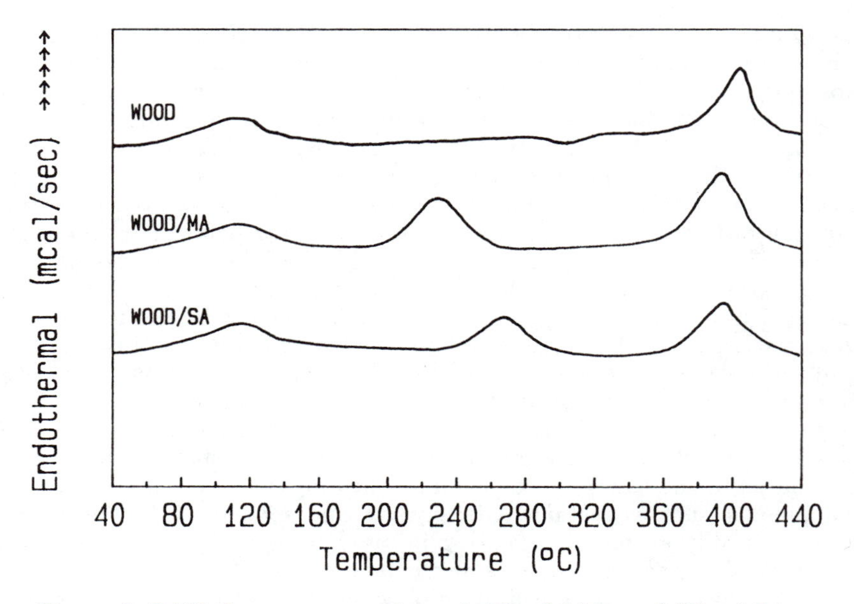

Figure 5. DSC thermograms of wood, Wood-MA, and Wood-SA.

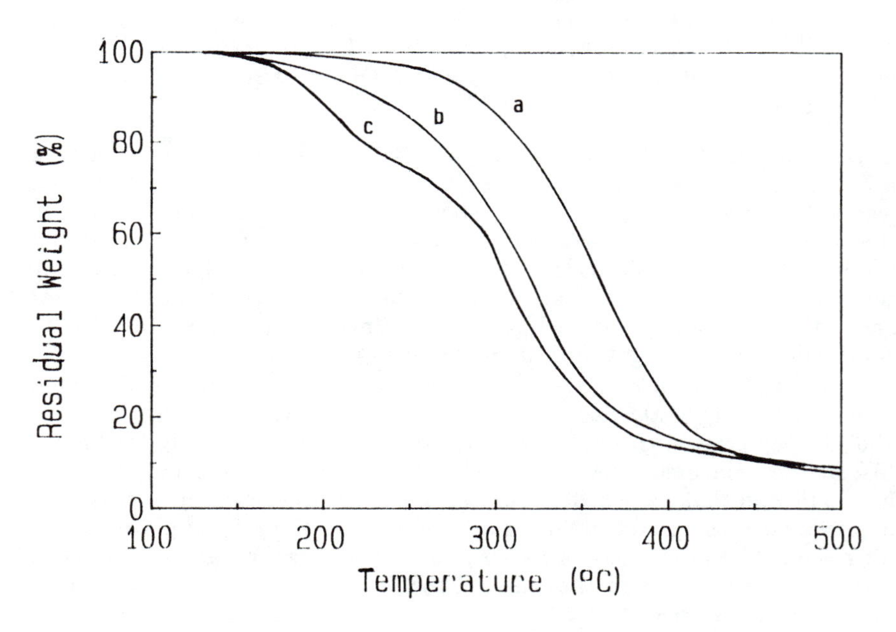

Figure 6. TG curves of wood (a), Wood-SA (b), and Wood-MA (c).

struments to evaluate this property. Dynamic mechanical thermal analysis senses any changes in molecular mobility in the sample as the temperature is raised or lowered. The dynamic modulus is one of the most important mechanical properties of materials for structural applications. Mechanical damping is often the most sensitive indicator in determining all kinds of molecular motions which are taking place in a polymeric material even in the solid state (19).

The storage modulus (E″) is defined as the stress in phase with the strain in a sinusoidal shear deformation divided by the strain. It is a measure of the energy stored and recovered per cycle when different systems are compared at the same strain amplitude. The loss modulus (E′) is deifned as the stress 90° out of phase with the strain divided by the strain. It is a measure of the energy dissipated or lost as heat per cycle of sinusoidal deformation. The angle which reflects the time lag between the applied stress and strain is δ. The tan δ (damping factor) is a measure of the ratio of energy lost to energy stored in the material during one cycle of oscillation.

Figure 7 shows the effects of temperature on the storage modulus and tan δ of untreated earlywood. Two noticeable peaks were found from the damping curve (tan δ). The peak at −80°C was attributed to the motion of side groups such as methyl groups which attached to lignin and hemicelluloses (18). Another contribution can be derived from the absorbed moisture which formed hydrogen bonds with the unesterified hydroxyl groups, particularly with the hydroxymethyl group in cellulose (20). The peak at 250°C is due to the glass transition of wood polymeric components. This transition resulted in a reduction of the storage modulus E′ from 4.47×10^{10} to 3.16×10^{9} Pa, which is small compared to many thermoplastic polymers (21). This may account for the highly crystalline structure of the wood cell wall which greatly restricts the motion of macromolecules and imparts good thermal stability to wood. Untreated latewood showed a slightly higher glass transition temperature, around 280°C, due to its denser and thicker cell walls which were more resistant to the diffusion of heat. This effect, however, did not change the storage modulus.

The loss modulus spectra of Wood-MA and Wood-SA with a weight gain of 104% and 109%, respectively, exhibited several transition or relaxation processes within the test temperature ranging from −150 to 200°C, as shown in Figure 8. Similar dispersions were observed from both samples. Primary transitions (the α transition) were detected at about 60°C. Other transitions were found in the glassy state on the lower temperature side of the primary transition, i.e., secondary transition, and were designated as β, γ and δ dispersion or relaxation.

The α transition process (glass transition) is attributed to the initiation of micro-Brownian motion of large segments of polymeric chains in wood. Below the glass transition temperature, the micro-Brownian motion of the main chain is frozen in, and only motion of small segments of polymeric chains can occur. This motion is largely of the local relaxation mode and can be the local mode relaxation of the main chain or the side chain relaxation.

The β dispersion did not occur in the spectrum of the untreated wood but occurred in Wood-MA and Wood-SA at about 0°C. Hence, the β dispersion is due to the torsional vibration of the propenoyl or propinoyl group around the ester bond attached to the backbone of cellulose, hemicelluloses and lignin.

The γ dispersion occurred at a temperature of about −80°C. Since this process was also found in the DMTA thermogram of the untreated wood, it can be assigned to the motion of methoxyl groups derived mostly from lignin. These methoxyl groups are able to rotate about their axes to the phenyl rings, resulting in the γ relaxation process. In addition, the γ process may also relate to the motion of methyl groups which are encountered extensively in lignin and hemicellulose structures. For instance, the large number of acetyl groups in hemicelluloses may be one of the origins of methyl groups. The involvement of water in this region may have some effect, as discussed earlier. An additional δ peak was observed at about −125°C. The weak δ relaxation process at this temperature is possibly derived from the local mode motion of chain molecules due to the cooperative torsional excursion about bands without change of configuration, i.e., within the same potential energy well. In contrast to the β and the relaxation processes due to the motion of the side chain, the local mode motions occurred on the main chain. These local relaxation modes are distinguished from the transition modes or the micro-Brownian motions since they are responsible for the establishment of the equilibrium distribution in the neighborhood of local equilibrium configurations of a polymer molecule. This so-called secondary dispersion phenomenon widely observed in amorphous polymers in the glassy state has been attributed to the local mode motions (22). This local mode relaxation was not observed in the untreated wood. This may be due to the high crystallinity of cellulose and the strong intermolecular interconnections among polymeric components in wood. Between Wood-MA and Wood-SA, it is obvious that the loss modulus is larger in Wood-SA. This implies that the Wood-SA is slightly more flexible than the Wood-MA.

Figure 9 shows the effect of temperature on the storage modulus (E′) of Wood-MA and Wood-SA. From these spectra, five regions may be defined. In region I, the polymer is glassy and brittle. Here, cooperative molecular motion along the chain is frozen, causing the material to respond like an elastic solid to the stress. A slight decrease in the storage modulus with rising temperature is noted. This may be attributed to the effects of the δ and γ relaxations. In this glassy state, molecular motions are largely restricted to vibrations and short-range rotational motions.

Region II corresponds to the β process. It is evident that the motion of propenoyl or propionyl side groups results in a remarkable reduction of storage modulus. Because of their long chain, the motion of propenoyl or propionyl groups exhibit a great effect on the storage modulus.

Region III is the glass transition region (α dispersion). The modulus drops by a factor of about one thousand over a 30-50°C range.

The behavior of Wood-MA and Wood-SA in this region is best de-

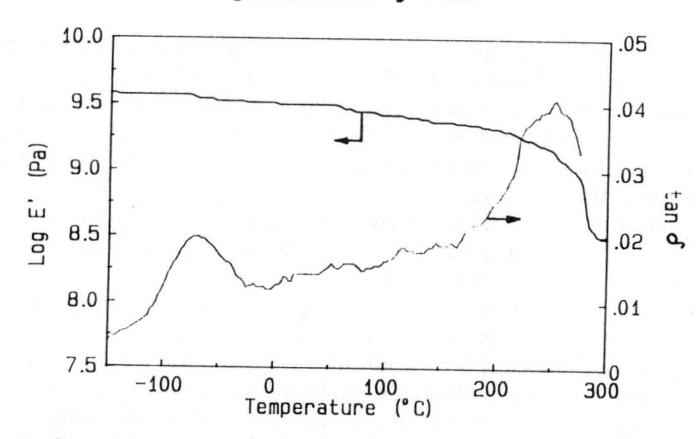

Figure 7. Storage modulus and damping factor of untreated earlywood.

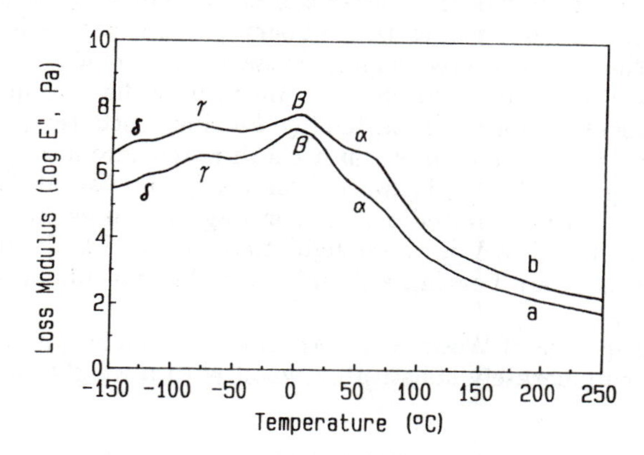

Figure 8. Change in loss modulus of Wood-MA (a) and Wood-SA (b) as the function of temperature.

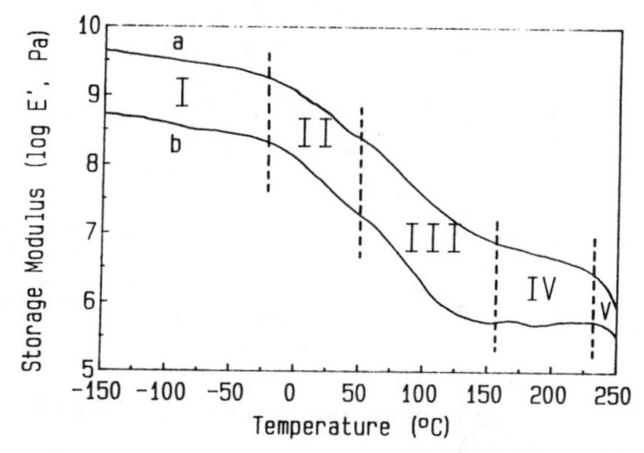

Figure 9. Change in temperature modulus of Wood-MA (a) and Wood-SA (b) as the function of temperature.

scribed as leathery, although a few degrees of temperature change will obviously affect the stiffness of the Wood-MA and Wood-SA. Qualitatively, the glass transition region can be interpreted as the onset of a large-range, coordinated molecular motion. While only side groups or small segments of macromolecules are involved in the small-scale motions below the glass transition temperature, large segments of macromolecules attain sufficient thermal energy to move in a coordinated manner in the glass transition region. The large segments can be derived from all polymeric components, cellulose, hemicelluloses, and lignin.

Region IV is the rubbery plateau region. After the sharp drop of the storage modulus, it becomes almost constant in the rubbery plateau region. In this region, the Wood-MA and Wood-SA exhibit long-range rubber elasticity. The width of the plateau depends on the diester values; the larger the diester values, the smaller plateau the esterified wood exhibits. This is due to the diester which functioned like a bridge to hold cellulose chains together. Hence, the higher the diester content, the more rigid the material.

As the temperature passes the rubbery plateau region for the esterified woods, the rubbery flow region, Region V, is reached. In this region, the Wood-MA and Wood-SA are marked by flow properties. The increased molecular motion imparted by the increased temperature permits assemblies of chains to move in a coordinated manner, and hence to flow or disintegrate. Again, between Wood-MA and Wood-SA, it is also clear that Wood-MA exhibited a higher storage modulus than Wood-SA, indicating that Wood-MA is more rigid than Wood-SA. It also suggests that a propenoyl group bearing a double bond has an influential effect on storage modulus.

Damping spectra of Wood-MA with different weight gain are shown in Figure 10. For moderately substituted wood with a weight gain of 35%, a

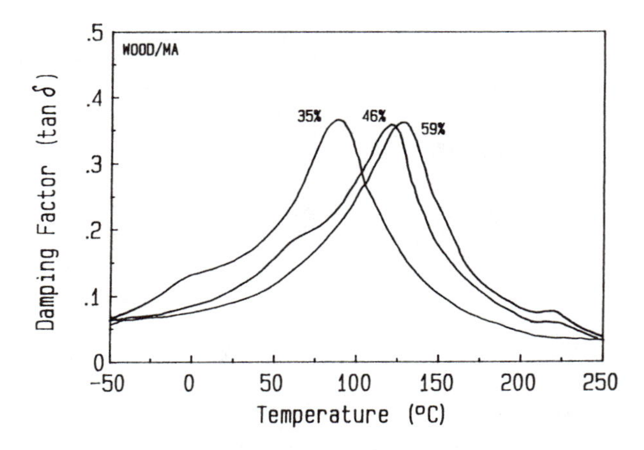

Figure 10. Effect of weight gain on damping factor of Wood-MA. Percentage represents weight gain (%).

broad transition peak at 85°C was observed. As the weight gain increased, the transition shifted towards a higher temperature range and the transition peak also became broad. The shift to higher temperature is probably due to the increase in diester content which in turn contributed to higher rigidity of the sample. The increase in the half-power width is considered to be the presence of an increased range of order (15). Similar damping spectra of Wood-MS were also obtained. They are shown in Figure 11. It is clear that Wood-SA exhibited a lower damping temperature than Wood-MA.

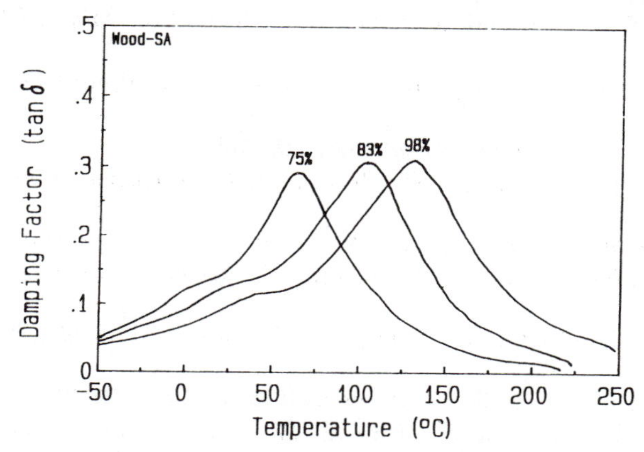

Figure 11. Effect of weight gain on damping factor of Wood-SA. Percentage represents weight gain (%).

Conclusions

Wood can be converted into a thermoplastic material by esterification with maleic and succinic anhydride. A DMTA study revealed that the Wood-MA and Wood-SA possessed viscoelastic properties. They exhibited four transition relaxations between −150 and 200°C. The α transition is attributed to the initiation of micro-Brownian motion; the β transition at 0°C is due to the torsional vibration of the propenoyl and propionyl groups around the ester bond attached to cellulose, hemicelluloses, and lignin. The γ transition at −80°C is due to the motion of methoxyl groups in lignin and methyl groups in lignin and hemicelluloses as well as due to the absorbed moisture reacting with hydroxy groups. The δ transition at −125°C is due to the local mode motion of the main chain molecules. From the change in storage modulus as a function of temperature, it is also noticed that esterified woods transformed from a glassy state to leathery, then to rubbery plateau and to rubbery flow states. Between Wood-MA and Wood-SA, the former always exhibited higher rigidity than the latter. The higher the diester content in either Wood-MA or Wood-SA, the more difficult it was to achieve flow property. A DSC study also showed that esterified woods decarboxylated between 200 and 300°C. TGA also showed that esterified woods lost their weight between 200 and 300°C.

Literature Cited

1. Worthy, W. *Chem. and Eng. News* 1990, **68**, 19.
2. Borman, S. *Chem. and Eng. News* 1990, **68**, 19.
3. Shiraishi, N. In *Wood and Cellulosic Chemistry;* Hon, D. N.-S.; Shiraishi, N., Eds.; Marcel Dekker: New York, 1990.
4. Hon, D. N.-S.; Ou, N. H. *J. Polym. Sci., Part A: Polym. Chem.* 1989, **27**, 2457.
5. Hon, D. N.-S.; San Luis, J. *J. Polym. Sci., Part A: Polym. Chem.* 1989, **27**, 4143.
6. Hon, D. N.-S. *Proc. Intl. Symp. on Wood and Pulping Chem.,* 1989, p 185-191.
7. Shiraishi, N. *Mokuzai Kogyo* 1980, **35**, 150.
8. Arora, M.; Rajawat, M. S.; Gupta, R. C. *Holzverwertung* 1980, **32**, 138.
9. Matsuda, H.; Ueda, M.; Hara, M. *Mokuzai Gakkaishi* 1984, **30**, 737.
10. Matsuda, H.; Ueda, M.; Murakami, K. *Mokuzai Gakkaishi* 1984, **30**, 1003.
11. Matsuda, H.; Ueda, M.; Murakami, K. *Mokuzai Gakkaishi* 1985, **31**, 103.
12. Matsuda, H.; Ueda, M. *Mokuzai Gakkaishi* 1985, **31**, 215.
13. Nguyen, T.; Zavarin, E.; Barrall, E. M. II. *J. Macromol. Sci. Rev. Macromol. Chem.* 1981, **20**, 1.
14. LeVan, S. L. Thermal Degradation. In *Concise Encyclopedia of Wood and Wood-Based Materials;* Scniewind, A. P., Ed.; Pergamon Press, 1989, pp 271-273.
15. Shafizadeh, V.; DeGroot, W. F. In *Thermal Uses and Properties of Carbohydrates and Lignins;* Shafizadeh, F.; Sarkanen, K. V.; Tillman, D. A., Eds.; Academic Press: New York, 1976; pp 1-17.
16. Gyrne, G. A.; Gardiner, D.; Holmes, F. H. *J. Appl. Chem.* 1966, **16**, 81.
17. Noller, C. R. *Chemistry of Organic Compounds;* Chapman: London, 1965.
18. Perry, J. D. *Viscoelastic Properties of Polymers;* 3rd ed.; Wiley: New York, 1980.
19. North, A. M. In *Molecular Behaviour and the Development of Polymeric Materials;* Ledwith, A.; North, A. M., Eds.; Chapman and Hall: London, 1975; pp 368-403.
20. Scandola, M.; Ceccorulli, G. *Polymer* 1985, **26**, 1963.
21. Murayama, T. *Dynamic Mechanical Analysis of Polymeric Materials;* Elsevier: New York; 1978.
22. Wetton, R. E.; Russell, G. S. *Molecular Relaxation Process;* Academic Press: London, 1966.

RECEIVED February 10, 1992

Chapter 9

Viscoelasticity of In Situ Lignin as Affected by Structure

Softwood vs. Hardwood

A-M. Olsson and L. Salmén

Swedish Pulp and Paper Research Institute, Stockholm S–114 86, Sweden

The viscoelastic properties of various wood species under water-saturated conditions have been studied by mechanical spectroscopy. Tests were made at temperatures from 20° to 140°C at frequencies ranging from 0.05 to 20 Hz. It is shown, contrary to earlier data, that hardwood lignins have lower softening temperatures than softwood lignins. For the hardwood lignins the softening process also has a lower apparent activation energy. Increased cross-linking of the lignin, achieved by heating in an acid environment, raises the softening temperature and increases the apparent activation energy of the softening process. In all cases the softening follows a WLF type of behavior, indicating that under wet conditions the viscoelastic properties of the lignin govern the properties of the wood fiber. The differences noted between softwood and hardwood lignin are discussed in terms of structural parameters.

The properties of wet wood reflect to a large extent the properties of the water-saturated lignin within the wood (1). This is due to the fact that the carbohydrates, both the hemicelluloses and the amorphous cellulose, are highly softened under water-saturated conditions already at 20°C, leaving only the cellulose crystals and the stiff lignin as load-transferring materials. The cellulose crystals have mainly an elastic response and may thus be viewed as an inert filler material in a lignin matrix. It has also been shown that the viscoelastic properties of water-saturated spruce wood follow a WLF type of behavior (1), indicating that the lignin behaves as a normal amorphous polymeric material under these circumstances. Thus, studies on water-saturated wood may make it possible to deduce something about the specific properties of the native lignin in that particular wood species.

0097–6156/92/0489–0133$06.00/0

Viscoelastic measurements on various wood species have earlier indicated that there is no difference in the softening behavior between softwood and hardwood species (2), which suggests that structural differences between hardwood and softwood lignins have no bearing on the viscoelastic properties of the lignin. More recent measurements with differential scanning calorimetry on moist lignin samples (3,4) have, however, indicated some differences in softening behavior between wet lignins from hardwood and softwood.

The present study was undertaken in order to clarify the influence of the native lignin structure on its viscoelastic properties and to see whether the structural differences between hardwood and softwood lignins have any effect on the mechanical properties of the wood. The viscoelastic properties were studied under water-saturated conditions between 20° and 140°C for spruce, pine, birch, and aspen, and also for a spruce sample deliberately cross-linked under acidic conditions.

Experimental Procedure

Materials. Four wood species have been used: two softwoods, Norwegian spruce (*Picea abies*) and Scandinavian pine (*Pinus silvestris*), and two hardwoods, Scandinavian birch (*Betula verrucosa*) and European aspen (*Populus tremula*). The wood samples were cut with the longitudinal direction across the grain with a length of 70 mm and a cross-section of 15 × 50 mm. Before being tested, the samples were saturated with water and then preconditioned with a steam treatment for 30 min at 135°C. One sample of spruce wood was instead impregnated with HCl at a pH of 1.8 and then heated for 30 min at 135°C in order to cross-link the lignin within the wood.

Mechanical Spectroscopy. Dynamic mechanical properties were measured at temperatures between 20° and 140°C, keeping the relative humidity at 100% in the testing autoclave. Forced sinusoidal vibrations with zero mean stress in the frequency range from 0.05 to 20 Hz were applied using a Material Testing System (MTS) servohydraulic testing machine. The deformation was measured with an extensometer attached to the specimen. In all cases the amplitude was kept within the viscoelastic range, up to 0.1% deformation (1), and data have been obtained by extrapolation to zero deformation.

Evaluation. Dynamic mechanical properties for each temperature and frequency were calculated at each of three amplitudes from the means of about 50 sinusoidal loops. The storage modulus E' and the loss coefficient $\tan\delta$ were determined from the complex modulus E^*, obtained from the maximum and minimum values of the stress and strain and the area of the hysteresis loop. The moduli are based on the macroscopic dimensions of the wood specimens at 20°C and are thus corrected for changes in specimen volume with temperature (1).

Master curves were constructed by shifting $\log E'$ –log frequency curves for a specific temperature horizontally with respect to the mean values of

Table I. The glass transition temperatures at different frequencies for the wood species tested

Frequency Hz	Softening Temperature, T_g, °C				
	Aspen	Birch	Pine	Spruce	Spruce Crosslinked
0.05	—	—	—	83	—
0.1	64	—	—	—	—
0.5	—	—	92.5	88.5	113
0.6	68	70	—	—	—
2.0	—	75	—	—	—
4.0	73.5	—	—	—	—
5.0	—	—	99	96	120.5
6.0	—	80	—	—	—
20	82	85	103	100	124

the linear regression curves for two consecutive temperatures, Figure 1. In order to apply the WLF concept [Williams, Landel and Ferry (5)] to the master curve and then more accurately evaluate the WLF-constants, these were determined separately by first fitting fifth order polynomial curves to the experimental log E′-temperature curves as illustrated in Figure 2. Shift factors log a_T were then determined from calculated log E′-frequency curves taken every fourth degree. The WLF constants, C_7 and C_2, were determined by rewriting the WLF equation in the form:

$$(T - T_{ref})/\log a_T = 1/C_1(T - T_{ref}) + C_2/C_1 \tag{1}$$

and then taking the linear part of the graph of $(T - T_{ref})/\log a_T$ *vs.* $T - T_{ref}$. The reference temperature, T_{ref}, is taken to correspond to the maximum value of the WLF-type behavior.

The WLF equation thus obtained is well in line with the experimentally determined shift factor as illustrated in Figure 3 where log a_T is given as a function of $T - T_{ref}$ based on the experimental determination, with the curve representing the WLF equation given as determined from the polynomeric fits.

Results

The softening temperature obtained by mechanical spectroscopy is most easily defined by the maximum in the mechanical loss coefficient, tanδ. Table I gives the glass transition temperatures of water-saturated native lignin, defined in this way, at the various frequencies for the wood species tested. These tests refer to measurements across the grain, but no major difference in glass transition temperature is noticed for measurements along the grain (6).

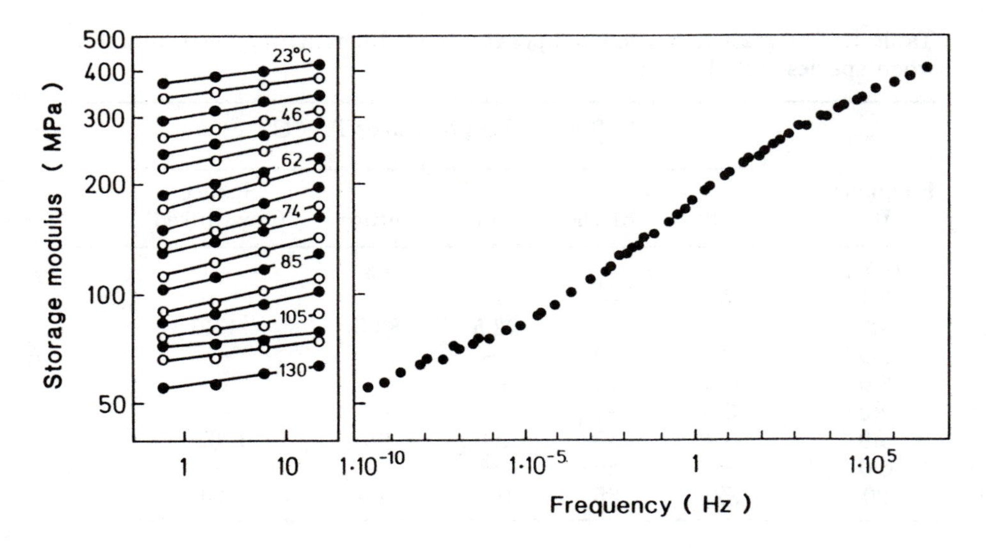

Figure 1. Master curve for the storage modulus of birch wood tested under water-saturated conditions in the temperature range from 23 to 130°C.

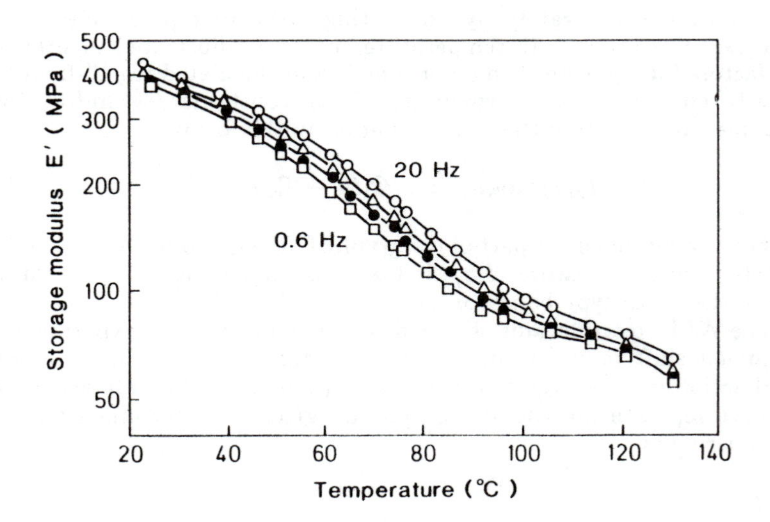

Figure 2. The storage modulus at different frequencies (0.6, 2.0, 6.0 and 20 Hz) for birch wood as a function of temperature. Fifth order polynomial curves are fitted to the experimental values.

It is obvious that softening in the hardwood species occurs at a lower temperature than in the softwood species, in contrast to the results of some measurements performed in torsion (2,7). The frequency dependence of the glass transition for the different wood species is illustrated in Figure 4, where the logarithmic frequency is plotted vs. the reciprocal glass transition temperature in an Arrhenius diagram. The slope of the lines is proportional to the apparent activation energy for the softening process, ΔH_a, which is evidently lower for the hardwood than for the softwood lignins. The apparent activation energies for the various lignins are given in Table II as determined by:

$$\Delta H_{a(\text{Arrhenius})} = 2.303 \bullet R(\Delta \log f)/\Delta(1/T_g) \qquad (2)$$

For the spruce wood in which the lignin had been cross-linked by the action of acid, both the softening temperature and the apparent activation energy of the softening process were higher than for all other samples tested.

Table II. Data for the softening process for the various wood species tested

	Aspen	Birch	Pine	Spruce	Spruce Cross-linked
$\Delta H_{a(\text{Arrh.})}$ (kJ/mol)	290	240	400	380	420
$\Delta H_{a(\text{WLF})}$ (kJ/mol)	330	360	440	390	460
T_{ref}(°C)	50	62	66	71	90
α_f(deg^{-1})	1.86×10^{-4}	0.79×10^{-4}	1.11×10^{-4}	0.82×10^{-4}	0.29×10^{-4}
f_g	0.0212	0.0142	0.0155	0.0144	0.0084

Figure 5 shows the relation between the softening temperature and the apparent activation energy of the softening process, ΔH_a, together with the empirical relations obtained by Lewis (8) for sterically restricted and unrestricted polymers. Apparently, the general trend found for polymers that ΔH_a increases with increasing softening temperature also holds for the lignin in different wood species. It is, however, difficult from this figure to discuss the structural restrictions of the lignin polymer, since at high temperatures the experimental data fall between the two equations even for synthetic polymers.

Earlier investigations (1,9) have shown the possibility of constructing time-temperature correspondence curves, master curves, according to the WLF concept for water-saturated wood samples. Such master curves show the relation for the stiffness over a larger frequency interval at a given temperature and have here been constructed from the measurements on the different wood species, as described earlier.

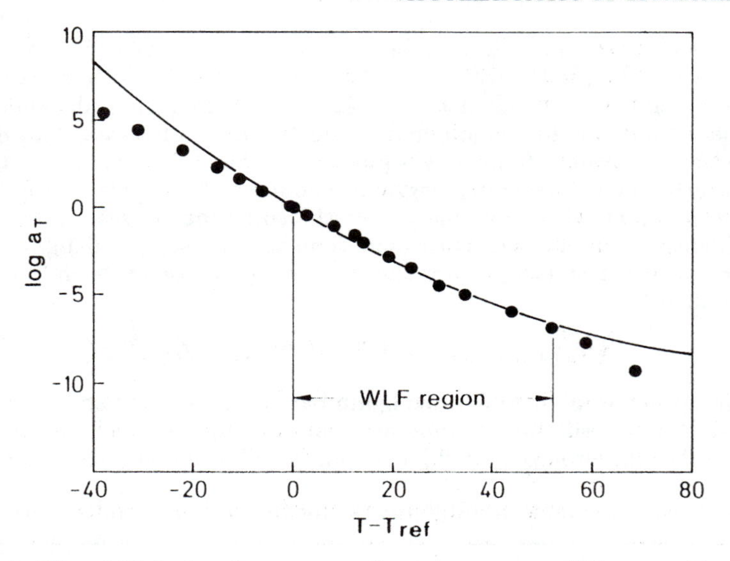

Figure 3. The shift factor, log a_T, vs. the temperature difference, $T - T_{ref}$, for the master curve of birch. The curve is calculated from the polynomial fit and the points are taken from experimentally shifted data.

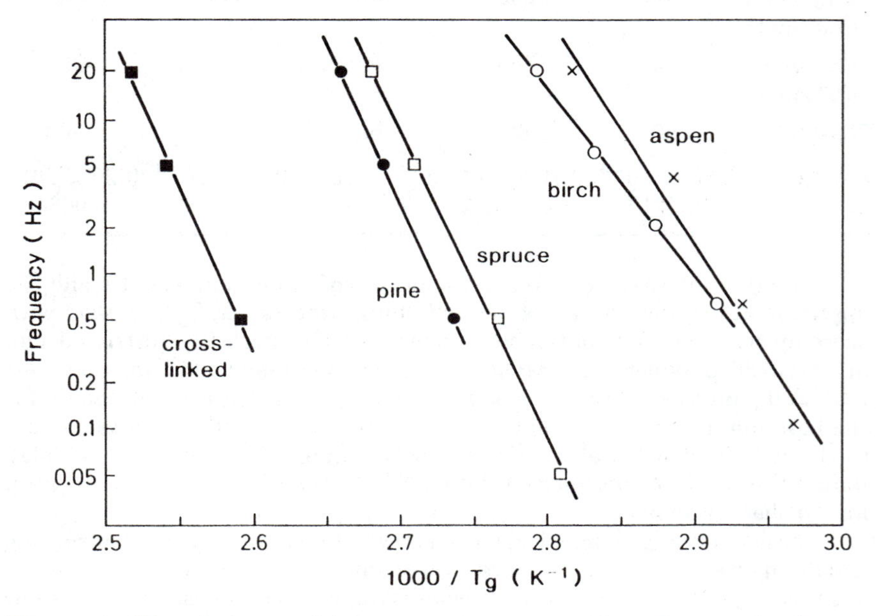

Figure 4. The glass transition temperature at various frequencies for the different wood species plotted in an Arrhenius diagram. The temperature decreases to the right in the diagram.

Figure 6 shows master curves for the spruce and aspen specimens at a reference temperature of 100°C. In order to compare the softening processes of the two samples, the curves have been shifted vertically to the same modulus at 20°C. These curves clearly show that the softening of the aspen occurs at higher frequencies, which correspond to lower temperatures, than the spruce sample. This is yet another confirmation that the glass transition temperature of hardwood lignins is lower than that of softwood lignins.

The modulus drop over the softening region in Figure 6, taken as the relative decrease E_{rubber}/E_{glass}, is about 72%, determined as the vertical difference between the tangents to the master curve at 10^{-5} and 10^{10} Hz, respectively. This decrease is somewhat larger than the drop calculated for wood across the grain using modulus values for isolated lignin (10) but, considering the approximations, it is quite a reasonable value for the type of cross-linked polymer in a composite here considered.

When the WLF equation is expressed in the linear form, the range of its validity and its parameters may be determined, as discussed under Experimental. The lowest temperature where the WLF equation is valid is considered to be equal to the glass transition temperature at very low frequencies and is given in Table II as T_{ref}. In all cases, the WLF equations are valid up to about 120°C. This is somewhat lower than the usual validity range of $T_g + 100°C$, which may be because degradation at high temperatures affects the measurements. The fractional free volume, f_g, and the volumetric expansion coefficient of the free volume, α_f, determined from the WLF constants, are also given in Table II. Both f_g and α_f decrease for both hardwood and softwood with increasing glass transition temperature, and there is only a small difference between the two types of wood species at about the same softening temperature. From the shift factors of the master curve, the limiting apparent activation energy, $\Delta H_{a(\mathrm{WLF})}$, at T_{ref} may also be determined as:

$$\Delta H_{a(\mathrm{WLF})} = 2.303 \bullet R(\delta \log a_T)/\delta(1/T) \qquad (3)$$

This value for the different wood species is given in Table II.

The relation between the apparent activation energy determined from the WLF equation and T_{ref} is given in Figure 7 together with the data based on the Arrhenius plot given in Figure 5. In this case there is also a general tendency for the apparent activation energy to increase with increasing softening temperature as given by T_{ref}. The increase with temperature is also very similar for the two determinations, whether by the WLF equation or by the Arrhenius plot. This is as expected, of course, for the behavior of synthetic polymers, as the apparent activation energy for the glass transition increases at lower frequencies.

The effect of cross-linking of the lignin on the viscoelastic properties of the wood is illustrated in Figure 8 with master curves for spruce samples, with the reference temperature taken as 100°C. It is obvious that cross-linking shifts the softening region towards lower frequencies, which implies a higher softening temperature. It is evident that neither the modulus value in the glassy region at high frequencies nor that in the rubbery region at low

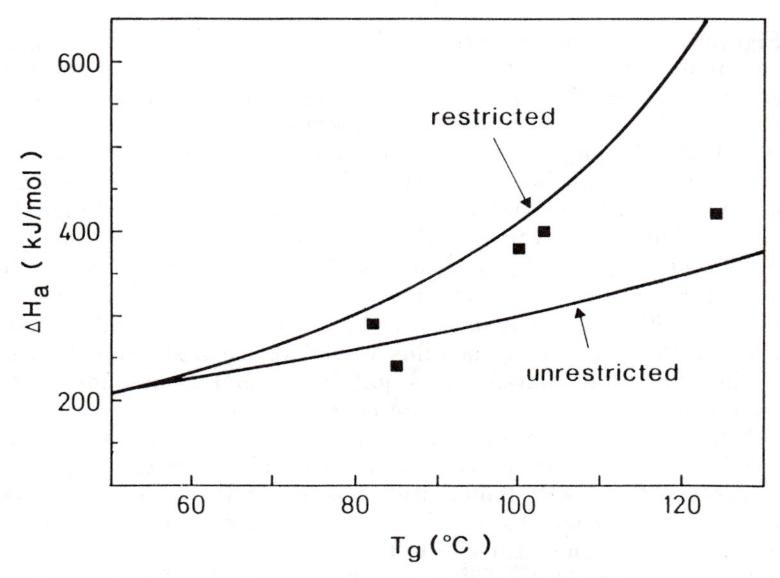

Figure 5. The apparent activation energy for the softening process as a function of T_g at 20 Hz for wood together with the empirical relations of Lewis (8) for synthetic polymers.

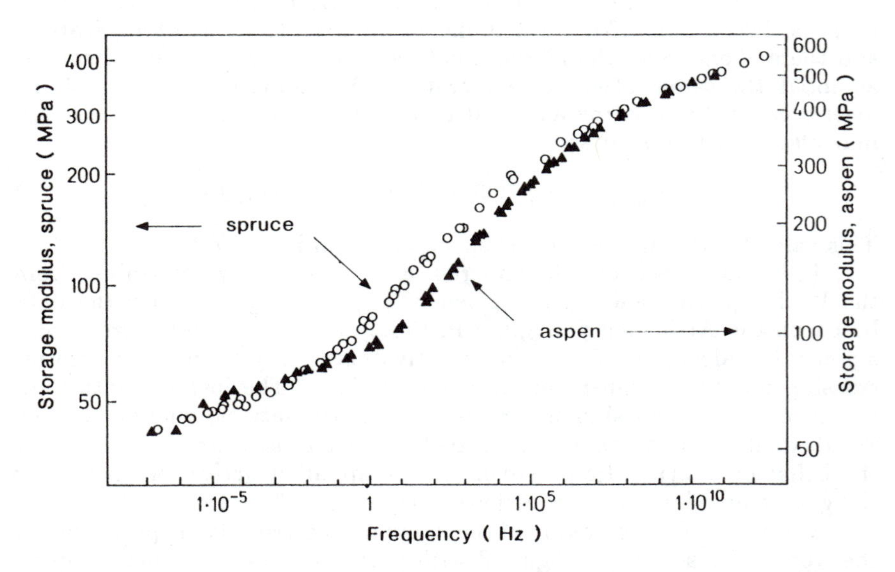

Figure 6. Master curves for spruce and aspen wood, shifted vertically to the same modulus at 20°C. The reference temperature is taken as 100°C and the frequency is obtained from the shift factor log a_T.

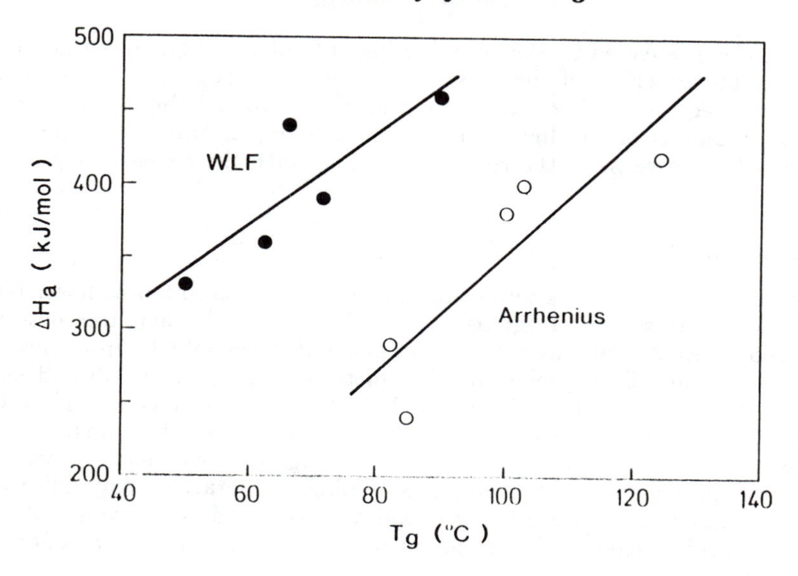

Figure 7. The apparent activation energy for the softening process, ΔH_a, as a function of the softening temperature, T_g.

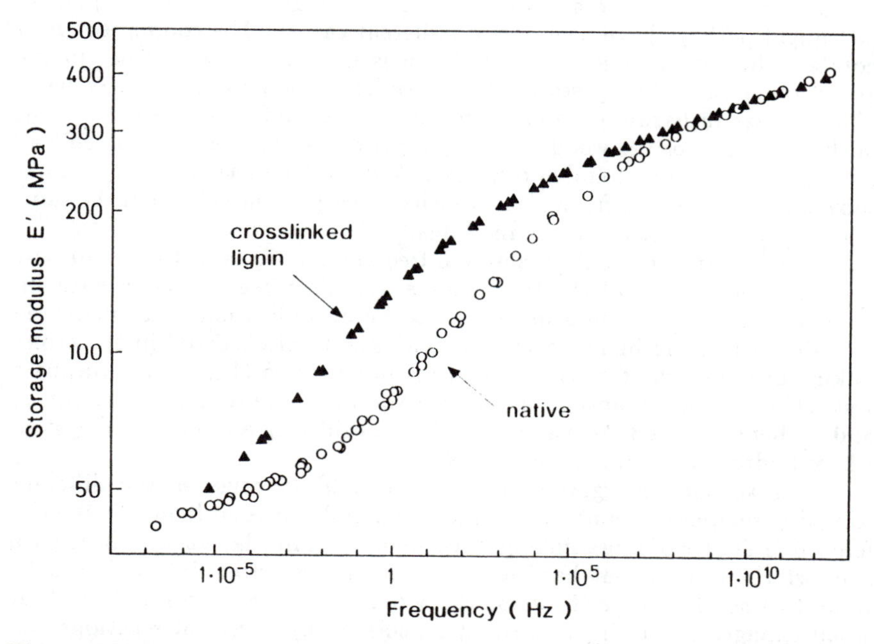

Figure 8. Master curves for spruce and a spruce wood sample in which the lignin has been cross-linked with acid. The reference temperature is taken as 100°C and the frequency is obtained from the shift factor log a_T.

frequencies is affected by the cross-linking of the lignin. The fact that there is no apparent effect of the cross-linking on the modulus in the rubbery region is very surprising. The modulus of the wood is highly dependent on the modulus of the lignin in the rubbery region and one might have expected an increase in the rubbery modulus with the cross-linking of the lignin.

Discussion

It is obvious from these results that hardwood lignins soften at lower temperatures than softwood lignins. The main structural differences between the two types of lignin are a lower content of free phenolic hydroxyl groups and a substantially higher content of methoxyl groups in hardwood than in softwood lignins (11). This can be interpreted as meaning that the hardwood lignins are less cross-linked than the softwood lignins. An increase in the degree of cross-linking of the lignin would, as is known for synthetic polymers, lead to a higher softening temperature. Any difference in molecular weight between hardwood and softwood lignin would, at the high molecular weights of these polymers, be too small to have an effect on T_g (12).

A deliberate cross-linking of the lignin, as with acidic treatment, increases the softening temperature. This is achieved apparently without affecting the modulus in the rubbery region, as is also evidently the case when comparing the hardwood and softwood species, Figure 6. This unexplained fact might perhaps have a number of different causes. The comparison may not be valid, as no strict plateau region is in fact reached, either in the glassy or in the rubbery state. The disordered cellulose zones may be of different size in hardwood and softwood and thus affect the reinforcement in the structure, or the wood as a complex composite system may itself lead to an unpredictable modulus response. Also, different types of cross-links show quite different shifts in glass transition temperature (13) and the same may apply to the effect on the modulus.

The fact that both the fractional free volume, f_g, and the volumetric expansion coefficient of the free volume, α_f, decrease markedly with increasing softening temperature may also be an indication of an increased cross-linking of the lignin. Such an effect has been observed in the cross-linking of unsaturated polyester and in epoxy resins (14,15) and means a restriction of the expansion of the free volume. The actual values for f_g and α_f here obtained are also reasonable, considering a cross-linked system where hydrogen bonding is possible (16).

It is therefore suggested that the most likely cause of the different softening temperatures noted for hardwood and softwood lignins is the difference in degree of cross-linking of the lignin within the wood. These data support the notion that hardwood lignins are less cross-linked than softwood lignins. The large effects on the softening temperatures indicate that small changes in the lignin structure achieved by chemical reactions can have a large impact on the softening temperature. This is well in line with the interpretation of the effect of sulphonation of lignin, which involves the breakage of α-ether bonds, thus breaking cross-links in the lignin network (17).

Acknowledgments

The authors thank Professor Anders Björkman, DTH, Denmark, for kindly providing the aspen wood sample, and Dr. Anthony Bristow for the linguistic revision.

Literature Cited

1. Salmén, L. *J. Mater. Sci.* 1984, **19**, 9:3090.
2. Becker, H.; Höglund, H.; Tistad, G. *Paperi ja Puu* 1977, **59**, 123.
3. LaPierre, C.; Monties, B. *J. Appl. Polym. Sci.* 1986, **32**, 4561.
4. Jordao, M. C. S.; Neves, J. M.; Otsuki, H. *Preprint, Fourth Intl. Symp. Wood and Pulping Chem.*; Paris, 1987; Vol. 2, p 81.
5. Williams, M. L. W.; Landel, R. F.; Ferry, J. D. *J. Amer. Chem. Soc.* 1955, **77**, 3701.
6. Salmén, L. *Progress and Trends in Rheology II*; Giesekus, H., Ed.; Darmstadt: Steinkopff Verlag, 1988; p 243.
7. Heitner, H.; Atack, D. *Preprint, Intl. Symp. Wood and Pulping Chem.*; Japan, 1983; Vol. 2, p 36.
8. Lewis, A. F. *J. Polym. Sci.* 1963, **B-1**, 649.
9. Kelley, S. S.; Rials, T. G.; Glasser, W. G. *J. Mater. Sci.* 1987, **22**, 617.
10. Salmén, L. In *Paper Structure and Properties*; Bristow, J. A.; Kolseth, P., Eds.; New York: Marcel Dekker, 1966; p 51.
11. Sjöström, E. *Wood Chemistry: Fundamentals and Applications*; Academic Press: London, 1983.
12. Hatakeyama, H.; Iwashita, K.; Meshitsuka, G.; Nakano, J. *Mokuzai Gakkaishi* 1975, **21**, 11:618.
13. McCrum, N. G.; Read, B. E.; Williams, G. *Anelastic and Dielectric Effects in Polymeric Solids*; New York: Wiley, 1968; p 394.
14. Shibayama, K.; Suzuki, Y. *J. Polym. Sci.* 1965, **A3**, 2637.
15. Kitoh, M.; Suzuki, K. *Kobunshi Ronbunshu* 1976, **5**, 1:26.
16. Kunugi, T.; Isobe, Y.; Kimura, K.; Asanuma, Y.; Hashimoto, M. *J. Appl. Polym. Sci.* 1979, **24**, 923.
17. Atack, D.; Heitner, C. *Trans. Tech. Sect. (Can. Pulp Pap. Ass.)* 1979, **5**, 4:TR99.

RECEIVED December 16, 1991

Chapter 10

Cellulose-Based Fibers from Liquid Crystalline Solutions

Solution Properties of Cellulose Esters

Vipul Davé and Wolfgang G. Glasser

Biobased Materials Center, Department of Wood Science and Forest Products, Virginia Polytechnic Institute and State University, Blacksburg, VA 24061

Solutions of cellulose esters with different concentrations in dimethyl acetamide (DMAc) and with different types of substituents, but with nearly constant total degree of substitution and constant molecular weights, were studied in relation to their liquid crystalline solution behavior. Observations made by dynamic mechanical spectrometry and by cross-polarized optical microscopy revealed classical liquid crystalline behavior for all solutions. Critical polymer concentration levels (V_p^c) were observed for all cellulose esters, and these were found to vary with substitution pattern. V_p^c is highest for cellulose acetate and lowest for the cellulose acetate butyrate with maximum degree of butyration. This is opposite to expected behavior based on the classical model by Flory, which predicts an increase in V_p^c with decreasing aspect ratio. Cellulose ester solutions are viscoelastic in nature.

Rigid rod-like polymers are recognized for their ability to form anisotropic liquid crystalline solutions (1,2). For rod-like species, the classical Flory equation of:

$$V_p^c \approx (8/X)(1 - 2/X) \quad (1)$$

relates the critical volume fraction, V_p^c, of the polymer in solution to the appearance of a stable anisotropic phase; where X is the aspect ratio (L/d) of the polymer; L is the contour length; and d is the average diameter of the polymer chain. Cellulose is a linear homopolymer of β-linked 1,4-anhydroglucose units. Liquid crystalline solution behavior has been observed with many cellulose and cellulose derivatives (3). Flory is credited with first commenting on the possibility of mesophase formation in cellulosic polymers (4). The most important parameter controlling the formation of a liquid crystalline phase appears to be chain stiffness for cellulose

0097–6156/92/0489–0144$06.50/0

derivatives (5). The solution conformations of cellulosic polymers studied so far indicate that the cellulose backbone is characterized as a semi-rigid polymer (6). Like other semi-rigid polymers, cellulose and its derivatives are better represented by the Kratky-Porod worm-like chain model rather than the random flight model. For worm-like chains, persistence length (q) is the measure of chain stiffness. It has been shown that the aspect ratio (X) in Equation 1 is related to the persistence length (q) through Equation 2 (3).

$$X = 2q/d \tag{2}$$

Very long worm-like chains behave as random flight models and the persistence length can be related to the equivalent Kuhn segment length (l_k) by Equation 3.

$$q = l_k/2 \tag{3}$$

Thus, the stiffness of a worm-like chain may also be characterized by the equivalent Kuhn segment length. Hence, the aspect ratio X in Equation 1 can be represented by Equation 4.

$$X_k = 1_k/d \tag{4}$$

$$\text{where } 1_k = < r^2 >_o /nlu \tag{4a}$$

$$\text{and } d = (Mu/Na_\rho lu)^{1/2} \tag{4b}$$

Mu is the molecular weight of the repeat unit, ρ is the density of the polymer, Na is the Avogadro number, and lu is the length of the repeat unit projected on the molecular axis (7).

Since persistence length, Kuhn segment length, and diameter are known quantities for a given polymer, values of $V_p{}^c$ can be both calculated and experimentally determined. There is, however, not an exact agreement between the experimental and theoretical values of $V_p{}^c$. For cellulose derivatives, minor deviations between theoretical and experimental values of $V_p{}^c$ have been reported (8,9). Attempts have been made to overcome these discrepancies by some appropriate modification of Flory's theory for specific solvents (10,11). Values of $V_p{}^c$ are influenced by solvent, temperature, molar mass and nature and degree of substitution mainly because these parameters affect chain conformation and stiffness (5,12). The change of $V_p{}^c$ with solvent is primarily due to a change in the persistence length of the polymer chain in solution. Solvents which help in forming intra-chain hydrogen bonds between anhydroglucose units are expected to increase stiffness and hence reduce $V_p{}^c$. Highly polar or acidic solvents generally favor mesophase separation at lower $V_p{}^c$ (13).

Persistence length decreases as temperature rises; and this influences $V_p{}^c$ as well (12).

It has been shown that values for $V_p{}^c$ increase with a decrease in molar mass of cellulose derivatives, and these observations are rationalized in terms of chain stiffness (10,11,14).

Unmodified cellulose is soluble only in selected solvents and solvent mixtures at the high concentrations required for the formation of liquid crystalline solutions (15,16). However, a myriad of cellulose derivatives can be formed due to the presence of three hydroxyl groups per anhydroglucose unit that may be available for modification. The presence of side chains; a high degree of substitution (DS); and a non-uniform pattern of substitution all increase the solubility of cellulose derivatives as crystallinity is repressed. The effective aspect ratio (X, Equation 1) will also vary with changes in the chain diameter caused by substitution. Long flexible side chains may behave like an internal plasticizer for the cellulose backbone and aid in allowing some mobility to form thermotropic liquid crystals in the absence of a solvent (3).

Much of the current interest in polymeric liquid crystals can be attributed to the extraordinary physical properties that are obtained when solution or melt anisotropy are retained during the solidification process. Although flexible chain polymers orient during processing, retention of this orientation becomes difficult due to fast relaxation. Instead, entanglements seem to control the structure and properties of flexible chain polymers. Rigid polymers, by contrast, have properties that result from macroscopic structures which are the consequence of the rheological properties of polymeric liquid crystals (17). The viscosity of the anisotropic solutions (lyotropic liquid crystalline systems) increases with concentration and passes through a maximum with the onset of liquid crystalline order before it decreases with further increase in concentration (18). The liquid crystalline phase exhibits unique birefringence when viewed with the polarized light microscope under static conditions; and this phenomenon corresponds very closely to the concentration at which maximum viscosity is observed (19).

Commercially, polymeric liquid crystals are important because they can be processed into ultrahigh modulus fibers. There is always a good possibility to convert polymer liquid crystalline solutions into high performance fibers, and to date cellulose triacetate and Kevlar have shown this behavior. In order to understand the processing of liquid crystalline solutions, it is essential to characterize their rheological properties. The objective of the present study is to examine the rheological and morphological properties of liquid crystalline solutions of cellulose esters (cellulose acetate and cellulose acetate butyrate) in dimethyl acetamide (DMAc) in relation to the substitution pattern of butyrate and acetate substituents on the cellulose backbone.

Experimental

Materials. Cellulose acetate (CA) and cellulose acetate butyrates (CAB) were obtained from Eastman Kodak, Kingsport, Tennessee, as either commercial or experimental samples. The samples obtained were CA 394-60 (CA), CAB 171-15 (CAB-1), CAB 381-20 (CAB-2), and CAB 500-5 (CAB-3). Their average acetyl contents were 39.4, 29.5, 13.5, and 4.0 wt.%, respectively; their butyryl contents were 0, 17.1, 38.1, and 50.0 wt.%, respectively; and their falling-ball viscosities were 60 (228 poise), 15 (57 poise), 20 (76 poise), and 5 (19 poise) seconds, respectively.

Reagent grade dimethyl acetamide (DMAc) was used as received.

Methods. Degree of substitution (DS) by acetate and butyrate groups on the cellulose backbone was determined using a JEOL Model GX-270 NMR spectrometer (operating at 270 MHz for ^1H) at 80°C (20,21). Deuterated dimethyl sulfoxide (DMSO) with a trace of trifluoroacetic acid served as the solvent.

The molecular parameters (molecular weight, Mark-Houwink-Sakurada constants, and intrinsic viscosity) of the cellulose esters were determined using gel permeation chromatography with a differential viscosity detector (Viscotek Model No. 100) and a differential refractive index (concentration) detector (Waters 410) in sequence. Solutions of the cellulose esters were prepared in HPLC-grade tetrahydrofuran (THF), and all the calculations were based on a universal calibration curve.

A complete range of solutions [5–50% (w/w)] were prepared by mechanically mixing known weights of the cellulose esters with DMAc at ambient temperature. The solutions were allowed to equilibrate for 1–2 weeks prior to analysis. Extreme care was taken to treat the solutions similarly so that the relative aging effect is constant for all of them.

Viscosity measurement of the solutions below 20% (w/w) concentration were made at 25°C using a Wells-Brookfield Cone/Plate Viscometer at different cone rotation speeds. A Rheometrics Mechanical Spectrometer (RMS 800) was used to determine the rheological properties of 20% (w/w) and higher concentrated solutions. The solutions were placed in a parallel-disk geometry. The dynamic mechanical properties were measured at 26°C using a strain amplitude of 25% of the value at which the respective sample showed viscoelasticity. The frequency ranged from 1.0 to 100.0 rad/sec.

Polarized light microscopy was performed with a Zeiss Axioplan Universal Microscope. Small portions of a cellulose solution were placed between microscope slide and cover slip, and this was examined for birefringence between the cross polarizers of the microscope at room temperature.

Results and Discussion

The chemical and molecular characteristics of the four cellulose ester derivatives employed in this study are summarized in Table I. The degree of substitution (DS) of acetyl (DS_{AC}) and butyryl (DS_{Bu}) was evaluated by proton NMR spectroscopy (20,21). DS_{OH} was determined by difference [*e.g.*, for CA; $DS_{OH} = 3 - 2.44 = 0.56$]. The results (Table I) indicate that DS_{OH} is almost constant for all four cellulose ester derivatives with the exception of CA; that molecular weights differ only by a factor of 1.5 to 2; that the Mark-Houwink-Sakurada (MHS) constant is almost constant by classifying all four cellulose esters as semi-flexible polymer chains ($a \approx 0.90$); and that intrinsic viscosity varies by less than a factor of 1.4. Although it is recognized that all factors, *i.e.*, DS_{OH} (22), molecular weight (10,11,14,22), chain flexibility (6), and intrinsic viscosity (10,11), influence liquid crystalline solution properties, the cellulose ester derivatives employed in this study differ mainly with regard to substitution pattern. No systematic study with respect to the effect of substitution pattern on liquid crystalline solution behavior of cellulose ester derivatives has been reported.

Table I. Chemical and Molecular Characteristics of Cellulose Esters

	CA	CAB-1	CAB-2	CAB-3
DS_{AC}[1]	2.44	2.03	0.96	0.29
DS_{Bu}[1]	0	0.8	1.77	2.57
DS_{OH}[2]	0.56	0.17	0.27	0.14
$\overline{M}_n(\times 10^3)$	41.5	66.8	60.0	52.6
$\overline{M}_v(\times 10^3)$	76.8	176.0	143.0	123.0
MHS - Constant (a)	0.90	0.92	0.98	0.95
Intrinsic Viscosity (dL/g)	1.88	1.63	2.07	1.51

[1] Determined by H-NMR spectroscopy.
[2] Determined by difference.

The relationship between dynamic shear viscosity, frequency or shear rate [applying the Cox-Merz rule (23)] and concentration of the cellulose ester derivatives is illustrated in Figure 1. All four cellulose ester samples exhibit an increase in viscosity with increasing concentration up to a maximum before decreasing with further increase in concentration (*i.e.*, after $V_p{}^c$). Such an anomalous viscosity behavior is typical for liquid crystalline solutions, and this has been shown earlier for cellulose acetate (22,24), ethyl cellulose (25), hydroxypropyl cellulose (24,26), cellulose triacetate (27), and cellulose acetate butyrate (12,28). This change in viscosity is due to a change in the structure of the solutions. Baird has proposed an explanation that as the concentration rises, the size and density of the ordered regions increases and a suspension of ordered clusters in an isotropic matrix is created (29). The energy dissipated by these clusters during flow is less than the total sum of the energy dissipated by each individual molecule; hence, the viscosity decreases with concentration. Another possible explanation for this anomalous viscosity behavior is that beyond $V_p{}^c$ there may be a competition between ordering due to shear and ordering due to packing of the molecules in solution. The viscosity initially decreases with concentration at levels exceeding $V_p{}^c$ as a consequence of shear ordering; and viscosity eventually increases again at higher concentrations as packing effects become important (24).

All four cellulose ester samples display a viscosity peak at constant concentration, irrespective of shear frequency. This indicates that unlike steady shear, dynamic shear is not responsible for driving the thermodynamic transition of the anisotropic phase to a different concentration (30). Previous investigators have made similar observations (22,24,26,30). The overall magnitude of the dynamic shear viscosity peak decreases as the frequency increases. This behavior has been justified by several mechanisms.

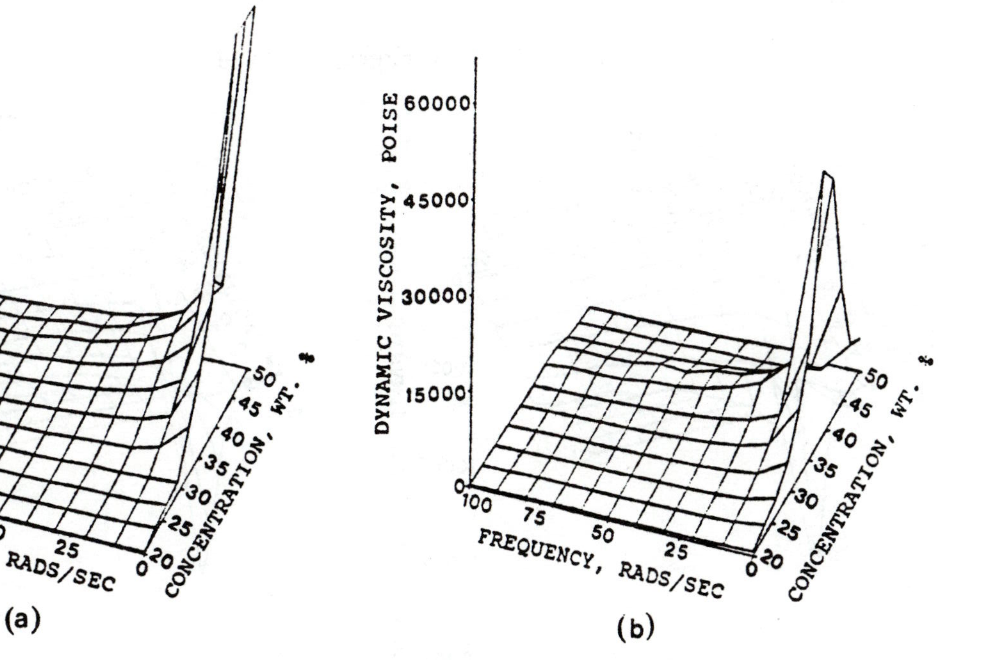

Figure 1. Dynamic viscosity vs. frequency for different concentration solutions of cellulose esters in dimethyl acetamide: (a) CA; (b) CAB-1; (c) CAB-2; (d) CAB-3.

Continued next page

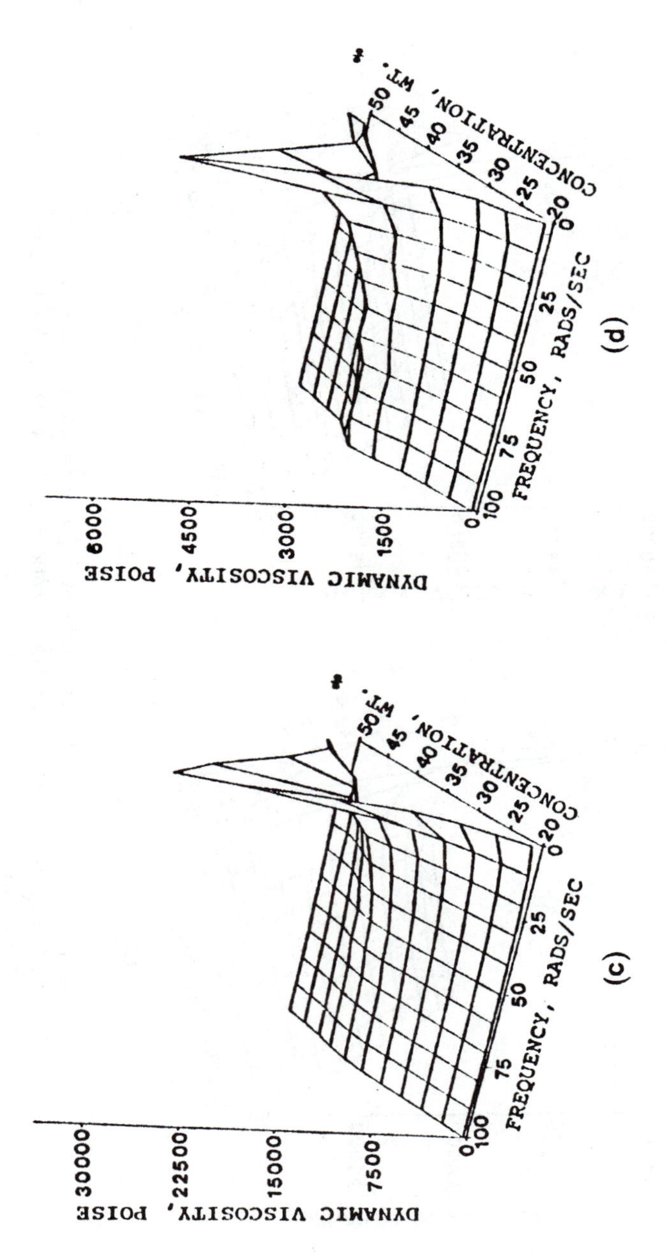

Figure 1. Continued.

Hermans (31) suspected that there might be a competition between the ordering produced by shear and the ordering produced due to thermodynamic reasons. At high shear rates, the former dominates over the latter one. It is also possible that with increasing shear rates there is orientation of the rigid segments of macromolecular chains under flow giving rise to macroscopic texture (32). This ultimately causes the viscosity to decline. At high frequencies, the viscosity values are almost the same for all concentrations.

The drop in viscosity with increase in shear rate is known as "shear thinning" or "pseudoplastic behavior," and this phenomenon begins several decades of shear rate lower for an anisotropic solution than for an isotropic one of the same viscosity (33). The dependence of viscosity on shear rate for polymer liquid crystals has been divided into a three-region flow curve by Onogi and Asada (34). In region I, the apparent viscosity decreases rapidly with increasing shear rate. Region II is the plateau region sometimes called Newtonian region, and region III is the power-law shear thinning region. According to the three-region flow curve, all four cellulose esters seem to represent region I or region III behavior. Assuming region III is applicable for the cellulose esters, shear thinning is more pronounced for CA and reduces with high DS_{Bu} (CAB's).

The rise in viscosity with polymer concentration is more pronounced for CA than for CAB's (Figure 1). The relationship between shear viscosity at 1 rad/sec frequency and concentration for the four cellulose esters (Figure 2) reveals that CA exhibits a greater increase in viscosity than CAB, and that the viscosity rise diminishes with butyryl content. Simultaneously, $V_p{}^c$ shifts to lower concentrations as DS_{Bu} rises (Figure 2). This shift in $V_p{}^c$ cannot be accounted for by hydrogen bonds or molecular weight, as both of these parameters are almost constant for all four cellulose esters. Although it is not surprising that $V_p{}^c$ is related to substituent size (and bulk), a decrease in $V_p{}^c$ with increase in substituent bulk is contrary to expectations. The experimental values of $V_p{}^c$ for CA and CAB-3, *i.e.*, 0.38 and 0.29, are in reasonable agreement with those obtained by Dayan *et al.* [0.49] (22) and Bheda *et al.* [0.30] (12), respectively. The CAB used by Bheda *et al.* had unspecified characteristics in terms of its DS, and so it becomes difficult to correlate their CAB with the ones used in this study.

In order to verify the experimental values of $V_p{}^c$, theoretical calculations of $V_p{}^c$ are made using Equations 1–4 (see Appendix A).[1] Diameter

---

[1] The following direct quotation is taken from Flory (7): "Tanner and Berry (35) had obtained $< r^2 >_o /n = 1080\AA^2$ for secondary acetate (CA) of degree of substitution 2.45 dissolved in trifluoroacetic acid or in a mixture of methylene chloride and methanol. They found larger values of 1350 or greater for cellulose triacetate (CTA); Kamide, Miyazaki and Abe (38), on the other hand, report lower values of ca. 600 for CTA. The disparity may reflect the difficulties caused by aggregation of CTA. Hence, we adopt the same value for CTA as for CA, on the plausible grounds that they should be similar in this respect." Tanner and Berry had obtained relatively narrow range of values of $< r^2 >_o /n$ for cellulose hexanoate, cellulose nitrate and

and molecular weight of the repeat unit increase from CA to CAB-3 (Table II). This observation is expected because the substituent bulk increases. The Kuhn segment length and persistence length of all the cellulose esters is constant. This implies that the chain stiffness does not change with an increase in substituent bulk and is in agreement with the unchanged value of MHS constant (a) for all cellulose esters. It can be inferred that the value of the axial ratio, X_k, is affected only by Mu. As Mu increases from 264.5 to 354.1, X_k diminishes from 25.77 to 21.40 and simultaneously, the theoretical value of V_p^c increases from 0.29 to 0.34. There is fairly good agreement between the absolute experimental and theoretical values of V_p^c (Table II), but there is poor correlation in the trend of change in V_p^c with Mu (Figure 3); *i.e.*, $(dV_p^c/dMu)_{expt} < 0$ and $(dV_p^c/dMu)_{Th.} > 0$. Such an opposite trend is unexpected and conflicts with Flory's theory. In the derivation of Equation 1, the intermolecular and intramolecular interactions were not considered and the value of V_p^c is based only on geometrical aspects of the molecules. The orientation-dependent or anisotropic interactions considered by Flory and Ronca (36) may play an important role in concentrated solutions exhibiting liquid crystalline behavior. As Mu increases, there is crowding on the backbone which may lead to intermolecular and intramolecular interactions causing sufficient stiffening to form liquid-crystalline solutions at lower V_p^c. The stiffness of the chains is much higher than predicted by the MHS equation. These reasons may explain the discrepancy in the experimental and theoretical values of V_p^c. Based on these arguments, the shift to lower V_p^c values with increasing substituent size may possibly be attributed to an increase in the chain stiffness caused by intermolecular and intramolecular interactions. This unexpected relationship has not yet been reported in the literature for cellulose esters. When the experimental V_p^c values are extrapolated to the molecular weight of anhydroglucose unit, *i.e.*, 162 (Figure 3), $(V_p^c)_{expt.}$ is approximately 0.48, and this is somewhat close to that predicted on theoretical grounds by Gilbert, *i.e.*, 0.70 (27).

The relationship of dynamic elastic modulus (G′) and dynamic loss

---

cellulose N-phenyl-carbamate in different solvent systems.

Light scattering measurements of some selected fractions of cellulose acetate butyrate (29.5% acetyl and 17% butyryl by weight) in nitromethane and in ethyl acetate were carried out (39). The results were discarded because the apparent molecular weight varied from solvent to solvent; there were inconsistent changes of the second virial coefficient and root-mean-square end-to-end distance with the nature of solvent; and the curvilinear nature of the scattered light intensity envelops made extrapolation inaccurate. These problems could be due to the fact that the cellulose acetate butyrate chains are heterogeneous in composition (40–42).

It seems difficult to experimentally obtain coherent $< r^2 >_o /n$ values for cellulose acetate butyrates. Hence, it is reasonable to assume the same $< r^2 >_o /n$ (*i.e.*, 1080 Å^2) value for the cellulose esters used in this study even though Tanner and Berry did not measure them.

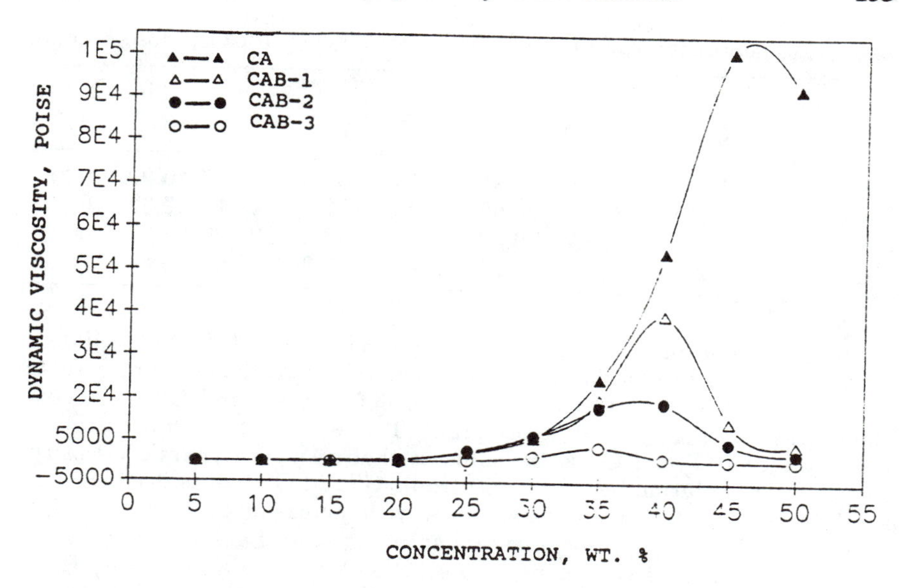

Figure 2. Low-frequency limit (1 rad/sec) of dynamic viscosity vs. concentration for cellulose esters in dimethyl acetamide.

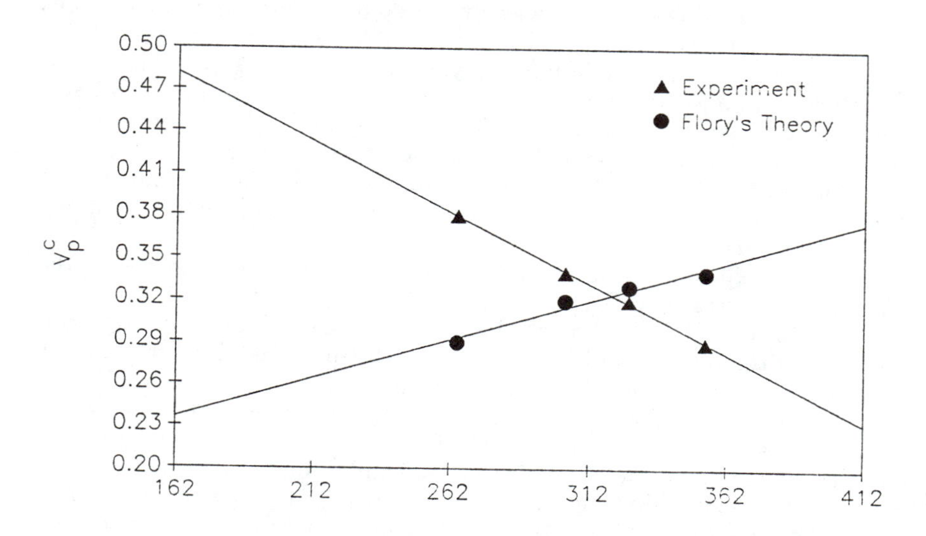

Figure 3. Experimental and theoretical variation of critical volume fraction of cellulose esters with molecular weight of the repeating unit.

Table II. Theoretical Values of Molecular Parameters for Cellulose Esters

Cellulose Esters	ρ, g/cm^3	Mu, g/mole	d, Å	l_k, Å	q, Å	X_k	Th. $V_p{}^c$	Expt. $V_p{}^c$
CA	1.3	264.5	8.06	207.7	103.85	25.77	0.29	0.38
CAB-1	1.2	303.3	8.98	207.7	103.85	23.13	0.32	0.34
CAB-2	1.2	326.2	9.32	207.7	103.85	22.29	0.33	0.32
CAB-3	1.2	354.1	9.71	207.7	103.85	21.40	0.34	0.29

modulus (G″) at different frequency and concentration of the four cellulose ester solutions is summarized in Figures 4 and 5, respectively. G′ represents the amount of energy stored per cycle of deformation and G″ represents the energy loss per cycle of deformation. The feature of the concentration dependence of G′ and G″ is the same as that seen for dynamic viscosity, *i.e.*, after the solution becomes anisotropic G′ and G″ values decline. The solutions are seen to become more elastic with increasing concentration and even in the liquid crystalline phase its nature is that of a liquid rather than of a crystal. This is so because the solution will be highly elastic if the continuous liquid crystal structure is perfectly preserved (37). The liquid crystalline solutions are less elastic and some of the elastic properties can be due to regions of isotropic fluid dispersed in anisotropic regions (18). Similar to any viscoelastic material, G′ increases with increasing frequency. As the DS_{Bu} increases, there is a gradual drop in G′ value and the solutions become less elastic.

G″ rises with increase in frequency. After reaching $V_p{}^c$, G″ falls showing that damping characteristics of liquid crystalline solutions are lower than those of isotropic solutions. The rise of damping behavior with frequency is most pronounced with CA and least with CAB-3.

These results reveal that liquid crystalline state is fluid in nature and these systems are viscoelastic, *i.e.*, they both store and dissipate energy during deformation.

During exposure to shear, such as during viscosity measurements, the phases are microscopically intermingled and they fail to reach a state of true equilibrium. In the absence of shear, phase equilibrium will be established, and the phases can be identified by cross-polarized light microscopy (19). Cross-polarization helps to distinguish isotropic (I), biphasic (B), and anisotropic (A) solution character empirically. Polarized light microscopy observations for solutions containing 5–50% (w/w) cellulose esters in DMAc are summarized in Table III. A distinctive band of biphasic behavior is recognized that is shifted to higher concentrations as butyryl content declines. The liquid crystalline phase appears at approximately the same concentration in this static experiment as the peak viscosity did in the dynamic test (Figure 2). A similar behavior had been observed for other lyotropic systems as well (19).

The photomicrographs of the four cellulose ester solutions illustrate the change in solution morphology in relation to concentration (Figures 6

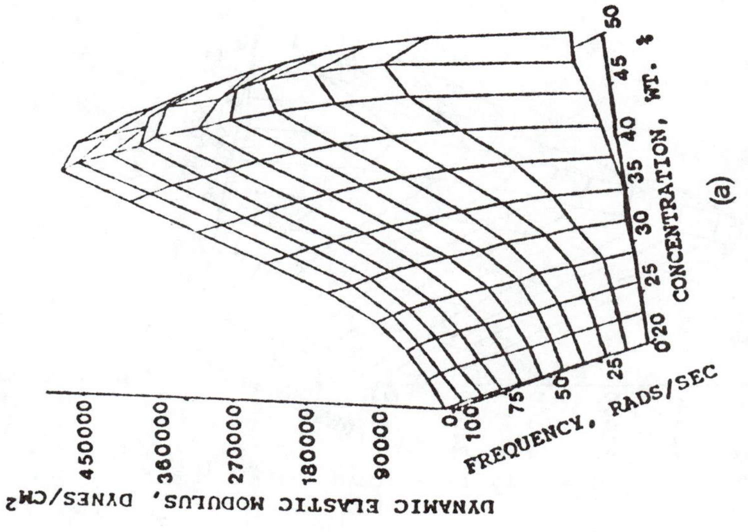

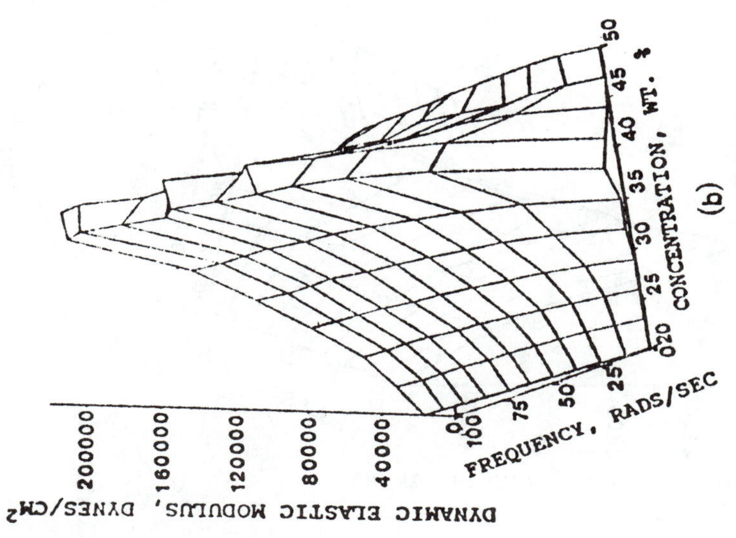

Figure 4. Dynamic elastic modulus vs. frequency for different concentration solutions of cellulose esters in dimethyl acetamide: (a) CA; (b) CAB-1; (c) CAB-2; (d) CAB-3.

Continued next page

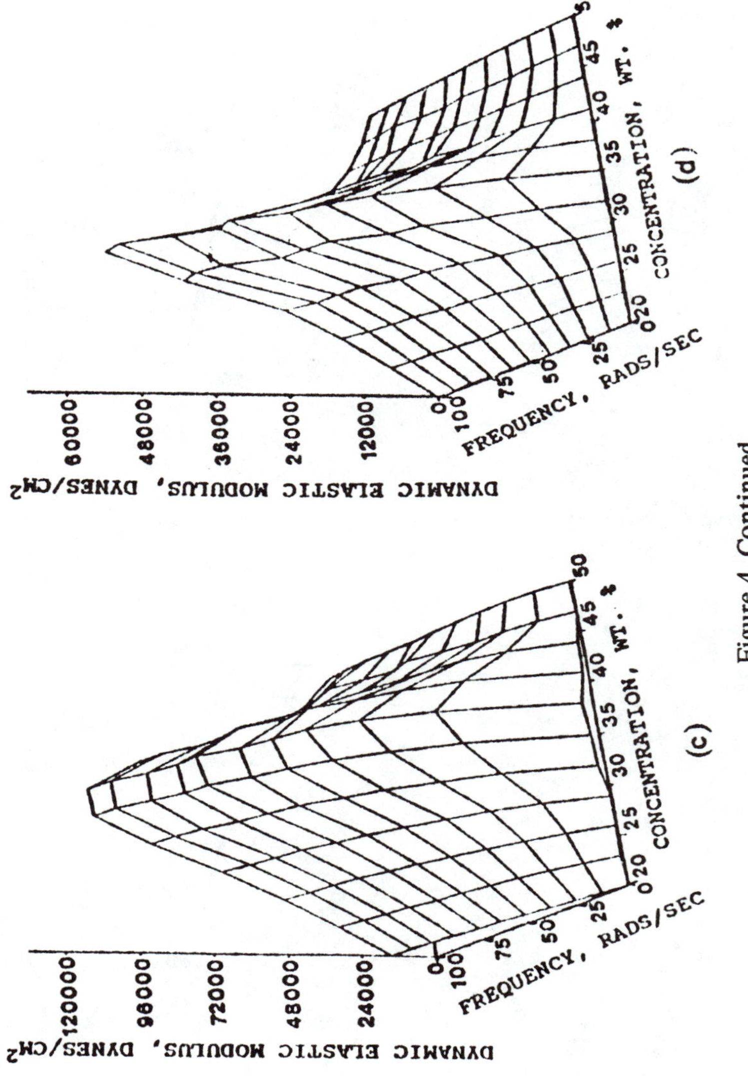

Figure 4. Continued.

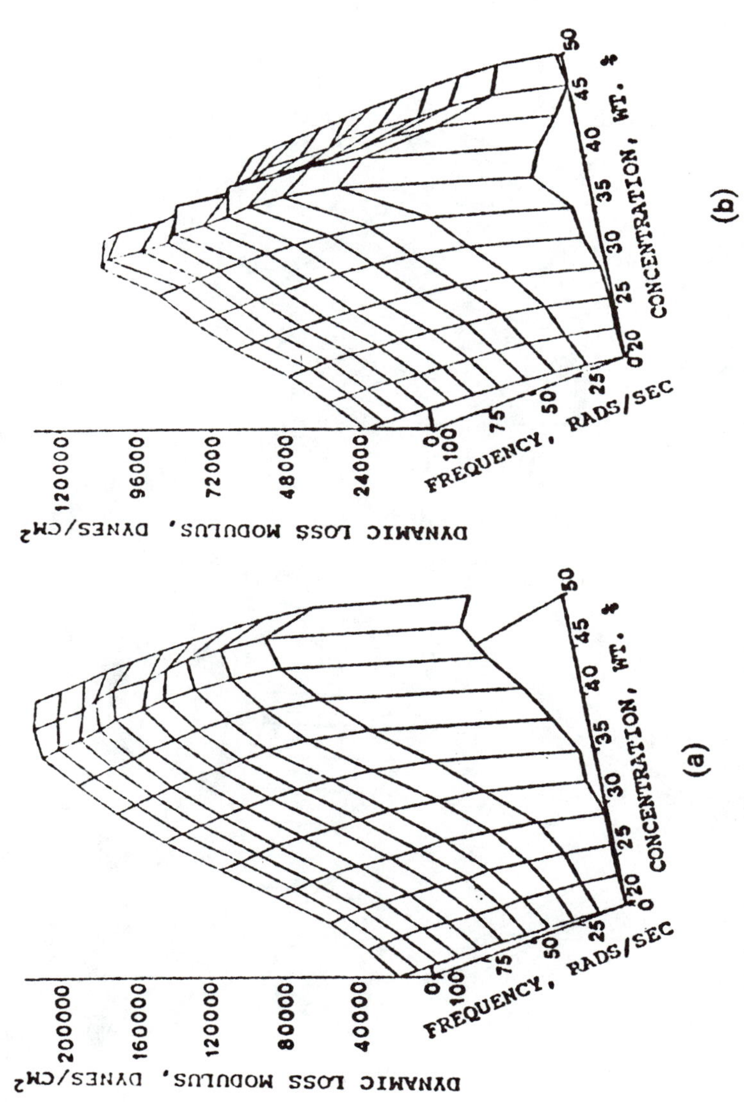

Figure 5. Dynamic loss modulus vs. frequency for different concentration solutions of cellulose esters in dimethyl acetamide: (a) CA; (b) CAB-1; (c) CAB-2; (d) CAB-3.

Continued next page

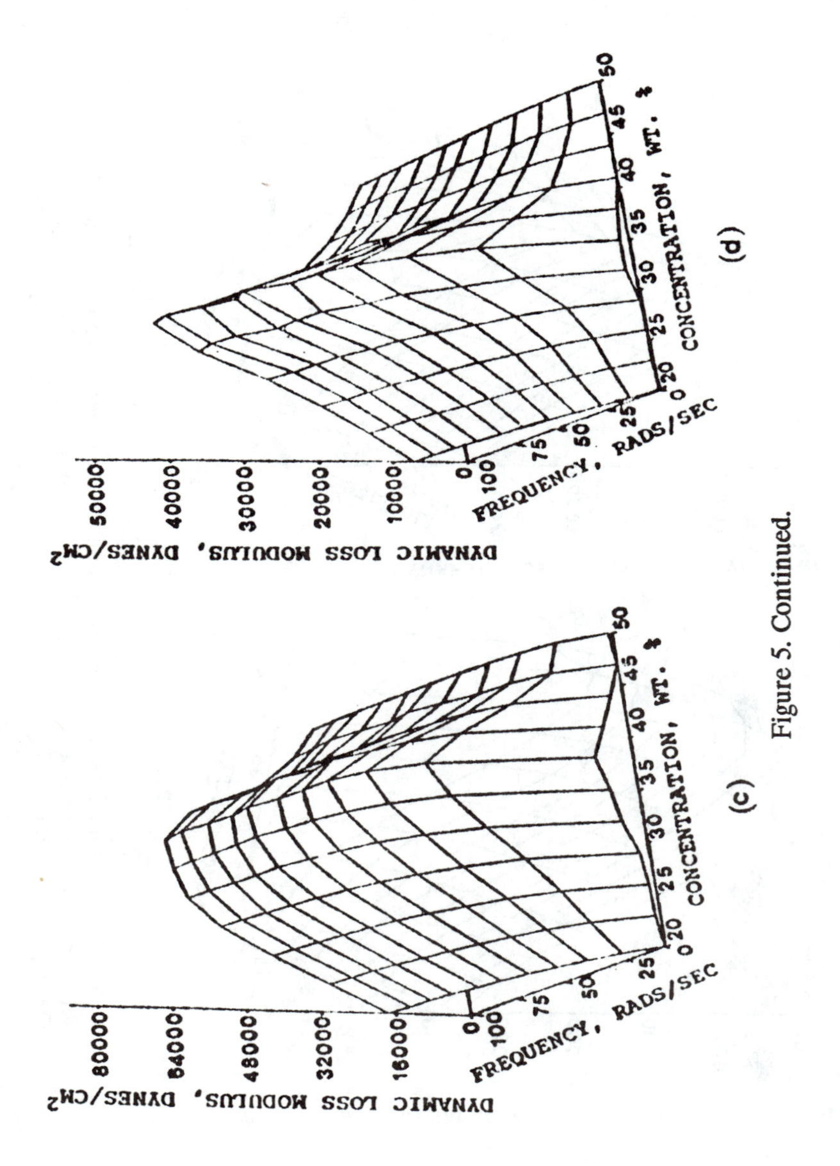

Figure 5. Continued.

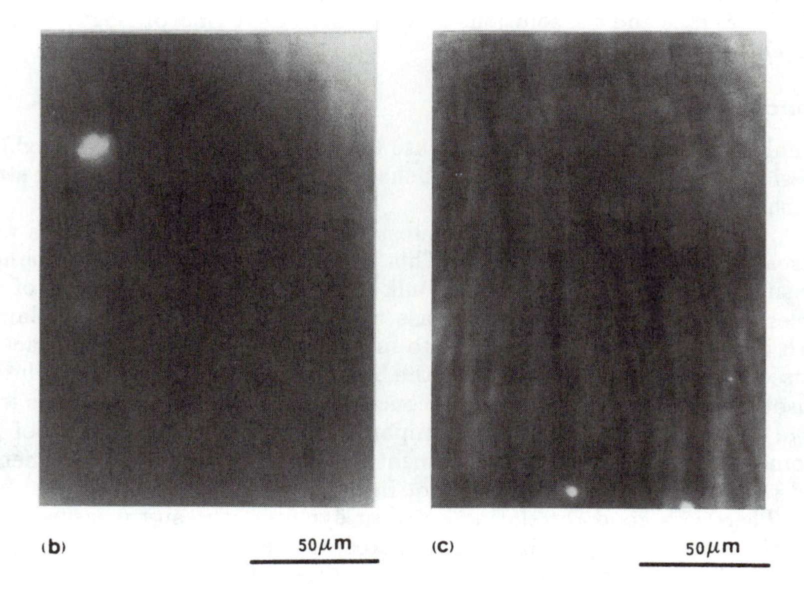

Figure 6. Polarized light optical micrographs of cellulose ester solutions in dimethyl acetamide: (a) CA, 50% (w/w) conc.; (b) CAB-1, 45% (w/w) conc.; (c) CAB-1, 50% (w/w) conc.

Table III. Summary of Polarized Light Microscopy Observations for Solutions Containing 5–50% (w/w) Cellulose Esters in Dimethyl Acetamide

Sample	5%	10%	15%	20%	25%	30%	35%	40%	45%	50%
CA	I	I	I	I	I	I	I	I	I	B
CAB-1	I	I	I	I	I	I	I	I	B	A
CAB-2	I	I	I	I	I	I	I	B	A	A
CAB-3	I	I	I	I	I	I	B	B	A	A

Note: I, B, and A designate isotropic, biphasic, and anisotropic, respectively.

and 7). Figure 6a shows that for CA, the solution becomes birefringent at approximately 50% concentration. The change from B to A for CAB-1 and CAB-2 is shown in Figures 6b and 6c and Figures 7a and 7b, respectively. The transformation from isotropic to anisotropic phase for CAB-3 is seen in Figures 7c and 7d.

A structure-property map (Figure 8) is prepared which summarizes the information from rheology, morphology, and chemistry data for the cellulose ester solutions. $V_p{}^c$ represents the peak viscosity values from Figure 2. The map shows that as the DS_{Bu} (chemistry) increases, the peak viscosity (rheology) falls and the solutions become anisotropic (morphology) at lower concentrations.

Conclusion

Cellulose acetate and cellulose acetate butyrate solutions in DMAc exhibit classical liquid crystalline solution behavior in agreement with earlier studies on cellulose derivatives.

$V_p{}^c$ is strongly related to the substitution pattern and it declines with increasing degree of butyration. This is contrary to the expected behavior since an increased substituent bulk (Mu) increases the diameter of the molecular chain and thereby increase the $V_p{}^c$. This anomaly is explained with increased chain stiffness due to interchain and intrachain interaction caused by substituent crowding on the backbone. This implies that solution anisotropy can be reached at lower concentrations with cellulose esters with large and bulky substituents as compared to smaller ones. In spite of the anomaly, there is fairly good agreement between the absolute experimental and theoretical values of $V_p{}^c$ but not in the trend of $dV_p{}^c/dMu$.

There is a good correlation between dynamic viscosity measurements and the static cross-polarized light microscopy results.

The liquid crystalline state is fluid and the cellulose ester solutions are viscoelastic in nature.

Acknowledgments

Financial support for this study was provided by the NSF Science and Technology Center on High Performance Polymers and Adhesives of Vir-

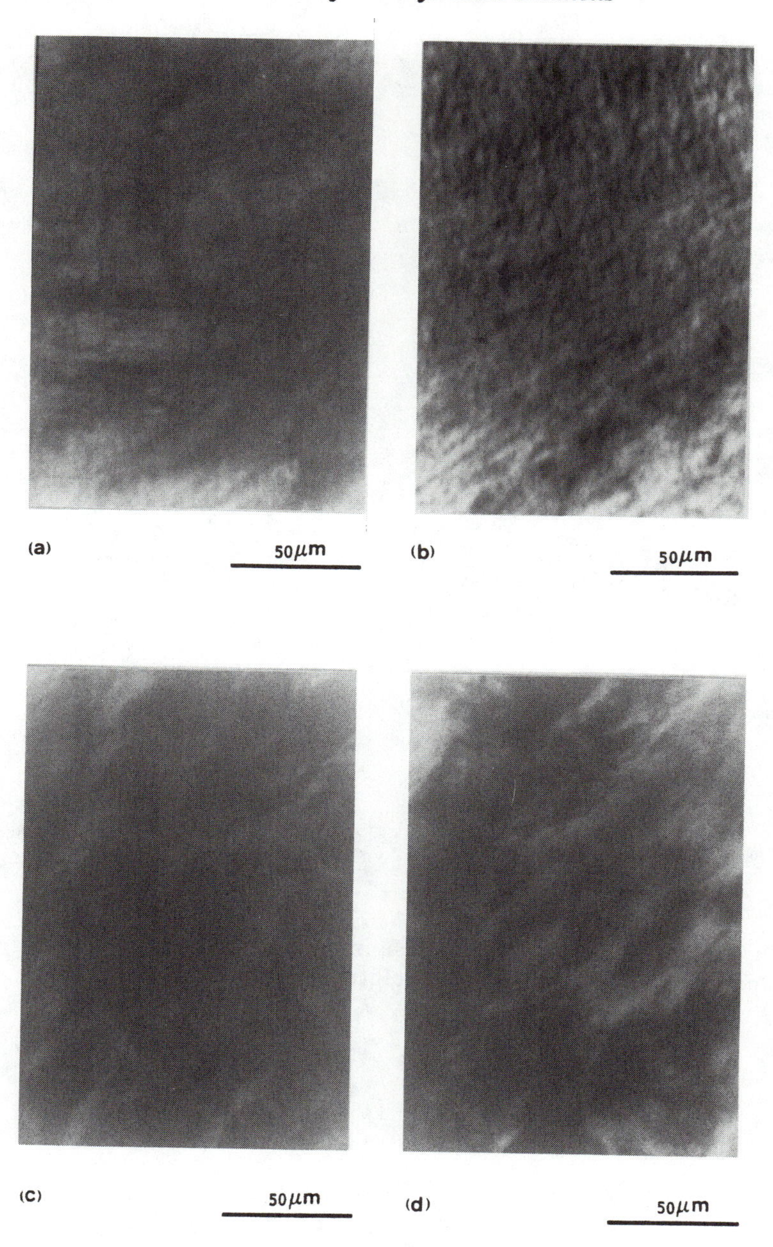

Figure 7. Polarized light optical micrographs of cellulose ester solutions in dimethyl acetamide: (a) CAB-2, 40% (w/w) conc.; (b) (CAB-2, 45% (w/w) conc.; (c) CAB-3, 30% (w/w) conc.; (d) CAB-3, 40% (w/w) conc.

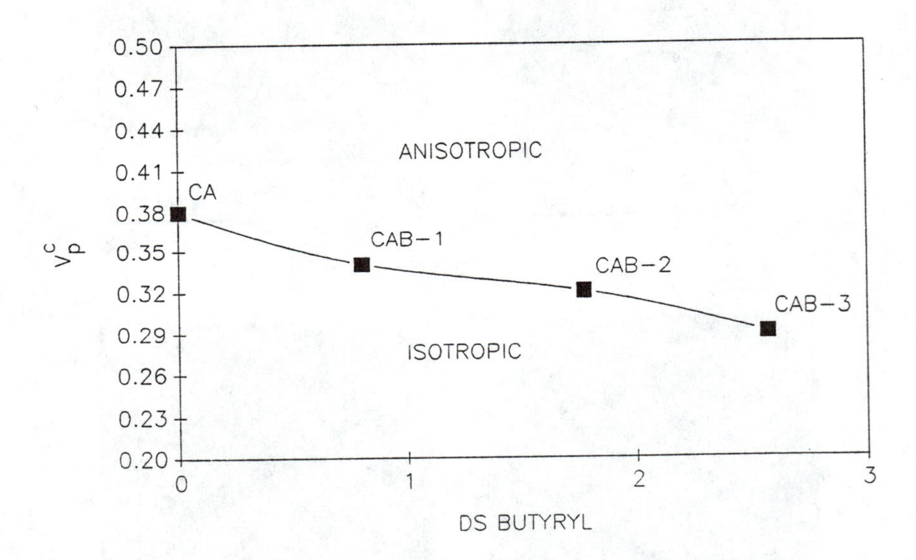

Figure 8. Structure-property map summarizing viscosity, morphology, and chemistry data for cellulose ester solutions in dimethyl acetamide.

Tech, and by the Biobased Materials Center of the Center for Innovative Technology at Virginia Tech. The authors would like to thank Dr. H. Marand (Chemistry) and Dr. D. G. Baird (Chemical Engineering) for their help with polarized light microscopy and RMS, respectively. Helpful counsel by Dr. Rick Davis (Chemical Engineering) is acknowledged with gratitude.

Appendix A

1. Calculation of molecular weights of repeat units, Mu:
   CA; $Mu = 2.44 \times 42 + 162 = 264.5$
   CAB-1; $Mu = 2.03 \times 42 + 0.8 \times 70 + 162 = 303.30$
   CAB-2; $Mu = 0.96 \times 42 + 1.77 \times 70 + 162 = 326.2$
   CAB-3; $Mu = 0.29 \times 42 + 2.57 \times 70 + 162 = 354.1$

2. Calculation of chain diameter, d: $d = (Mu/N_{A\rho} lu)^{1/2}$

   $lu = 5.2\text{Å}(7); N_A = 6.02 \times 10^{23}$ molecules/mole.

   $$d_{CA} = \left(\frac{264.5}{6.02 \times 10^{23} \times 1.3 \times 10^{-24} \times 5.2}\right)^{1/2} = 8.06\text{Å}$$

3. Calculation of Kuhn segment length, l_k: $l_k = < r^2 >_o /nlu$;

   From light-scattering [Tanner and Berry, (35)];

   $< r^2 >_o /n = 1080\text{Å}^2$;

   A narrow range of values of $< r^2 >_o /n$ have been reported for diverse cellulose esters (35); therefore, 1080Å^2 is assumed for all the cellulose esters in this study (see footnote in text).
   $l_k = 1080/5.2 = 207.70\text{Å}$

4. Calculation of persistence length, q: $q = l_k/2 = 207.70/2 = 103.85\text{Å}$

5. Calculation of aspect ratio, X_k: $X_k = l_k/d_{CA} = \frac{207.70}{8.06} = 25.77$

6. Calculation of critical concentration, $V_p{}^c$: $V_p{}^c = (8/X_k)(1 - 2/X_k)$;
   $V_p{}^c{}_{CA} = 8/25.77(1 - 2/25.77) = 0.29$

Literature Cited

1. Papkov, S. P.; Kulichikhin, V. G.; Kalmkoya, V. D.; Malkin, A. Ya. *J. Polym. Sci.: Polym. Phys. Ed.* 1974, **12**, 1753-1770.
2. Morgan, P. W. *Macromolecules* 1977, **10**, 1381-1390.
3. Gray, D. G. *J. Appl. Polym. Sci.: Appl. Polym. Symp.* 1983, **37**, 179-192.
4. Flory, P. J. *Proc. R. Soc. London, Ser. A* 1956, **234**, 60.
5. Gray, D. G. *Faraday Discuss. Chem. Soc.* 1985, **79**, 257-264.
6. Gilbert, R. D.; Patton, P. A. *Prog. Polym. Sci.* 1983, **9**, 115-131.
7. Flory, P. J. *Adv. in Polym. Sci.* 1984, **59**, 2-36.
8. Werbowyj, R. S.; Gray, D. G. *Macromolecules* 1980, **13**(1), 69.

9. Werbowyj, R. S.; Gray, D. G. *Mol. Cryst. Liq. Cryst.* 1976, **34**, 97.
10. Conio, G.; Bianchi, E.; Ciferri, A.; Tealdi, A.; Aden, M. A. *Macromolecules* 1983, **16**, 1264-1270.
11. Aden, M. A.; Bianchi, E.; Ciferri, A.; Conio, G.; Tealdi, A. *Macromolecules* 1984, **17**, 2010-2015.
12. Bheda, J.; Fellers, J. F.; White, J. L. *Colloid and Polymer Sci.* 1980, **258**, 1335-1342.
13. Aharoni, S. M. *Mol. Cryst. Liq. Cryst.* 1980, **56**, 237.
14. Bhadani, S. N.; Tseng, S. L.; Gray, D. G. *Makromol. Chem.* 1983, **184**, 1727-1740.
15. Chanzy, H.; Peguy, A.; Chaunis, S.; Monzie, P. *J. Polym. Sci.: Polym. Phys. Ed.* 1980, **18**, 1137.
16. Patel, D. L.; Gilbert, R. D. *J. Polym. Sci.: Polym. Phys. Ed.* 1981, **19**, 1231.
17. Baird, D. G. In *Polymeric Liquid Crystals*; Blumstein, A., Ed.; New York: Plenum Press, 1985; pp 119-143.
18. Baird, D. G. In *Rheology*; Astarita, G.; Marrucci, G.; Nicolais, L., Eds.; New York: Plenum Press, 1980; Vol. 3, pp 647-658.
19. Aharoni, S. M. *Polymer* 1980, **21**, 1413-1422.
20. Buchanan, C. M.; Hyatt, J. A.; Lowman, D. W. *Macromolecules* 1987, **20**, 2750-2754.
21. Buchanan, C. M.; Hyatt, J. A.; Lowman, D. W. *Carbohydr. Res.* 1988, **177**, 228-234.
22. Dayan, S.; Maissa, P.; Vellutini, M. J.; Sixou, P. *J. Polym. Sci.: Polym. Lett. Ed.* 1982, **20**, 33-43.
23. Bird, R. B.; Armstrong, R. C.; Hassager, O. In *Dynamics of Polymeric Liquids*; 1987; Vol. 1, p 150.
24. Dayan, S.; Gilli, J. M.; Sixou, P. *J. Appl. Polym. Sci.* 1983, **28**, 1527-1534.
25. Suto, S. *J. Polym. Sci.: Polym. Phys. Ed.* 1984, **22**, 637-646.
26. Navard, P.; Haudin, J. M. *J. Polym. Sci.: Polym. Phys. Ed.* 1986, **35**, 189-201.
27. Hong, Y. K.; Hawkinson, D. E.; Kohout, E.; Garrard, A.; Fornes, R. E.; Gilbert, R. D. In *Polymer Association Structure*; El-Nokaly, M. A., Ed.; ACS Symp. Ser. 384, 1989; Ch. 12, pp 184-203.
28. Bheda, J.; Fellers, J. G.; White, J. L. *J. Appl. Polym. Sci.* 1981, **26**, 3955-3961.
29. Baird, D. G. *J. Rheology* 1980, **24**(4), 465-482.
30. Kiss, G.; Porter, R. S. *J. Polym. Sci.: Polym. Symp. #65* 1978, 193-211.
31. Hermans, J. *J. Colloid Sci.* 1962, **17**, 638.
32. Kiss, G.; Porter, R. S. *J. Polym. Sci.: Polym. Phys. Ed.* 1980, **18**, 361.
33. Wissbrun, K. F. *J. Rheology* 1981, **25**(6), 619-662.
34. Onogi, S.; Asada, T. In *Rheology*; Astarita, G.; Marrucci, G.; Nicolais, L., Eds.; New York: Plenum Press, 1980; Vol. I, pp 127-147.
35. Tanner, D. W.; Berry, G. C. *J. Polym. Sci.: Polym. Phys. Ed.* 1974, **12**, 941.

36. Flory, P. J.; Ronca, G. *Mol. Cryst. Liq. Cryst.* 1979, **54**, 311.
37. Iizuka, E. *Mol. Cryst. Liq. Cryst.* 1974, **25**, 287-298.
38. Kamide, K.; Miyazaki, Y.; Abe, T. *Polym. J.* 1979, **11**, 523.
39. Vyas, N. G.; Shashikant, S.; Patel, C. K.; Patel, R. D. *J. Polym. Sci.: Polym. Phys. Ed.* 1979, **17**, 2021.
40. Goodlett, V. W.; Dougherty, J. T.; Patton, H. W. *J. Polym. Sci.* 1971, **A-1, 9**, 155.
41. Vasilév, B. V.; Grishin, E. P.; Zhegalova, N. N.; Malinine, L. P.; Pogosov, Yu. L. *Vysokomol. Soedin.* 1974, **Ser. A, 16**, 136.
42. Baker, W. O. *Ind. Eng. Chem.* 1945, **37**, 246.

RECEIVED December 16, 1991

Biogels and Gelation

Chapter 11

Molecular Transformations in Connective Tissue Hyaluronic Acid

Glyn O. Phillips

North East Wales Institute of Higher Education, Connah's Quay, Clwyd CH5 4BR, Wales

Free radicals, either induced by the action of ionizing radiations or produced by metal ion induced electron transfer reactions *in situ*, can initiate a marked reduction in the viscoelasticity of the connective tissue matrix. This paper examines the dominant role of hyaluronic acid in controlling this behavior at molecular level. Our results indicate that after a dose of 5Gy (500 rads), the average molecular weight of hyaluronic acid in skin would be reduced by a factor of 4, which would lead to a 60-fold reduction in viscosity of the glycosaminoglycan. Shorter chains so produced would further inhibit hyperentanglement and chain-chain interactions which are responsible for the viscoelasticity of the hyaluronic acid-polymer network.

This paper is concerned with the molecular properties which govern the viscoelasticity of the connective tissue, with particular emphasis on the effect of free radical systems on the integrity of the matrix. The subject is directly relevant to Human Tissue Banking, when the viscoelastic and mechanical properties of the membraneous allografts (skin, fascia lata, dura mater, amnion, cartilage) need to be retained following processing and radiation sterilization. More than 500,000 such allografts are now annually used in the USA alone. These complement other chemically modified biomaterials (1). *In vivo* also there is required a better understanding at the molecular level of the extreme vulnerability of connective tissue to low doses of radiation. Our studies are directed to examining the major role of hyaluronic acid, in controlling the viscoelastic properties of the bulk connective tissue.

0097–6156/92/0489–0168$06.00/0

Hyaluronic Acid in Connective Tissue

Balazs (2,3) has devoted a lifetime of research to identifying the relationship between chemical structure, molecular shape and viscosity of hyaluronic acid within the intercellular matrix. This matrix is the material which fills the space between cells in such diverse tissues as skin, tendons, muscles, and cartilage. A major chemical component of connective tissue is hyaluronic acid, which is a glycosaminoglycan. The molecular configuration and structure of the repeating disaccharide unit of hyaluronic acid is shown in Figure 1. The long linear chain is unbranched, and at physiological pH, due to its polyanionic character, exhibits considerable stiffness, which is responsible for its large hydrated volume. Light scattering shows that the long unbranched chain can form a random coil occupying a large water-containing molecular domain. The hyaluronic acid needs to have a molecular weight in excess of 2×10^6 to form such a large hydrated molecular volume of diameter approximately 300 nm. Balazs (4) has illustrated this behavior in Figure 1, when the individual molecular domains form an interpenetrating molecular network ("hydrated molecular sponge"). There is considerable evidence for entanglement between neighboring molecules. From measurements of its sedimentation constant and viscosity, the hyaluronic acid from synovial fluid has a MW of 8-9 $\times 10^6$ and a hydrodynamic volume which is 10^3 larger than the space occupied by the unhydrated polysaccharide chain. For this Random-Coil-Solvated Sphere Model, the radius of gyration approximates to the radius of the solvated sphere, given by hydrodynamic measurements (5). The ability to trap water by hyaluronic acid is considerably greater than other polysaccharides. A 2% solution of pure hyaluronic acid holds the water so tightly that it can be picked up as though it were a gel. Yet it is not a gel, since it is a true liquid capable of being diluted and exhibiting viscous flow, elastic and pseudoelastic properties. The combination of high molecular weight, and large molecular volume forces the overlap between individual hyaluronic and molecular domains, resulting in extensive chain entanglement and chain-chain interaction. The other glycosaminoglycans of connective tissue (heparin, chondroitin sulphates, keratan sulphate) do not exhibit this viscoelastic polymer network behavior. The resulting intertwined polymeric network acts as a jelly-like milieu supporting and influencing tissue function (6,7). In Figure 2 Balazs has schematically depicted the entangled hyaluronic acid chains filling the interfibrillar spaces of a collagen gel, holding the water like a "molecular sponge."

Thus hyaluronic acid solutions show quite characteristic rheological properties. It is the molecular dimensions of the individual molecules and their potential for intermolecular interactions that determines whether a particular hyaluronic acid preparation will form an elastoviscous matrix under a given set of conditions (7). For random coil solutions the limiting viscosity $[\eta]$ is related to the molecular weight M_r by the Flory-Fox relationship:

$$[\eta] = \Phi(r^2)^{3/2}/M_r$$

where Φ is a constant, and $(r^2)^{1/2}$ is the root mean square end to end distance.

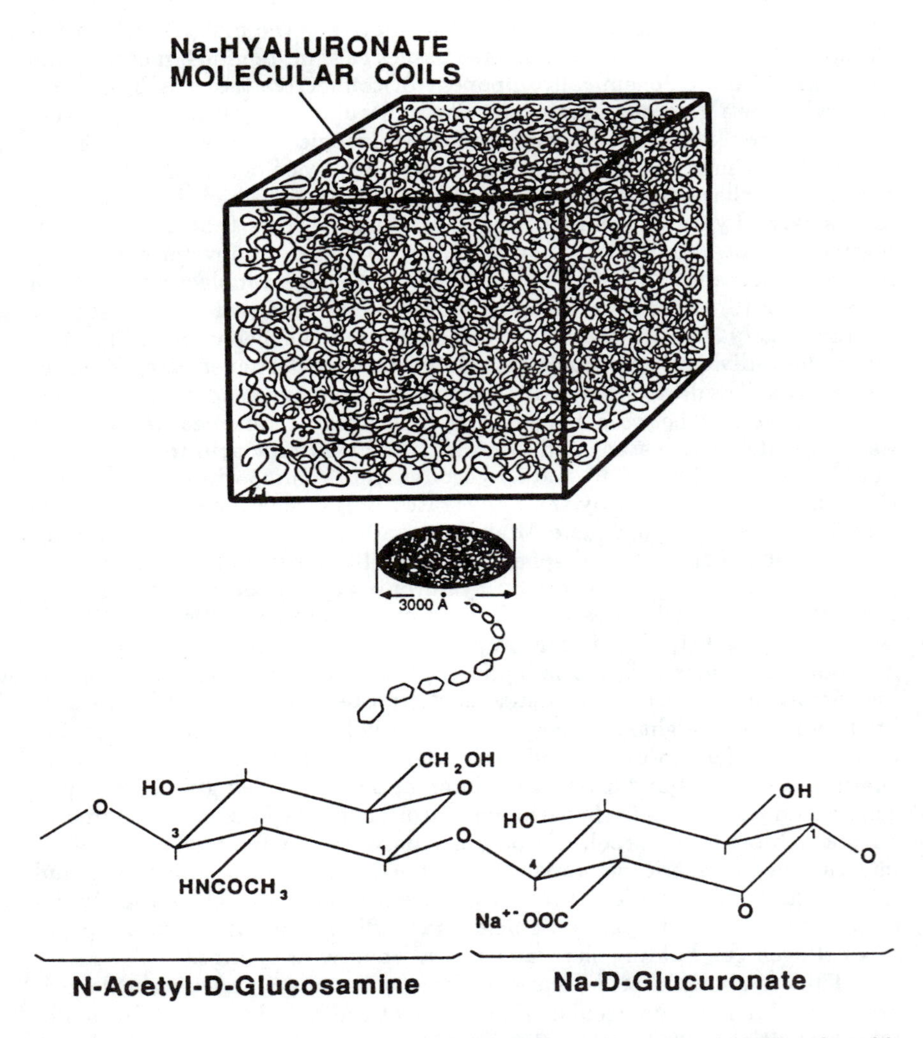

Figure 1. The Balazs illustration of the molecular dimensions of hyaluronic acid. (Reproduced with permission from reference 3. Copyright 1983 Wiley.)

Thus treating each coil as a sphere of radius R, the hydrodynamic volume is proportional to $[\eta][M_r]$. Since the total number of coils is proportional to C/M_r where C is the concentration, the degree of occupancy in space is characterized by $C[\eta]$, the coil overlap parameter, a dimensionless quantity. Based on the Flory-Fox relationship, hyaluronic acid molecules should begin their entanglement behaviour at a coil-overlap of $C[\eta]$ equal to 1.5 (8). Such molecular entanglement can be observed near ($C[\eta] = 2$) this theoretical value (Figure 3). The onset of entanglement and subsequent concentration dependence of the specific viscosity has been compared for a number of polysaccharides which include carboxymethyl cellulose, dextran, alginate, carrageenan, guar gum, locust bean gum and hyaluronic acid (8). Generally for polysaccharides, entanglement started at $C[\eta] = 4$ and $\eta_{sp} = 10$. The 'zero shear' specific viscosity η_{sp} varies as $C^{1.4}$ for dilute solutions and $C^{3.3}$ for concentrated solutions. The only deviations noted were guar, locust bean and hyaluronic acid. These show an even greater concentration dependence, and entanglement occurs at a lower concentration. The "normal" polysaccharides exhibit identical behaviour to non-associative polymer systems, such as polystyrene in toluene. The difference has been attributed to the formation of specific junctions between stiff, structurally related sequences. Such specific chain-chain association in hyaluronic acid increases with increasing concentration, thus in effect raising the average molecular weight, hence viscosity. However, on addition of smaller hyaluronic segments of $\sim$ 4 or 400 disaccharide units, all evidence of coupling is lost. These "hyper-entanglement" interactions have a specificity and different lifetime to the usual entanglement and are in our view of considerable significance in interpreting the behavior of the connective tissue matrix when attacked by free radical systems.

Elasticity is also dependent on matrix formation, and is related to $C[\eta]$. Figure 4 shows the linear dependence between dynamic storage modulus (G′) and coil overlap parameter $C[\eta]$. If the hyaluronic acid is present as isolated molecules, not a matrix, either because of low concentration or low molecular weight, the system exhibits no elastic behavior. Using the limiting viscosity it is possible to calculate the minimum concentration for matrix formation as a function of molecular weight (Figure 5). It demonstrates that the higher the molecular weight of a hyaluronic acid preparation, the lower the concentration necessary for matrix formation to occur. It also demonstrates why two different hyaluronic acid solutions can have completely different physical behavior even if identical in concentration and composition. It has been shown (8) that the elasticity (G′) of the hyaluronic acid matrix is also reduced by an order of magnitude by addition of shorter hyaluronic acid segments. They appear to eliminate network structure and reduce solution elasticity compared with that expected for isolated chains. This behavior is incompatible with entanglement being the sole mechanism of intermolecular coupling.

It is evident, therefore, that degradation of the hyaluronic acid will result in a dramatic effect on the structure and viscoelasticity of the connective tissue matrix.

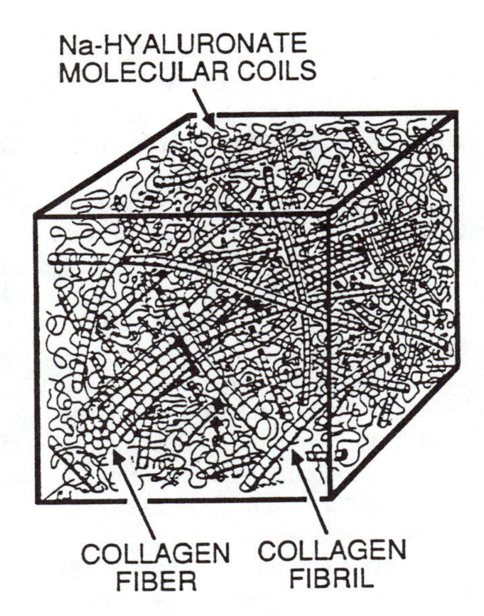

Figure 2. The Balazs illustration of a hyaluronic acid collagen gel. (Reproduced with permission from reference 3. Copyright 1983 Wiley.)

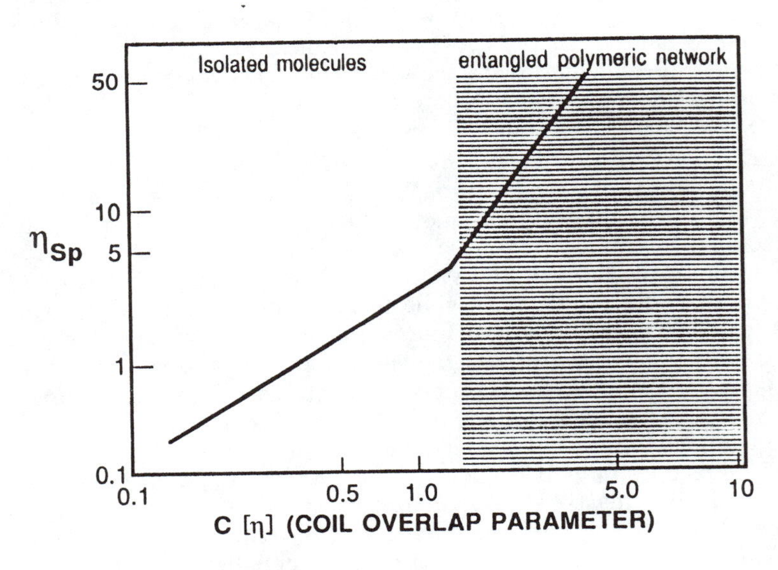

Figure 3. Abrupt discontinuity in the specific viscosity of an HA solution at the point of chain entanglements. (Reproduced with permission from reference 7. Copyright 1985.)

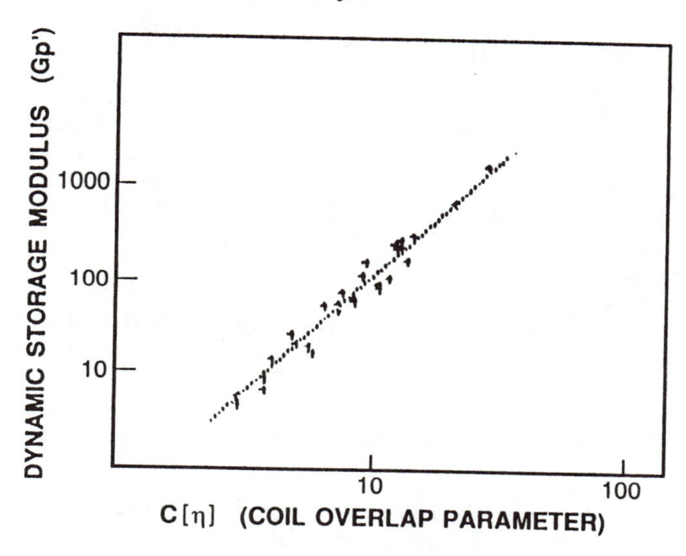

Figure 4. Dependence of matrix elasticity on HA chain entanglement, as measured by the coil overlap parameter. (Reproduced with permission from reference 7. Copyright 1985.)

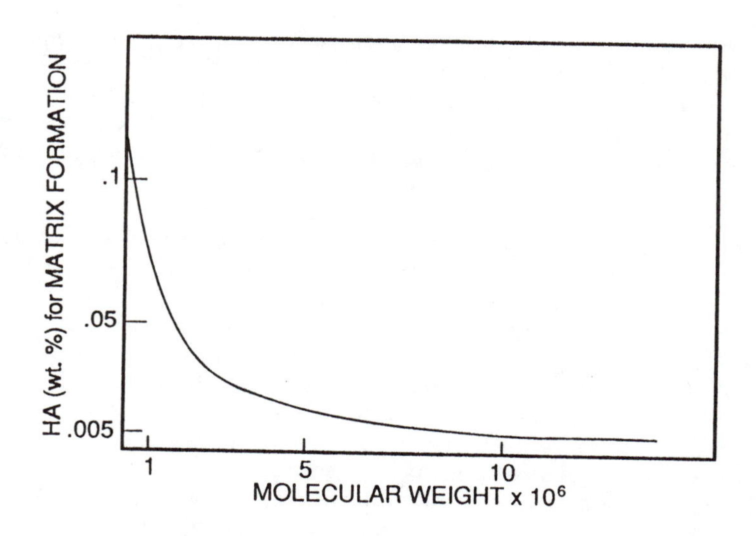

Figure 5. Molecular weight dependence of HA matrix formation. (Reproduced with permission from reference 7. Copyright 1985.)

Radiation Susceptibility of Hyaluronic Acid Systems

The Vitreous and Collagen Gels. The vitreous in most animals and humans is a collagen gel, which contains hyaluronic acid, and proteins. When exposed to ionizing radiations, the hyaluronic acid in the enucleated eyeball or isolated vitreous gel is degraded. The stability of the matrix, moreover, is correspondingly decreased (Figure 6), and this destabilization can be observed at doses well below 1000 rads (9). Collagen gels containing various concentrations of hyaluronic acid also show a dependence of stability on radiation dose as well as on the concentration of hyaluronic acid present (Figure 7). Irradiation leads to partial collapse of the gel, with separation of the liquid. However, irradiation does not alter liquid formation in gels which contain no hyaluronic acid. It is the stabilizing effect of the undegraded hyaluronic acid which is responsible for retaining the integrity of the gel. Conversely, when the hyaluronic acid is degraded there is collapse of both the vitreous body and reconstituted gels in an extracellular matrix.

Whole Connective Tissue. The classical work of Brinkman and Lamberts (10) on the radiation effects on connective tissue can now be related to these observations and are highly relevant to our present study.

They measured the effects of radiation on the water permeability of subepidermal peeled-off skin membranes of the rat. Such membranes are built up of several separate layers of fine interlaced fibrils bedded in the glycosaminoglycan matrix. Figure 8 shows the immediate increase in water permeability with 30 rads or less. Radical scavengers can provide radiation protection (Figure 9).

A similar phenomenon can be illustrated using another technique in human skin. Brinkman and Lamberts (10) measured the injection pressure in the subephithelial skin tissue, using a fine needle carefully inserted into the thin dermal layer. Saline solution is injected at a slow and constant rate and the pressure built up by this infusion can be measured. Irradiation of the spot around the needle tip causes an immediate fall in pressure. Figure 10 shows that a dose as low as 80 rads causes an immediate fall in pressure. If the irradiation precedes the injection, a much higher dose is needed to prevent the pressure from reaching a value of ±100 min Hg, the original injection pressure before irradiation.

Our own *in vivo* results (11) support the view that chemical degradation of the hyaluronic acid occurs in the connective tissue when exposed to doses of up to 1000 rads. Preferential degradation of the hexosamine moiety occurs, as shown by the reduction in hexosamine to hydroxyproline ratio after extraction with acetic acid. The enhanced effect when acetic acid is present points also to greater accessibility of the chemically modified hexosamine moiety after irradiation. Histologically, after staining with the cationic dyes, toluidine blue or methylene blue, reduced metachromasia is evident after 500 rads irradiation, indicative of depolymerization of the glycosaminoglycan component.

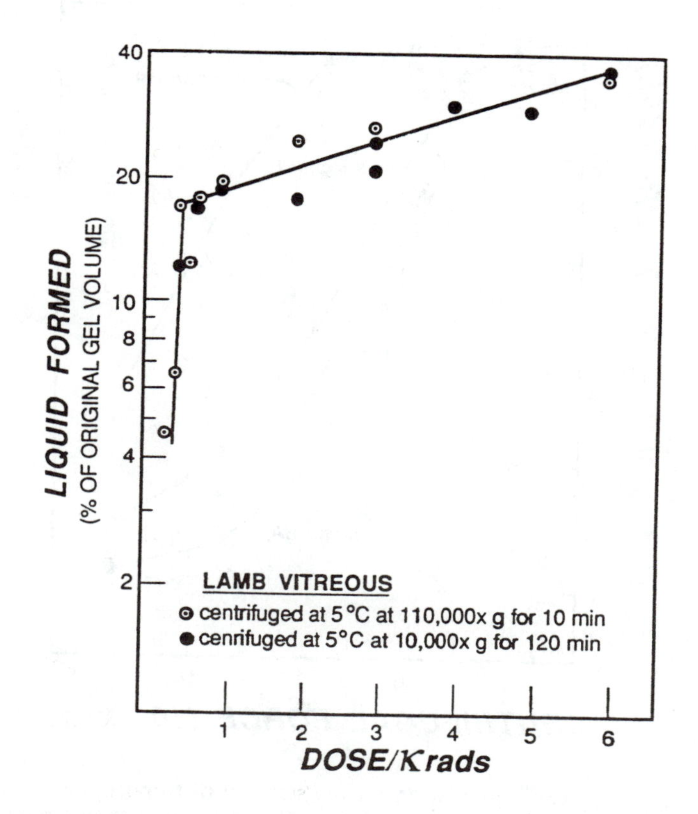

Figure 6. Radiation sensitivity of ovine vitreous irradiated with Co^{60} γ-radiation. (Reproduced with permission from reference 9. Copyright 1966 Academic.)

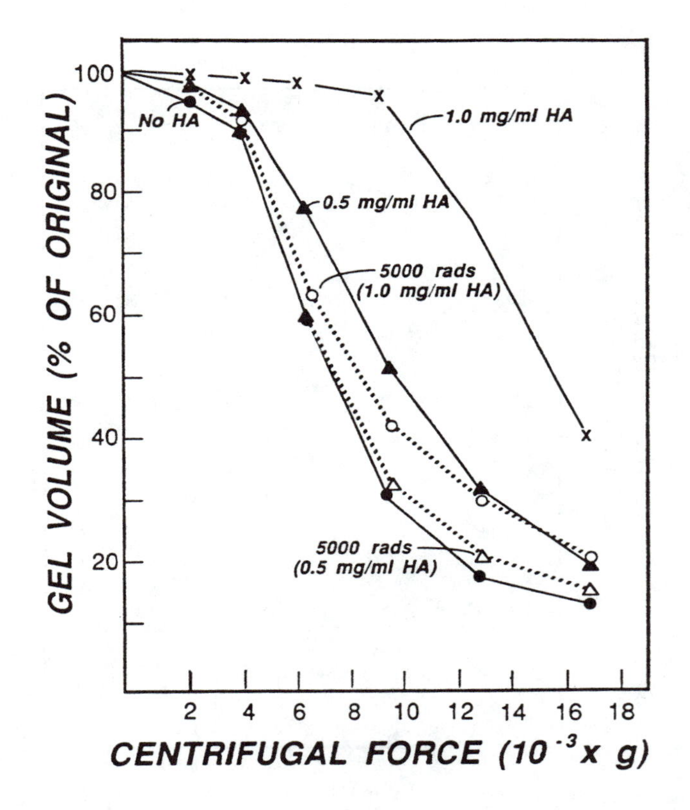

Figure 7. Effect of Co^{60} γ-radiation on the stability of tropocollagen gels in the presence and absence of hyaluronic acid. (Reproduced with permission from reference 9. Copyright 1966 Academic.)

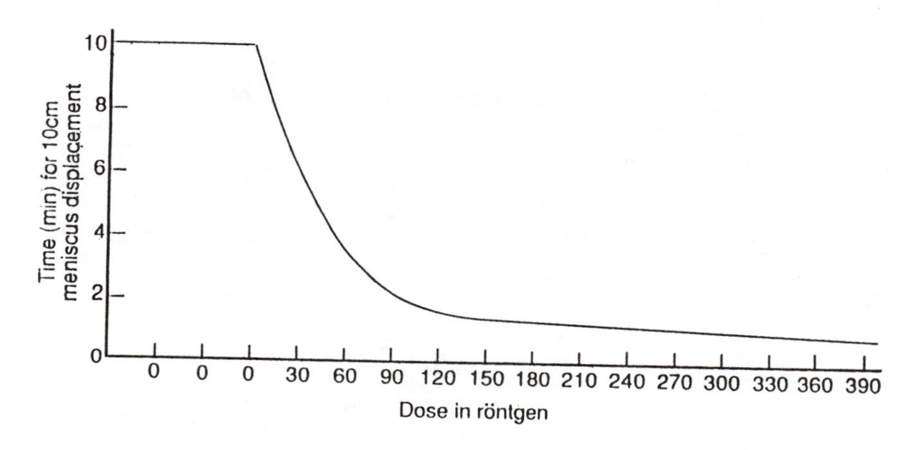

Figure 8. Increased water permeability of a connective tissue membrane by X-irradiation. (Reproduced with permission from reference 10. Copyright 1968 Academic.)

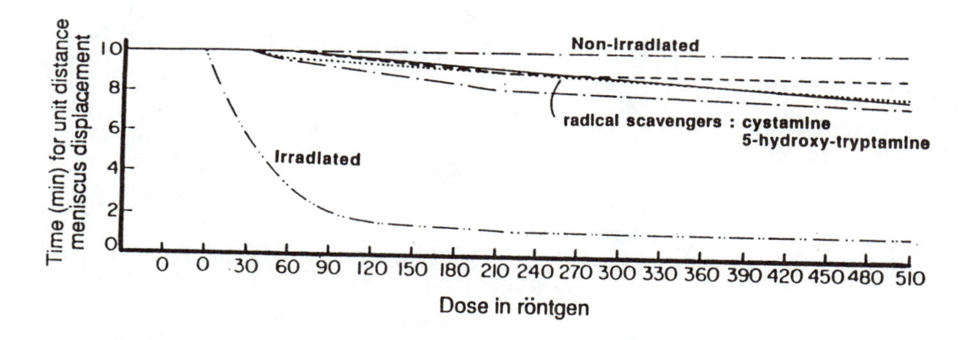

Figure 9. Chemoprotection against the radiation-induced increase of water permeability of connective tissue membranes. (Reproduced with permission from reference 10. Copyright 1968 Academic.)

The question now is whether we can account for these phenomena at the molecular level. How does hyaluronic acid break down on irradiation? Can we quantitatively account for the collapse in viscoelasticity of such connective matrices by such low doses? Collagen itself is highly stable to 3.5 Mrads, since irradiation is now the method of choice for sterilizing catgut sutures.

Hyaluronic Acid. The degradation of hyaluronic acid in the synovial fluid of rheumatoid arthritis patients is now well established (12,13). During the disease there is an increase in iron deposition, which accumulates in the membrane, synovial tissue and synovia (13, 14). This classical work of Pigman *et al.* into the role of metal ions and oxygen derived free radicals (16) suggested that the OH radical participated in the degradation. Our own studies have identified how metal ions can catalyze the degradation through the formation of superoxide.

$$O_{2}- + Fe^{3+} \longrightarrow Fe^{2+} + O_2$$

$$Fe^{2+} + H_2O_2 \longrightarrow Fe^{3+} + OH + OH^-$$

Copper ions can similarly lead to an OH-like species, although the mechanism is more complex (17).

Using pulse radiolysis and steady state irradiation studies we have quantified the reactivity of hyaluronic acid and related carbohydrates towards free radicals derived from water (18). OH radical reacts rapidly with hyaluronic acid ($k_2 = 8.8 \times 10^8 M^{-1} s^{-1}$). The radiation chemistry of hyaluronic acid has been reported in detail by us elsewhere (19-21). Here we summarize the findings relevant to the factors which influence the viscoelasticity of the connective tissue.

Random Chain Scission. The effect of γ-irradiation is to induce degradation in hyaluronic acid, and the effect is linearly dependent on dose (Figure 11). Viscosity average (M_v) and weight average molecular weight (M_n) changes for a series of hyaluronic acid samples degraded to different degrees by OH radicals are linearly related (Figure 12). This is symptomatic of random scission. Of the hyaluronic acid radicals formed by OH radical abstraction reactions, 60% lead to chain scission (G chain breaks = 3.5). The presence of oxygen reduces the yield of breaks by about 40%. Here, as in all radiation chemical yields quoted in this paper, G value refers to the yield per 100 eV. This chain scission process was studied quantitatively based on the pulse-conductivity technique of Bothe and Schulte-Frohlinde (22)

Mechanism. OH radicals interact with the polysaccharide by H-abstraction ($k_2 = 8.8 \times 10^8 M^{-1} s^{-1}$):

$$R - H + \bullet OH \longrightarrow R\bullet + H_2O$$

Of these radicals, 60% lead to chain breaks, and the remainder to chemical modification of the monomeric units. Our unpublished results with

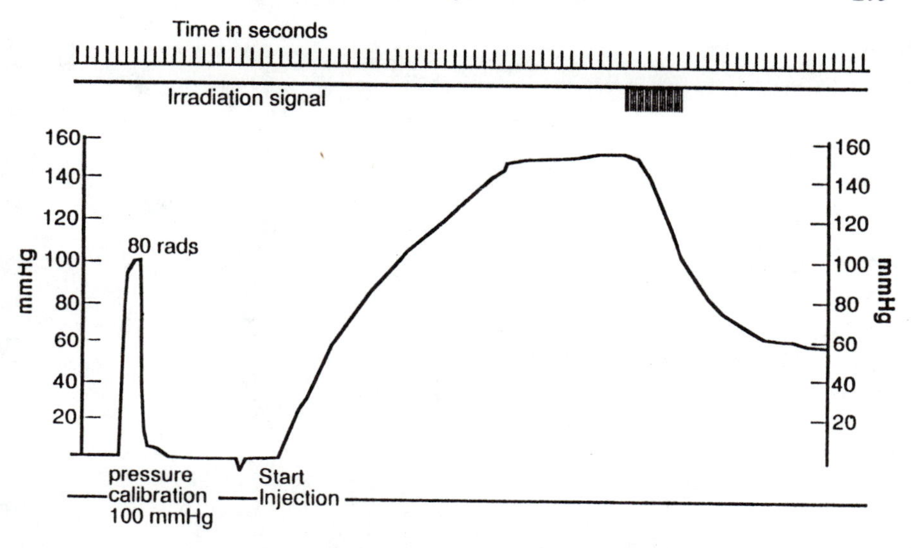

Figure 10. Effect of X-irradiation on the injection pressure in the human skin. (Reproduced with permission from reference 10. Copyright 1968 Academic.)

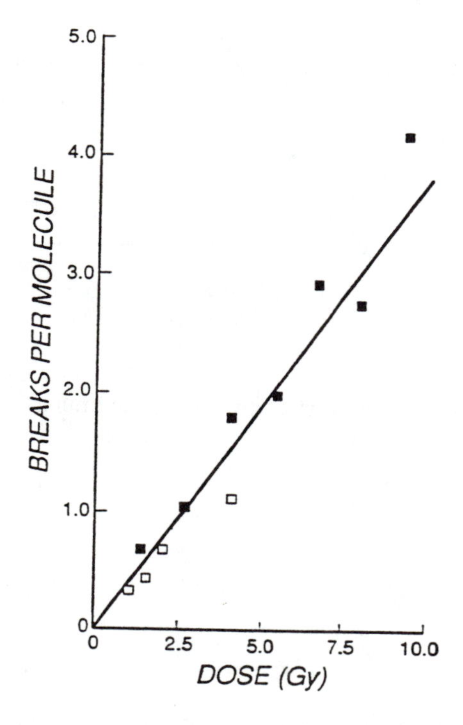

Figure 11. Average number of breaks per molecule as a function of dose.

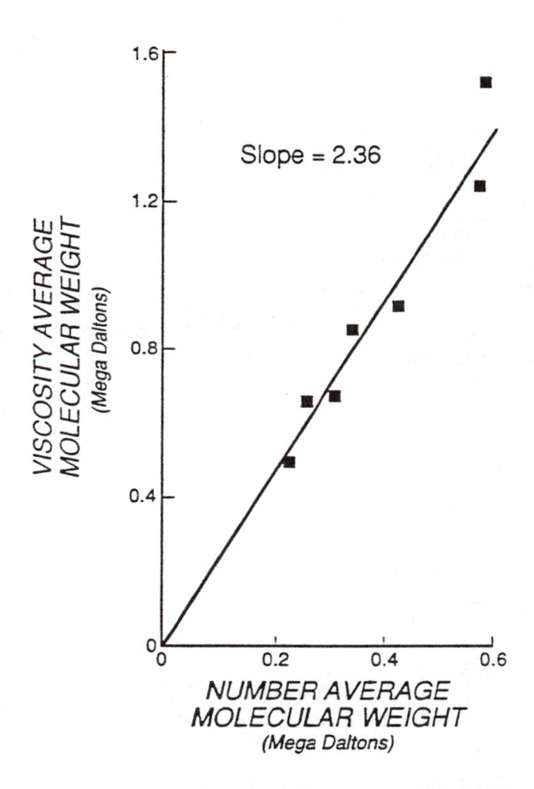

Figure 12. Relationship between viscosity average and number average molecular weight of HA: An indication of random chain cleavage.

the cycloamyloses indicate that there is spontaneous ring scission at the initially formed α-alkoxy radicals with a rate constant of ~ $80s^{-1}$. Many factors influence the behavior of the initially formed radicals. Both acid and alkaline pH's catalyze the rate of strand breakage. Particularly significant is the doubling of G value for chain breaks (pulse radiolysis, in the presence of O_2) at pH 10.6 compared with pH 9.7 and a seven-fold increase compared with pH 7. The kinetics, moreover, point clearly to second-order radical-radical reactions leading to enhanced chain breakage.

Stability of the Matrix to Free Radicals

Our observations on the chain scission reactions of hyaluronic acid induced by OH radicals were undertaken in dilute aqueous solution. Nevertheless, they provide a basis for understanding the dramatic changes in viscoelastic properties of the connective tissue matrix on exposure to ionizing radiation. The stability towards free radicals is clearly dependent on the hyaluronic acid component, which in itself is responsible for the viscoelasticity of the polymer network. It is the chain entanglement and chain-chain interactions which are responsible for the jelly-like milieu, characteristic of tissue. Extremely small doses (~ 30 rads) can disrupt this structure. The protection afforded by radical scavengers confirmed the free radical nature of the degradative action. Of the free radicals formed by ionizing radiations in water (H, OH and e^-_{aq}), we have shown that the OH radical is the most influential in initiating chain breaks.

In the matrix, the mobility of the polysaccharide based radicals produced after H-atom abstraction by OH radicals is reduced and the second order reactions between these radicals will be considerably retarded. The longer lifetime of these radicals combined with a favorable stereochemistry could permit the occurrence of chain reactions by their abstracting an H-atom from a neighboring molecule. Previously, such highly efficient chain reactions have been demonstrated in crystalline sugars, leading to G values of > 650 in 2-deoxy D-erythro pentose, for example (23). Once initiated, these processes proceed with insignificant activation energy (< 10 kcal mol^{-1}).

In the presence of oxygen, hyaluronic acid radicals produce $O_2{\cdot}^-$, which can diffuse freely through the tissue matrix and react with other radicals or with metal ions. Additional chain breakage may result from such reactions. Even without taking such 'extra' chain breaks in the matrix, our results on the gamma irradiation of pure and dilute hyaluronic acid solutions predict that after a dose of 5 Gy (500 rads) the average molecular weight of hyaluronic acid in skin would be reduced by a factor of 4. This alone, based on the quantitative data we have presented, would lead to a 60-fold reduction in viscosity. The effect of introducing short chains produced by free radical attack in disrupting the specific "hyper-entanglement" interactions must also be considered additionally. On these bases, the dramatic collapse of the matrix with such low radiation doses becomes more explicable. The estimated sixty-fold viscosity reduction takes no account of the influence of the OH radical induced short chain segments on the short range 'hyper-entanglement' effects. On the basis of all these factors the disintegration of the gel-like hyaluronic acid matrix would be anticipated. The radiation-induced collapse of the hyaluronic acid quasi-solid state matrix observed for connective tissue systems and related gel systems represents possibly the most sensitive chemical and biological effects which have been observed. The initial effect is chain-scission. However, the consequent physical disruption of the viscoelastic properties can now be understood in terms of the overall behavior of concentrated high molecular weight hyaluronic acid solutions which show both conventional and hyper-entanglement behavior.

Acknowledgments

The collaboration of my colleagues in studying the radiation chemistry of carbohydrates is gratefully acknowledged, particularly that of C. von Sonntag, D. J. Deeble, E. Bothe, and H.-P. Schuchmann. I have drawn heavily on my partnership over the years with Endre A. Balazs for expertise on hyaluronic acid.

Literature Cited

1. Hongu, T.; Phillips, G. O. In *New Fibres;* Ellis Horwood; ISBN 0-13-613266-9, 1990, p 221.
2. Balazs, E. A., Ed. *Chemistry and Molecular Biology of the Intercellular Matrix, Vols. 1, 2;* Academic Press, 1970.
3. Balazs, E. A. In *Healon: A Guide to its Use in Ophthalmic Surgery;* Miller, D.; Stegmann, R., Eds. John Wiley: New York, 1983; p 5.
4. Balazs, E. A.; Band, P. *Cosmetics and Toiletries* 1984, **99**, 5.
5. Laurent, T. C. In *Chemistry and Molecular Biology of the Intercellular Matrix, Vol. 2;* Balazs, E. A., Ed. Academic Press, 1970; p 703.
6. Pearce, R. H.; Grimmer, B. J. In *Advances in the Biology of Skin, Vol. II;* Montagna, W.; Bently, J. P.; Dobson, R. L., Eds.; Appleton: New York, 1970; p 89.
7. Band, P. *Drug and Cosmetic Industry* 1985, **4**.
8. Morris, E. R.; Rees, D. A.; Welsh, E. J. *J. Mol. Biol.* 1980, **138**, 375-382; 383-400; *Carbohydrate Polymers* 1981, 5-21.
9. Sundblad, L.; Balazs, E. A. In *The Amino Sugars. The Chemistry and Biology of Compounds Containing Amino Sugars*; Balazs, E. A.; Jeanloz, R. W., Eds.; Academic Press: New York, 1966, **2-B**, 229-250.
10. Lamberts, H. B. In *Energetics and Mechanisms in Radiation Biology*; Phillips, G. O., Ed.; Academic Press: New York and London, 1968; pp 387-396.
11. Phillips, G. O.; Morgan, R. E. Unpublished results.
12. Balazs, E. A.; Watson, D.; Duff, I.; Roseman, S. *Ann. Rheum. Dis.* 1956, **15**, 357.
13. Kofoed, J.; Barcelo, S. *Experienta* 1978, **34**, 1545.
14. Wong, S.; Halliwell, B.; Richmond, R.; Skowroneck, W. *J. Inorganic Biochem.* 1981, **14**, 127.
15. Carlin, G.; Djursater, R. *EBS Letters* 1984, **177**, 27.

16. Pigman, W.; Hawkins, W.; Gramlings, E.; Rizvi, S.; Holley, R. H. *Arch. Biochem. Biophys.* 1960, **89**, 184.
17. Deeble, D. J.; Parsons, B. J.; Phillips, G. O.; Myint, P.; Beaumont, P. C.; Blake, S. M. In *Free Radicals, Metal Ions and Biopolymers;* Beaumont, P. C., Deeble, D. J.; Parsons, B. J.; Evans, Catherine Rice, Eds.; Richelieu Press: London, 1989; p 159.
18. Myint, Pe; Deeble, D. J.; Beaumont, P. C.; Blake, S. M.; Phillips, G. O. *Biochimica et Biophysica Acta* 1987, **925**, 194.
19. Myint, P.; Deeble, D. J.; Phillips, G. O. In *IAEA-TECDOC-527, New Trends and Developments in Radiation Chemistry;* 1989; pp 105-116.
20. Deeble, D. J.; Phillips, G. O.; Bothe, E.; Schuchmann, H.-P.; von Sonntag, C. *Proc. Intl. Symp. Radiation of Polymers,* Tokyo, Japan, 1989. The Radiation Induced Degradation of Hyaluronic Acid. *Radiat. Phys. Chem.* 1990, in press.
21. Deeble, D. J.; Bothe, E.; Schuchmann, H.-P.; Phillips, G. O.; von Sonntag, C. *Z. Naturforsch*, submitted.
22. Bothe, E.; Schulte-Frohlinde. *Z. Naturforsch* 1982, **37C**, 1191-1204.
23. von Sonntag, C.; Neuwald, K.; Dizdaroglu, M. *Radiat. Res.* 1974, **58**, 1-8.

RECEIVED December 16, 1991

Chapter 12

Formation and Viscoelastic Properties of Cellulose Gels in Aqueous Alkali

K. Kamide, M. Saito, and K. Yasuda

Fundamental Research Laboratory of Fiber and Fiber-Forming Polymers, Asahi Chemical Industry Company, Ltd., 11–7, Hacchonawate Takatsuki, Osaka 569, Japan

Detailed gelling conditions of cellulose–9 wt.% aqueous NaOH solution system were investigated by kinematic viscosity and light scattering methods. The system has two gelation temperatures in the room temperature region. The gel generated at a higher temperature region (HTG) is thermally irreversible, while the gel formed at a lower temperature shows thermal reversibility. Sol-gel transition cannot be explained in terms of liquid-liquid phase separation. The viscoelastic behavior accompanied with sol-gel transition, and the thermal properties of the HTG, such as swelling in the solvent medium and syneresis, were also examined. Storage modulus in the transition increases monotonically with time over 100 hr and more, and the syneresis continues for at least one month when kept above 30°C. From these findings, HTG is concluded to be in a pseudo-equilibrium state in the temperature range of 30–50°C.

Recently polysaccharide gel has been receiving considerable attention because of its wide potential applicability to industrial fields such as medicine and civil engineering. Gelation phenomena of cellulose derivative solutions, for example, the methylcellulose-water system on cooling (1), and the cellulose nitrate-organic solvent system on warming (2), have been studied so far, and especially for the former system, detailed conditions of gelation and some thermal properties of these gels have emerged. With respect to cellulose solutions, it has been well known for a long time that the gel arises by vaporization of ammonia from a concentrated cellulose cuprammonium solution. However, contrary to cellulose derivatives, a systematic investigation on the gelation behavior of cellulose solutions has never before been performed. The fact that a stable and simple solvent, in which cellulose

0097–6156/92/0489–0184$06.00/0

can be dissolved molecularly without formation of any complex, has been scarcely found, seems to prevent us from studying this subject at advanced levels.

Very recently, Kamide *et al.* (3) found that native and regenerated celluloses, prepared under specific conditions, dissolved without forming alcoholate in aqueous alkali (sodium- and lithium-hydroxide) solution, regardless of the molecular weight of the cellulose samples, and the solubility in the solvents was closely correlated with the relative amount of the local region where intramolecular hydrogen bonds are at least partly broken. Successively, it was disclosed that the second virial coefficient, A_2, obtained by the light scattering method of dilute cellulose solutions in 8 wt.% aqueous NaOH solution, shows a significant negative temperature dependence near room temperature (r.t.), and A_2 becomes zero at ca. 40°C, *i.e.*, at which 8 wt.% aqueous NaOH solution becomes a Flory theta solvent for cellulose (4). In the course of these studies it was also observed that when the temperature of the cellulose solution elevates or falls from r.t. (in other words, with a concomitant decrease of the solvent power), gelation occurs at specific temperatures. The cellulose gels thus obtained are shown in Figure 1. While various systems which form gels by cooling or heating are already known, no polymer solution which has two gelation points near r.t. has been found so far, except for the cellulose–aqueous NaOH solution system.

This paper is principally devoted to disclosing, for regenerated cellulose in aqueous alkaline solutions, the detailed gelling conditions and the viscoelastic and thermal properties of the cellulose gels formed.

Experimental

Preparation of Cellulose Samples and their Solutions. Hydrolysis of purified cotton linter proceeded in a cuprammonium solution (Cu, 11; NH_3, 200; water, 1000 grams) by storing in a dark place at 20°C for ca. 1 week (4,5). Addition of a 20 wt.% aqueous H_2SO_4 solution to the above solution brought about a regenerated cellulose sample (sample code C1) as a precipitate. Nine regenerated cellulose samples (code C2–C10) with different molecular weights were prepared from the sample C1 slurry, including 25 wt.% aqueous H_2SO_4 solution at 30°C, by altering storage time. The viscosity-average molecular weight (M_v) of these samples, as listed in Table I, was estimated through the use of the Mark-Houwink-Sakurada equation (Equation 1) determined by Brown-Wikström (6) using cadoxen as a solvent.

$$[\eta] = 3.85 \times 10^{-2} M_w{}^{0.76} (cm^3 g^{-1}) \qquad (25°C) \qquad (1)$$

where $[\eta]$ is the limiting viscosity number and M_w is the weight-average molecular weight.

We dissolved the cellulose sample flake in 9 wt.% aqueous NaOH solution at 5°C (dissolving temperature T_p) by agitating for 1 minute with a home mixer, followed by additional agitation for the same period 15 hours later. In order to eliminate undissolved residue, the solution was centrifuged under 1.5×10^4 gravity for 1 hr, and finally transparent and clean cellulose

Table I. Molecular characteristics of cellulose samples used in this study

Sample Code	Molecular Weight $\times 10^{-4}$	Experimental Methods Utilized
C1	12.0 (M_w[a])	Starting sample, LS[c]
C2	4.0 (M_v[b])	LS, KV[d], BD[e], VE[f], SY[g]
C3	5.2 (M_v)	KV, VE, SW[h], SY
C4	7.3 (M_v)	KV, VE, SW
C5	7.8 (M_v)	LS, SW, SY
C6	1.9 (M_v)	LS, SW
C7	5.9 (M_v)	BD, VE
C8	2.4 (M_v)	SY
C9	6.4 (M_v)	SY
C10	9.7 (M_v)	SY

a, weight-average molecular weight; b, viscosity-average molecular weight; c, light scattering; d, kinematic viscosity; e, ball drop method; f, viscoelasticity; g, syneresis; h, swelling.

solutions were obtained. The solutions were stocked in a dark place maintained at 5°C, where the solutions were ascertained to be stable for at least several months.

Measurements of Gelation Temperature and Thermal Hysteresis Detected by the Temperature Jumping Method. Gelation temperature of the cellulose solution was determined by investigating the time dependence of kinematic viscosity ν ($\equiv \eta/\rho$; η, the viscosity; ρ, the density) for the system. At about 30 min after the centrifugative filtration at 5°C, modified Ubbelohde viscometers, each of which contained the sample C2–C4 solutions, respectively, were immersed in the water bath controlled at an annealing temperature, T_a, ranging from -5–35 ± 0.05°C. ν of these solutions was estimated through the equation:

$$\nu = at_v \qquad \text{(cSt)} \tag{2}$$

where a is a constant of the viscometer used and t_v is time required for a given amount of the solution to flow through the capillary of the viscometer. Here, constant a was evaluated using a standard liquid supplied by Showa Shell Oil Company Limited (Tokyo).

We denoted the maximum or minimum temperature, in the region of T_a in the temperature jumping method, in which any slight increase in ν was observed for the system within 24 hr, as gelation temperature $T^G{}_J$.

For the purpose of investigating the thermal hysteresis effect on cellulose (sample C3)–aqueous NaOH solution system, the solution prepared at $T_p = 5$°C was quickly heat-treated at T_a equal to 30°C and -2°C for various periods of time t_a ($t_a = 1, 3, 6$ hr at $T_a = 30$°C and 4, 4.5 hr at $T_a = -2$°C). After the heat treatment at $T_a = 30$°C, the solution was quickly cooled down to and kept at the temperature $T_j = 5°C$. In the case of $T_a = -2°C$, the solution was heated up to and kept at $T_j = 20°C$. We intermittently measured ν of the sample C3 solution maintained at T_a and T_j.

Measurement of Gelation Temperature by the Light Scattering Method when the System is Slowly Heated. A mother solution of each sample (C1, C2, C5, and C6), stored at 5°C for 24 hr after centrifugation, was diluted with 9 wt.% aqueous NaOH solution. Each of the diluted solutions was directly poured into a cylindrical light scattering cell, and then the cell was sealed to minimize the possible effect of carbon dioxide gas on the gelation behavior of the cellulose solutions, although this kind of effect has never been reported. In the light scattering apparatus, model DLS 700 manufactured by Otsuka Electronics (Osaka), temperature of the solution was elevated at the rate of 4°C/hr from 5°C to the measuring point. The difference between the intensity ratio of scattering intensity at an angle 90° to that of the incident beam for the solution and that for the solvent ($\equiv \Delta I_{90}$) was estimated. In the light scattering measurements, the following operating conditions were employed: wave length of incident beam, 633 nm; gate time, 160 msec; accumulation time, 100.

Measurements of Melting Point. A test tube with a diameter of 13.5 mm containing 10 grams of one of the sample C8 and C9 solutions was placed vertically in the water-ethylene glycol bath kept at $-9.5\pm0.2°C$ for 16 hr, to generate and develop a cellulose gel. A stainless steel ball with a diameter of 3.2 mm was put on the surface of the gel. The gel was then heated at the rate of 4°C/hr and the location of the ball was read by a cathetometer. The temperature at which the ball began to drop was defined as the melting point T^m of the gel involved.

Storage Modulus in Sol-Gel Transition. The viscoelastic parameters, such as storage modulus G' and loss modulus G'' of the cellulose solutions at the gelation process in the higher temperature region, were determined by a rotation-vibration rheometer, Auto Viscometer L-III, manufactured by Iwamoto Company (Kyoto). The cellulose solutions, prepared at 5°C, of samples C2–C4 and C7, were isothermally annealed in the cell of the rheometer controlled at 33, 35, 40, 45, and 50°C, to give the gels. From the results of preliminary measurements, G' of these gels was found to be almost independent of frequency in the range of 0.001 to 0.1 Hz. In addition, G'' was on the order of less than 1 dyne/cm^2, which was too small to be determined accurately by the rheometer used in this study. Then G' was determined at the constant frequency of 0.05 Hz.

Swelling Behavior and Syneresis of Cellulose Gels. The cellulose C3 solutions with the polymer concentration w of 6 wt.% were annealed at 50°C for three days and resulted in gels. To one part (in volume) of the gels five parts (in volume) of 9 wt.% aqueous NaOH solution was added as a medium. The cellulose gels suspended in 9 wt.% aqueous NaOH solution medium were annealed at various temperatures ranging from 4 to 50°C for

1, 3, and 7 days. As a measure of swelling or contraction of the gels, the ratio of the volume of the annealed gels to that of the unannealed gels, Q_v, was estimated.

In order to find conditions of occurrence of syneresis, the cellulose solutions (samples C2, C3, C5, and C8–10) were annealed at 30, 37, and 50°C for 5, 10, and 30 days.

Results and Discussion

Figures 2a and 2b show time (t) dependences of the kinematic viscosity of sample C3 ($M_v = 5.2 \times 10^4$) solution with $w = 5$ wt.% · ν of the solution treated (annealed before quenching) above 25 or below −2°C monotonically increased after the solution temperature was changed. In the figure treatment temperature is shown. In the case of the solution maintained at 2 and 0°C, we observed a slight increase of ν at $t \geq 22$ and $t \geq 21$ hr, respectively. Moreover, at $t = 24$ hr, all of the solutions examined except for those held at 5–20°C exhibited a significant increase in turbidity. Judging from these facts it can be said that cellulose chains in the solution treated below 2 or above 25°C tend to aggregate together at least within 24 hr, and as more time elapses, all these solutions are eventually supposed to approach gel, which is defined by Ferry (7) as the solution which exhibits no steady state flow, *i.e.,* $\nu \rightarrow \infty$. Following this definition, the cellulose-9 wt.% aqueous NaOH solution system has two gelation points, both very close to room temperature. We define the gel formed on warming and that formed on cooling as higher temperature gel (HTG) and lower temperature gel (LTG), respectively.

Figure 3 depicts the time dependence of ν of the sample C3 solutions with w ranging from 3 to 6 wt.% at 30°C (a) and 0°C (b). Below w = 4 wt.%, ν does not show any remarkable change with storage time at either temperature, indicating that there exists a critical concentration above which the solution transforms to the gel.

Figure 4 shows a liquid-gel phase diagram determined by the kinematic viscosity method in the temperature jumping method for the solutions of three cellulose samples (M_v = 4.0, 5.2, and 7.3×10^4). As expected, the gel region becomes narrower as the molecular weight and the polymer concentration increase. The relationship between $T^G{}_J$ and polymer concentration for a given cellulose sample can be well represented by two straight lines (HTG, solid line; LTG, broken line), of which the slopes are practically independent of the molecular weight of cellulose. The experimental finding that the upper and lower gelation lines cross at 5°C suggests that the cellulose-aqueous NaOH solution system with relatively high polymer concentration (w = 6–7 wt.%) cannot exist as the liquid state. In Figure 5, iso-kinematic viscosity curves of the sample C3 solution are plotted against w. The value of ν in the figure corresponds to that measured at 10 minutes after the viscometer was immersed in a water bath. Every iso-kinematic viscosity curve traverses two gelation lines, corresponding to HTG and LTG of the same sample solution, suggesting strongly that the absolute ν value

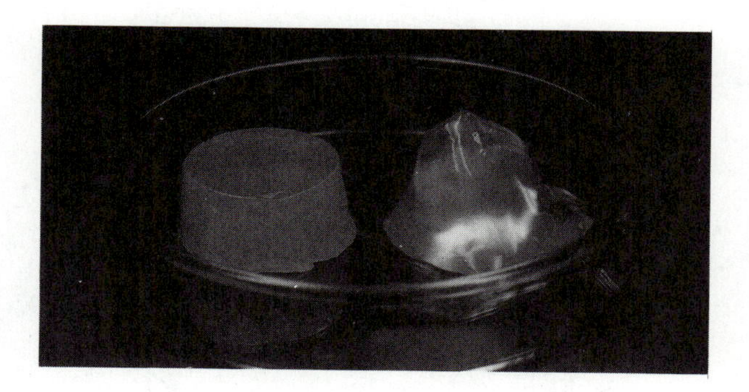

Figure 1. Gels of cellulose-9 wt.% aqueous NaOH solution system at 20°C. Right side gel, which is partially melting, generated at −9°C for 70 hr and left side gel generated at 40°C for the same period as the right side gel.

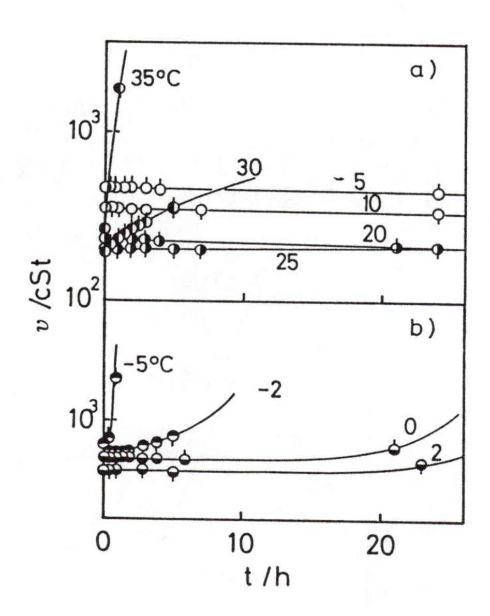

Figure 2. Time t dependence of the kinematic viscosity ν of cellulose (sample code C3, the viscosity-average molecular weight = 5.2×10^4) solution (the polymer concentration = 5 wt.%; dissolving temperature T_p = 5°C) at 5 (○), 10 (○), 20 (◐), 25 (◐), 30 (◐), and 35°C (◐) (a); and 2 (◒), 0 (◒), −2 (◒), and −5°C (◒) (b).

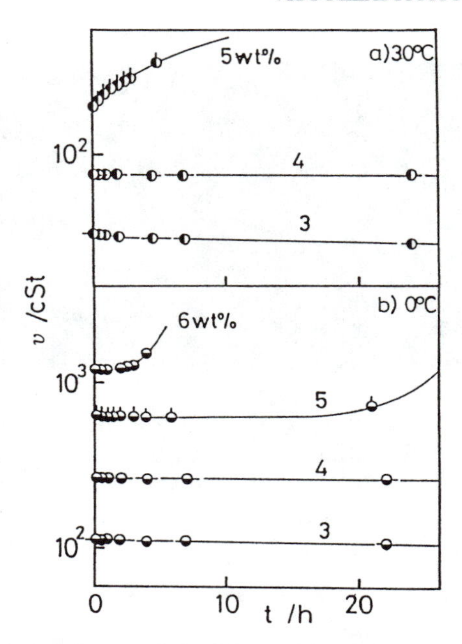

Figure 3. Time t dependence of the kinematic viscosity ν of cellulose (sample C3) solution with various polymer concentrations (the dissolving temperature T_p = 5°C) at 30°C (a) and at 0°C (b). (a)-◐, 3; ◐, 4; ◐, 5 wt.%; and (b)-◒, 3; ◒, 4; ◒, 5; ◒, 6 wt.%.

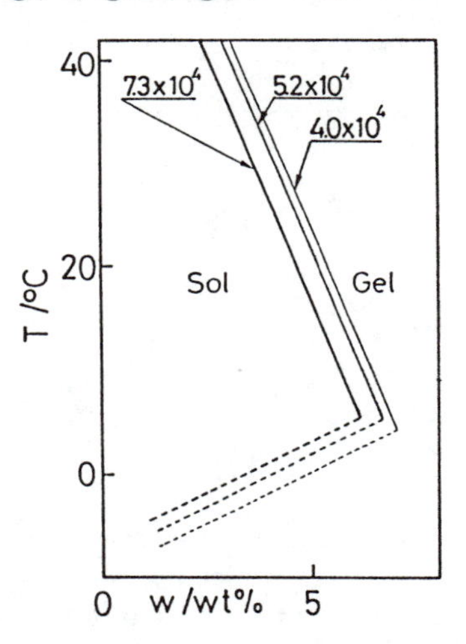

Figure 4. Upper gelation temperature (solid line) and lower gelation temperature (broken line) of cellulose-9 wt.% aqueous NaOH solution systems plotted as a function of the cellulose concentration. The numbers attached to the line denote the viscosity-average molecular weight of the cellulose samples.

of the solution does not serve effectively as a measure of generating the gel from the cellulose solution.

Figure 6a shows the change in ν of the cellulose C3 solution with w of 4.6 wt.% during the heat treatment. For the sake of comparison, data of untreated solutions were also plotted. Irrespective of t_a, ν of the solution (T_p = 5°C, T_a = 30°C) at T_j = 5°C declined rapidly for about the initial 2 hr, attaining a constant value which is apparently higher for longer annealing time. It became clear that the aggregation structure formed at 30°C remains partially undestroyed even at 5°C, which is sufficiently below the upper transition temperature. In other words, the gel formed at the higher temperature region is thermally irreversible. In the previous paper (4), we suggested that in the dilute solution region of the cellulose in aqueous NaOH solution there possibly occurs a specific transition of cellulose chain conformation, such as coil-globule transition. We suspect this transition occurs from the fact that when the cellulose-aqueous NaOH solution system is heated the radius of gyration of the chain decreases abruptly at 40°C. For the aggregated structure in the gel of naturally occurring polysaccharides, several models have been proposed, *e.g.*, the egg-box type for polyelectrolytes such as poly-L-guluronate (alginate) and its optical isomer, in water containing alkali metal ions (8), and the double helical form for carrageenan and agarose in water (9). However, with respect to cellulose gel, cellulose chains must not aggregate in any regular form, such as egg-box or helix, because cellulose chains in the solution or in the solid state have never been reported to be conformed in helical strands. But, unfortunately, detailed information on the microstructure of cellulose gels is still lacking. Considering the fact that in the relatively limited range of around room temperature solvent power gets larger as the temperature falls (4), an initial declination of ν as shown in Figure 6a is most likely caused by the partial melting of the gel.

Figure 6b is another example of the change in ν for the cellulose C3 solution (w = 4.6 wt.%, T_p = 5°C, T_a = −2°C, t_a = 4 and 4.5 hr, T_j = 20°C). In this case too, just after the temperature reached T_j, ν of the system immediately falls on the line of the untreated solution. It is interesting to note that the LTG unquestionably has thermal reversibility, contrary to HTG.

Figure 7 shows the relationship between ΔI_{90} of the C1 solution with w ranging from 0.6 to 3 wt.%, and the temperature, which was raised at the rate of 4°C/min. Every line in the figure has an upward bending point. An equation relating ΔI_{90} of semidilute and concentrated polymer solution with the polymer concentration c (g·cm^{-3}) was first derived by Benoit-Benmouma (10) in the form:

$$\Delta I_{90} = Kc[1/(M \cdot P(90)) + 2A_2 c]^{-1} \quad (3)$$

Here, K is an optical constant; M, molecular weight of the polymer chain; $P(90)$, scattering function at scattering angle 90; A_2, the second virial coefficient. Equation 3 has an analogous form to the Zimm's equation (11) for dilute solution. Over the temperature range of 4–35°C, K, A_2,

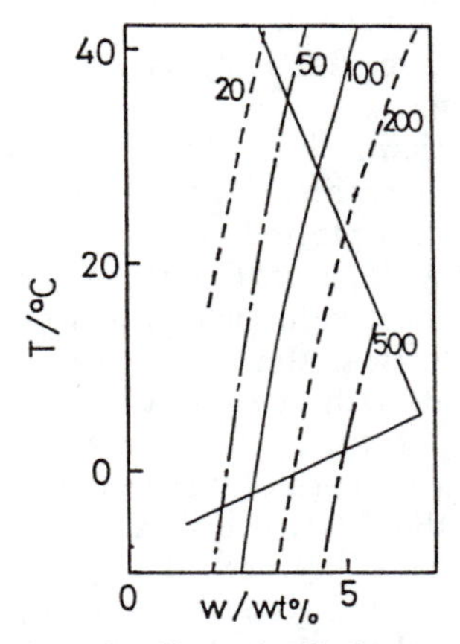

Figure 5. Iso-kinematic viscosity line of cellulose sample C3-9 wt.% aqueous NaOH solution system. Attached numbers to the lines denote the kinematic viscosity (cSt). Heavy solid lines, upper and lower gelation temperature lines.

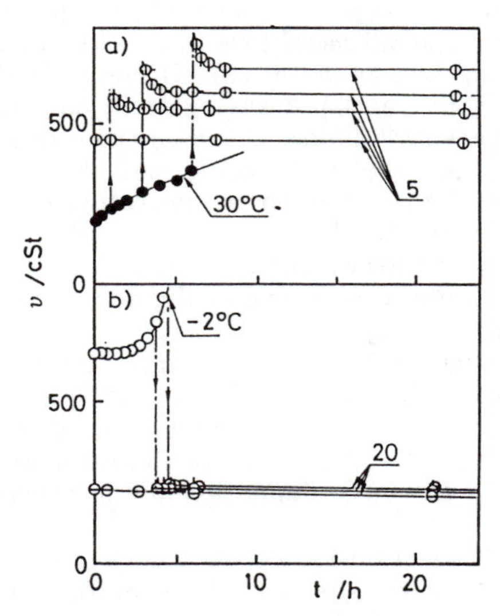

Figure 6. (a): ⦶, the untreated solution (the dissolving temperature $T_p = 5°C$) at 5°C; ●, the heat-treated solution ($T_p = 5°C$, heat-treated temperature $T_a = 30°C$) at 30°C; ⦶, the quenched solution after heat treatment ($T_p = 5°C$, $T_a = 30°C$, heat treatment time $t_a = 1$ hr, quenching temperature $T_j = 5°C$) at 5°C; ⦶, the quenched solution after heat treatment ($T_p = 5°C$, $T_a = 30°C$, $t_a = 3$ hr, $T_j = 5°C$) at 5°C; ⦶, the quenched solution after heat treatment ($T_p = 5°C$, $T_a = 30°C$, $t_a = 6$ hr, $T_j = 5°C$) at 5°C. (b): ⊖, the untreated solution ($T_p = 5°C$) at 20°C (*i.e.*, $T_j = 20°C$); ○, the treated solution ($T_p = 5°C$, $T_a = -2°C$) at $-2°C$; ⊖, the treated solution ($T_p = 5°C$, $T_a = -2°C$, $t_a = 4$ hr, $T_j = 20°C$) at 20°C; ⊖, the treated solution ($T_p = 4°C$, $T_a = -2°C$, $t_a = 4.5$ hr, $T_j = 20°C$) at 20°C.

and $P(90)$ of the dilute solution of the C1 sample had no inflection point and the drastic change in the volume of the cellulose solution was also not observed. Therefore, the upward inclination of ΔI_{90} curves in the figure may be reasonably attributed to an increase in apparent molecular weight with an onset of association of the polymer chains. The temperature at which ΔI_{90} curves bend upward (indicated with arrows) is defined as the gelation point at slow heating process, T^G. In Figure 8, the results on C1 solution with $w = 1.9$ wt.% held at 30 ± 0.1°C were demonstrated and, in general, ΔI_{90} of the solution held below T^G remained almost constant for at least one day.

In Figure 9, the concentration dependence of T^G for the samples with $M_v = 1.9 \times 10^4$ to 1.2×10^5 is demonstrated (solid line), and, in the figure for comparison, $T^G{}_J$ (see Figure 4) of HTG for the sample with $M_v = 4.0 \times 10^4$ was redrawn (broken line). While the concentration dependence of T^G is similar to that of $T^G{}_J$, the difference between T^G and $T^G{}_J$ is found especially in the concentrated region. The dependence of gelation temperature on the heating rate may be explained by the following mechanism: When the temperature of the relatively concentrated solution is elevated stepwise, the entanglement of the polymer chains develops into tight crosslinkage by hydrogen bond formation. Contrary to this, when the solution is heated at a slow rate, a shrinkage of cellulose chain sphere, resulting from coil-globule transition, reduces the probability of forming entanglements of the polymer chains, to yield a higher gelation temperature than that obtained by the temperature jumping method (*i.e.*, $T^G > T^G{}_J$). In the dilute region, the number of entanglements existing in the solution are so few that it is relatively hard to form a gel by the temperature jumping method. However, cellulose chains embedded in the poor solvent for hours (the slow heating process) tend to associate together with a high probability of collisions which may bring about a lower gelation temperature.

In order to clarify the relationship between the gelation phenomena in the higher temperature range and liquid-liquid phase separation, we attempted to determine a theoretical cloud point curve of the cellulose-aqueous NaOH solution system using some thermodynamic parameters, such as A_2, evaluated through the study on dilute solution properties of this system (4). According to the Flory-Huggins theory (12), the chemical potential of the solvent $\Delta\mu_0$ is represented as a function of the polymer volume fraction ϕ_1;

$$\Delta\mu_0 = kT[ln(1-\phi_1) + (1 - 1/X_n) + \chi\phi_1^2] \tag{4}$$

Here, k is the Boltzmann constant; T, the absolute temperature; X_n, the number-average degree of polymerization; χ, the polymer-solvent interaction parameter, which is semi-empirically expressed by (12):

$$\chi = [(1/2 - \psi_0) + \Theta\psi_0/T](1 + p_1\phi_1) \tag{5}$$

where ψ_0 is the entropy parameter; Θ, Flory's Θ temperature; p_1, the first order concentration-dependence parameter. From the temperature dependence of A_2 of the cellulose-aqueous NaOH solution system (see Figure 6

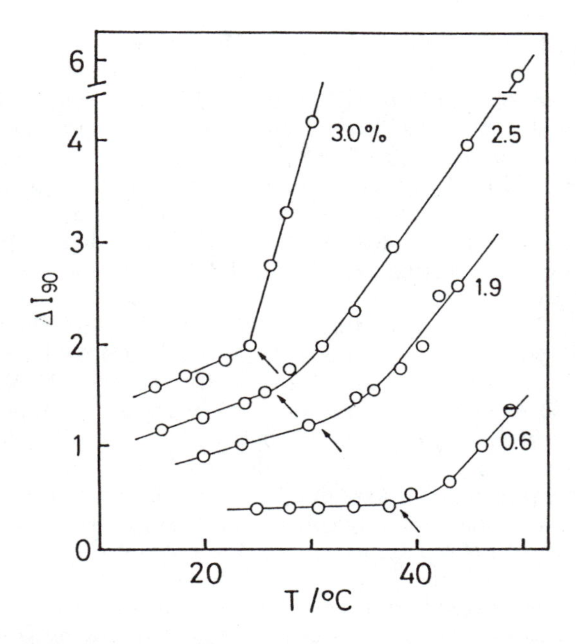

Figure 7. Scattered light intensity difference at scattering angle 90°, ΔI_{90}, of cellulose sample C1 solutions plotted against the temperature T, when the solutions were heated at constant rate of 4°C/hr. Numbers in the figure denote the cellulose concentration (wt.%) and arrows indicate the gelation temperature (T^G(°C)).

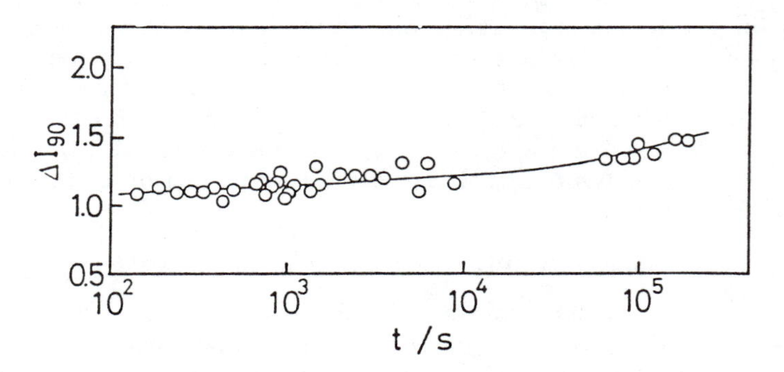

Figure 8. Time dependence of scattered light intensity difference at scattering angle 90°, ΔI_{90} of a cellulose sample (C1) solution (dissolving temperature $T_p = 5$°C) at 30°C.

in ref. 4), ψ_0 and Θ are estimated to be −0.77 and 313 K, respectively. Substitution of these values into Equation 5 enables us to calculate χ as a function of T and ϕ_1 if p_1 is assumed in advance. Then, we can calculate the cloud point curve according to the method proposed by Kamide *et al.* (12–14) under the following assumptions: The aqueous NaOH solution is a single component solvent. The molecular weight distribution of the cellulose samples can be represented by the Schulz-Zimm type distribution with X_w/X_n (X_w, the weight-average degree of polymerization) = 2. The results are shown as a chain line in Figure 9 for the hypothetical solution. Here, $p_1 = 0$ and 0.67 (theoretical value predicted for non-polar polymer–non-polar solvent system (12)) were chosen. Obviously, gelation points of the sample with the same M_w, experimentally observed, are far below, except for the range of $w \leq 1$ wt.%. The cloud point curves even if $p_1 = 0.67$ is assumed, so that gelation of this system seems not to be correlated with the phase separation of the solution. This is a reasonable explanation why the cloud point is not experimentally observed for the cellulose-aqueous NaOH solution system. However, it is not yet clear to what kind of thermodynamic transition the onset of gelation of cellulose solution at the higher temperature region corresponds.

Figure 10 shows the melting point, as determined by the ball drop method, of the LTG of samples C2 and C7, plotted against the polymer concentration. As expected, the melting of LTG occurs at a higher temperature as the concentration and the molecular weight increase. The heat of fusion (ΔH^G) per mole of cross linkages in these LTG's can be roughly estimated using the Eldridge-Ferry (EF) equation (15) (Equation 6), derived by assuming that only a binary association of polymer chains is responsible for the gelation of the solution:

$$H^G = kN_A T(\partial lnw/\partial(1/T^m))_M$$

where N_A is Avogadro's number. Plots of w vs. the reciprocal T^m for each sample, as demonstrated in Figure 11, can be approximately expressed by the straight line and the slopes of the lines, obtained for two cellulose samples, are practically equal, making ΔH^G = 31 kJ/mol. In Table II, the literature data on ΔH^G, determined through the EF equation, of various polymer solvent systems, are summarized. ΔH^G of cellulose LTG is very close to those of atactic vinyl type polymer gels (see Table II), including polyvinylalcohol (16), polyvinylchloride (17), and polystyrene (18). When we take into account the possible contribution to ΔH^G of the intramolecular hydrogen bond existing in the cellulose gel, the intermolecular association energy should not be larger than 31 kJ/mol. This value is small enough to allow the phase transition of LTG to sol.

Figure 12 shows the time course of G′ for the C7 solution (w = 5 wt.%) at five different annealing temperatures (T_a). At a given T_a, G′ shows a remarkable increase over the initial state. According to Kuhn's theory (21) on the viscoelasticity of the polymer network, G′ is directly related to the number of crosslinkages in unit volume of the polymer solutions. Inspection of Figure 12 leads us to the conclusion that the number of clusters, which

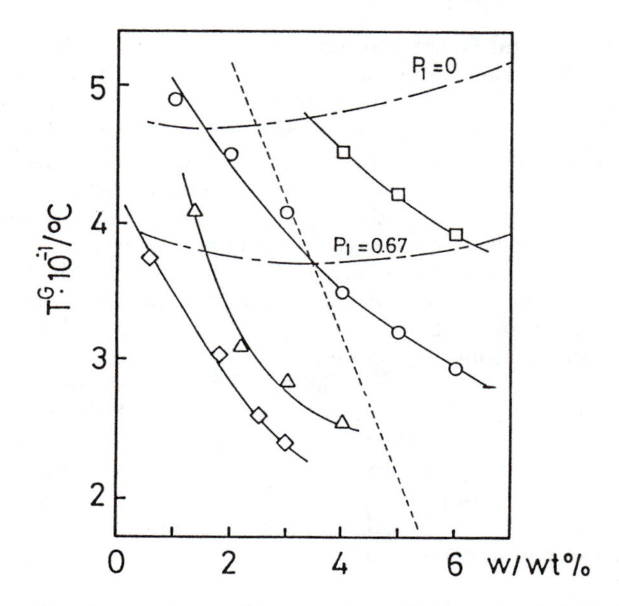

Figure 9. Concentration w dependence of gelation temperature T^G determined through the light scattering measurements (solid line) of cellulose-9 wt.% aqueous NaOH solution systems which are heated at a low rate. Broken line, gelation temperature T^G, which is determined through the kinematic viscometric measurements on a cellulose sample C2-9 wt.% aqueous NaOH solution system heated by the temperature jumping method (see Figure 3). ◇, the weight average-molecular weight = 1.2×10^5; △, the viscosity-average molecular weight $M_v = 7.8 \times 10^4$; ○, $M_v = 4.0 \times 10^4$; □, $M_v = 1.9 \times 10^4$. Attached numbers are the viscosity-average molecular weights of the cellulose samples. Chain line, theoretical cloud point curve calculated by Kamide *et al.*'s theory (12–14).

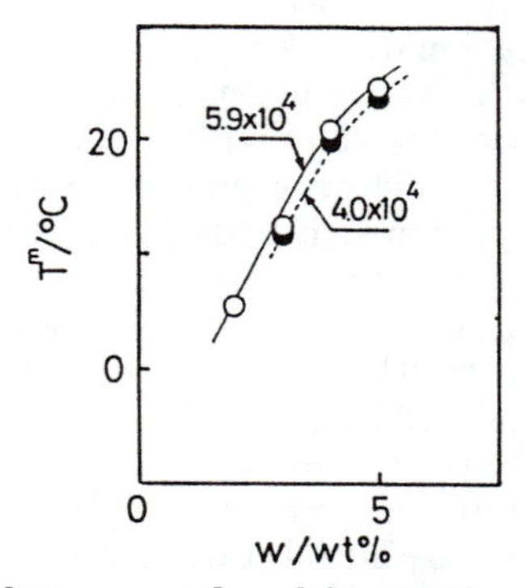

Figure 10. Relationship between gel melting temperature T^m determined by the ball drop method for cellulose gel (sample C2, ●; sample C7, ○), and the polymer concentration w. The viscosity-average molecular weight of cellulose samples is shown on curves.

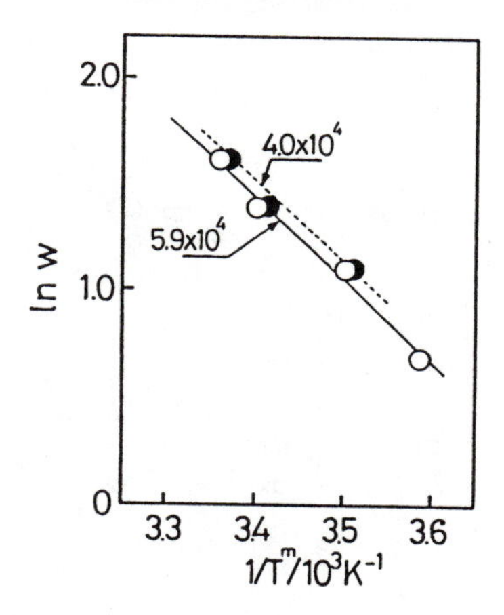

Figure 11. Relationship between the polymer concentration w and reciprocal of the gel melting temperature T^m as determined by the ball drop method. Attached numbers denote the viscosity-average molecular weight of cellulose samples.

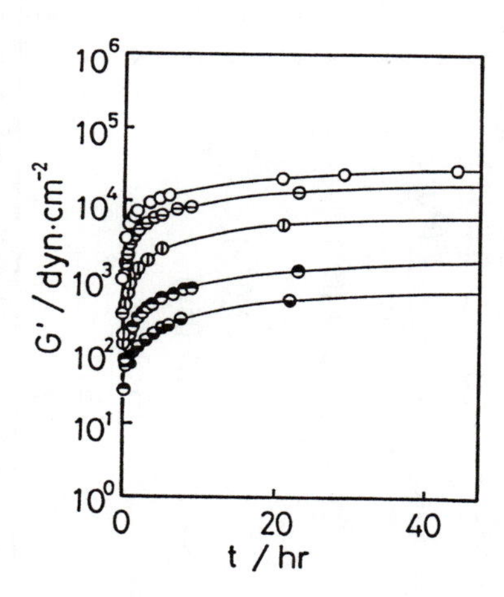

Figure 12. Time course of the storage modulus G′ of cellulose (M_v = 5.9 × 10^4) solutions with w = 5 wt.%. ◒, heat-treatment temperature T_a = 33; ◓, 35; ⦶, 40; ⊖, 45; ○, 50°C.

Table II. Heat of fusion of gel (ΔH^G) of various polymer-solvent systems

Polymer	Solvent	Molecular Weight $\times 10^{-4}$	Conc. wt %	Measuring Method	ΔH^G kJ/mol	Ref.
Cellulose	9%NaOH aq.	4.0 ~ 5.9 (Mv[h])	2 ~ 6	B D[k]	31	this work
Gelatin	0.9%NaCl aq.	3.3 ~ 7.2 (Mw[i])	2 ~ 6	S I[m]	209 ~ 920	15
a-PVA[a]	water	7.1	11 ~ 21	O F[n]	37	16
s-PVA[b]	water	5.1 (Mv)	2.9 ~ 7	S I	159 ~ 256	19
			2.2 ~ 2.9	S I	9.6 ~ 13	19
a-PVC[c]	diaxan	6 ~ 13 (Mw)	2.5 ~ 14	S F	35	17
s-PVC[d]	anisole	13.1 (Mv)	1.5 ~ 2.8[j]	O F	ca.54	20
	dioxan	13.1	2.0 ~ 4.3[j]	O F	ca.42	
	EDC[f]	13.1	2.0 ~ 4.3[j]	O F	ca.38	
a-PS[e]	CDS[g]	67	5.3 ~ 17	B D	32	18

a, atactic polyvinylalcohol (PVA); b, syndiotactic PVA; c, atactic polyvinylchloride (PVC); d, syndiotactic PVC; e, atactic polystyrene; f, ethylene dichrolide; g, carbon disulfide; h, viscosity-average molecular weight; i, weight-average molecular weight; j, volume fraction of polymer; k, ball drop method; m, solution inverse method; n, onset of flowing.

consist of the crosslinkings of the cellulose chains, is larger at higher T_a. All of the curves in the figure can be superimposed on a single master curve as indicated in Figure 13, in which log G′ is plotted against log t, taking the curve of $T_a = 40°C$ as a reference.

Effects of the cellulose concentration and the cellulose molecular weight on the aging time dependence of G′ are demonstrated in Figures 14 and 15, respectively. In both cases, as aging time proceeds, G′ increases in a very similar fashion, as in the case of the temperature effect as shown in Figure 12. All curves in Figure 14 or 15 can be readily superimposed with respect to the polymer concentration or the molecular weight. Evidently, these curves almost coincide (Figure 16), indicating that the number of clusters in the solution is larger in more concentrated polymer solutions with higher molecular weight, and the aggregation rate, evaluated from the slope of the master curves, is quite independent of these three factors.

Figure 17 shows the annealing temperature dependence of the volume expansion factor Q_v of the sample C3 gel suspended in 9 wt.% aqeous NaOH solution medium at annealing time of 1 (chain line), 3 (broken line), and 7 days (dotted line). After annealing for 1 day, the cellulose gel maintained at a lower temperature region tends to swell, because of the stronger solvent power at lower temperature. At about 32°C, the gel starts to shrink drastically. Above 35°C, the degree of shrinkage does not markedly depend on the annealing time. These decreases of Q_v seem to be accompanied by chain conformation changes, such as coil-globule transition, because the radius of gyration, determined by the light scattering method in dilute regime, decreased remarkably in almost the same temperature range (4). At annealing temperatures below 30°C, in which 9 wt.% aqueous NaOH solution behaves as a good solvent toward cellulose, gel swelling occurs initially, followed by volume contraction.

Figures 18a-f illustrate schematically the phase diagram of the cellulose-aqueous NaOH solution system as a function of polymer concentration, molecular weight, annealing temperature, and aging time. In the figures, the unfilled part is the undissolved state (*i.e.*, two phases of cellulose solid and aqueous NaOH solution), and the filled part denotes the area in which syneresis of the gel occurs. Evidently, longer annealing at a higher temperature brings about syneresis in the area of lower polymer concentration and lower molecular weight. In addition, cellulose gel, formed from its solution in aqueous NaOH solution, never attains an equilibrium state even if kept longer than one month. In this sense, HTG is observed here as only in a thermally pseudo-equilibrium state.

Conclusions

Now we can summarize briefly the conclusions reached from this study:

1. The cellulose-aqueous NaOH solution system has two gelation temperatures near room temperature.
2. While higher temperature gel (HTG) is thermally irreversible, lower temperature gel (LTG) has thermal reversibility.
3. Cellulose gel suspended in the solvent shrinks drastically at specific temperatures.
4. Cellulose gel is in a thermally pseudo-equilibrium state.

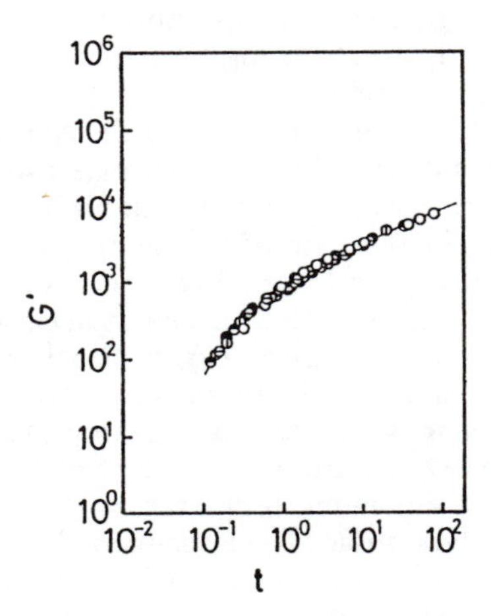

Figure 13. Master curve obtained by superposition of the data shown in Figure 11. The results on $T_a = 40°C$ is taken as a reference.

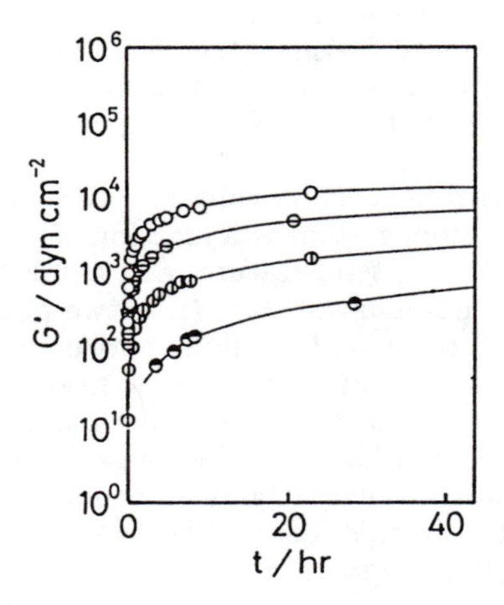

Figure 14. Time course of storage modulus G′ of cellulose ($M_v = 5.9 \times 10^4$) solutions with various polymer concentrations w at constant T_a (=40°C). ◓, w=4; ◑, 4.5; ⊖, 5.0; ○, 5.5 wt.%.

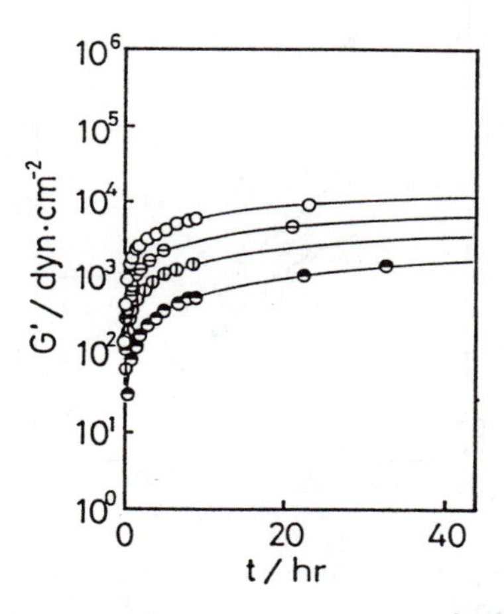

Figure 15. Time course of storage modulus G′ of cellulose solutions with polymer concentration w = 5 wt.% at constant T_a (=40°C). ◓, the viscosity-average molecular weight $M_v = 3.9\times10^4$; ⏀, 5.2×10^4; ⊖, 5.9×10^4; ○, 7.3×10^4.

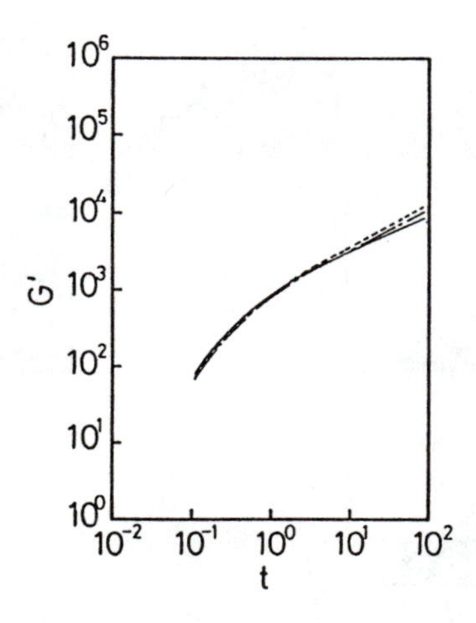

Figure 16. Three master curves obtained by superposition with respect to the heat-treatment temperature (solid line), the polymer concentration (chain line), and the molecular weight (broken line).

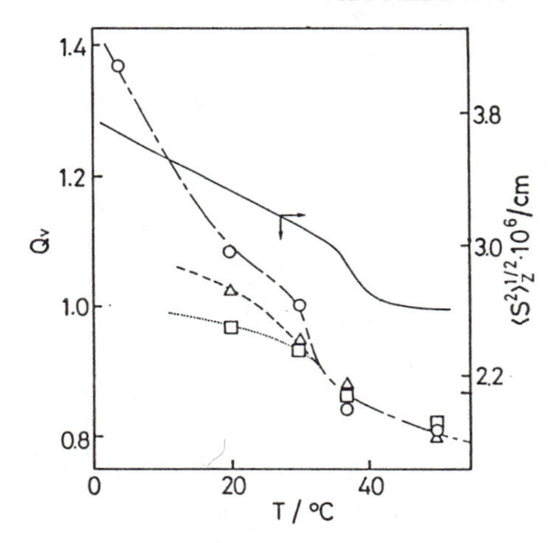

Figure 17. Aging temperature dependence of the volume expansion ratio Q_v of higher temperature gel (HTG) ($M_v = 5.2 \times 10^4, w = 6.0$ wt.%) in 9 wt.% aqueous NaOH solution medium. ○, aging time 1 day; △, 3 days; □, 7 days. Solid line, the z-average radius of gyration $< S^2 >_z {}^{1/2}$ cited from Figure 5 of ref. 4.

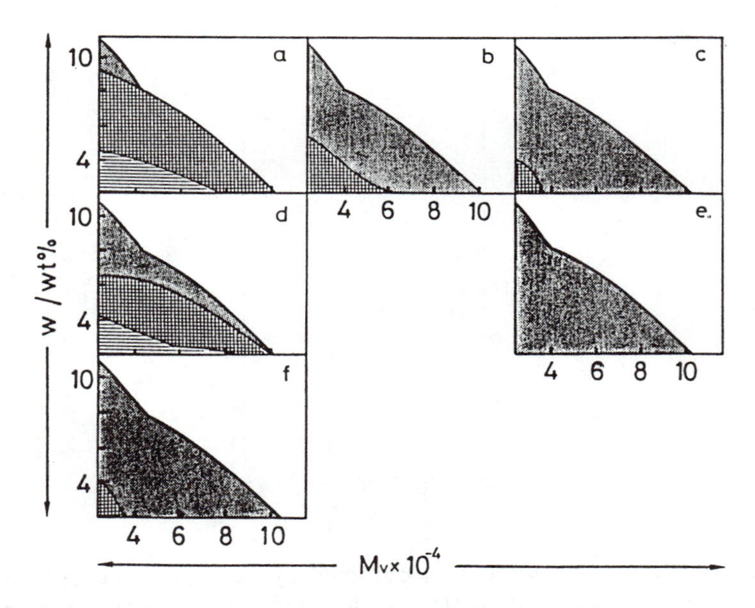

Figure 18. Phase diagrams of cellulose-9 wt.% aqueous NaOH solution system as a function of polymer concentration, molecular weight, annealing temperature, and aging time. Unfilled part, undissolved state; hatched part, solution; cross-hatched part, gel; filled part, syneresis.

Literature Cited

1. Heymann, E. *Trans. Faraday Soc.* 1935, **31**, 846.
   Savage, A. B. *Ind. Eng. Chem.* 1957, **49**, 99.
   Kuhn, W.; Moser, P.; Mijer, H. *Helv. Chim. Acta* 1961, **44**, 770.
   Neerly, W. B. *J. Polym. Sci., C* 1963, **1**, 311.
   Rees, D. A. *Adv. Carbohydr. Chem. Biochem.* 1969, **26**, 267.
   Klug, E. D. *J. Polym. Sci., C* 1971, **36**, 491.
   Kagemoto, A.; Baba, Y.; Fujishiro, R. *Makromol. Chem.* 1972, **154**, 105.
   Baba, Y,; Kagemoto, A. *Kohbunshi Runbunshu* 1974, **31**, 446.
   Baba, Y.; Kagemoto, A. *Kohbunshi Runbunshu* 1974, **31**, 528.
   Kato, T.; Yokoyama, M.; Takahashi, A. *Colloid Polym. Sci.* 1978, **256**, 15.
   Nakura, S.; Nakamura, S.; Onda, Y. *Kohbunshi Runbunshu* 1981, **38**, 133.
2. Newman, S.; Krigbaum, W. R.; Carpenter, D. K. *J. Phys. Chem.* 1956, **60**, 648.
3. Kamide, K.; Okajima, K.; Matsui, T.; Kowsaka, K. *Polym. J.* 1984, **16**, 85.
4. Kamide, K.; Saito, M.; Kowsaka, K. *Polym. J.* 1987, **19**, 1173.
5. Kamide, K.; Saito, M. *Polym. J.* 1986, **18**, 569.
6. Brown, W.; Wikström, R. *Eur. Polym. J.* 1965, **1**, 1.
7. Ferry, J. D. *Viscoelastic Properties of Polymers*, 3rd Ed.; New York: John Wiley, 1980; 537pp.
8. Rees, D. A.; Welsh, E. J. *Angew. Chem. Int. Ed.* 1977, **16**, 214.
   Kohn, R. *Pure Appl. Chem.* 1975, **30**, 371.
   Morris, E. R.; Rees, D. A.; Thom, D. *Chem. Comm.* 1969, **701**.
9. Rees, D. A.; Steele, I. W.; Williamson, F. B. *J. Polym. Sci. C* 1969, **29**, 261.
10. Benoit, H.; Benmouma, M. *Macromolecules* 1984, **17**, 535.
11. Zimm, B. *J. Chem. Phys.* 1948, **16**, 1093.
12. Kamide, K. In *Thermodynamics of Polymer Solutions*; Amsterdam: Elsevier, 1990; Ch. 2.
13. Kamide, K.; Matsuda, S.; Dobashi, T.; Kaneko, M. *Polym. J.* 1984, **16**, 839.
14. Kamide, K.; Matsuda, S.; Saito, M. *Polym. J.* 1985, **17**, 1013.
15. Eldridge, J. E.; Ferry, J. D. *J. Phys. Chem.* 1954, **58**, 992.
16. Maeda, H.; Kawai, T.; Kashiwagi, R. *Kohbunshi Kagaku* 1956, **58**, 992.
17. Hanson, M. A.; Morgan, P. H.; Park, G. S. *Eur. Polym. J.* 1972, **8**, 1361.
18. Tan, H. M.; Moet, A.; Hiltner, A.; Baer, E. *Macromolecules* 1983, **16**, 28.
19. Matsuzawa, S.; Yamaura, K.; Maeda, R.; Ogasawara, K. *Makromol. Chem.* 1979, **180**, 229, and related refs. cited therein.
20. Takahashi, A.; Nakamura, T.; Kagawa, I. *Polym. J.* 1972, **3**, 207.
21. Kuhn, W. *Kolloid Zeits* 1936, **76**, 258.

RECEIVED February 10, 1992

Chapter 13

Physical Gelation of Biopolymers

Simon B. Ross-Murphy[1]

Cavendish Laboratory, University of Cambridge, Cambridge CB3 0HE, United Kingdom

Polymer networks can be divided into two main classes, those formed at high segment concentrations by the topological entanglement of polymer chains and those formed formally by the covalent cross-linking of preformed linear chains. The former are networks at frequencies higher than some typical entanglement lifetime, and are viscoelastic liquids, whereas the latter are of infinite molecular weight, and have an equilibrium modulus—they are viscoelastic solids. However, there are many intermediate systems of chains which are "physically" cross-linked, the cross-links themselves being of small but finite energy, and/or of finite lifetime; these are called physical networks. Specific systems include both biological and synthetic polymers. Amongst the former are gelatin, and the seaweed polysaccharides such as agarose and the carrageenans. Such biopolymers form physical gels involving junction zones of known, ordered secondary structure. Models are presented that relate the modulus and geltime kinetics to properties of the macromolecular chains.

Covalent and Entanglement Networks

Polymer gels or networks can be divided into two main classes, chemically cross-linked materials (including bulk elastomers), and "entanglement networks". The covalently cross-linked materials are formed by a variety of routes including cross-linking high molecular weight linear chains, either chemically or by radiation, by end-linking reactant chains with a branching unit, or by step-addition polymerisation of oligomeric multi-functional precursors. They are true macromolecules, where the molecular weight is nominally infinite, and they therefore possess an infinite relaxation time

[1]Current address: Biomolecular Sciences Division, King's College London, West Kensington, London W8 7AH, United Kingdom

0097–6156/92/0489–0204$06.00/0

and an equilibrium modulus (1). By contrast, entanglement networks are formed by the topological interaction of polymer chains, either in the melt or in solution when the product of concentration and molecular weight becomes greater than some critical molecular weight M_e (1, 2). In this case they behave as "pseudo-gels" at frequencies higher (timescales shorter) than the lifetime of the topological entanglements. This depends for linear chains on $M_r{}^3$, where M_r is molecular weight.

Rheological discrimination between these two classes of networks can be made by the technique of dynamic mechanical analysis ("mechanical spectroscopy") (1). In this, a small deformation, oscillatory strain (usually a shear strain) of frequency ω is applied to the material, and the real and imaginary parts, G' and G'', of the complex shear modulus are measured. Typical spectra are illustrated in Figure 1; for the entanglement networks, as the frequency is decreased there is a "cross-over" in G' and G''. At very low frequencies, in the "terminal zone" they flow as high viscosity liquids. The precise response of the covalently cross-linked system depends upon whether or not CM_r for the system *before cross-linking* was above or below M_e. In the case $CM_r < M_e$, the cross-linked networks generally show no entanglement effects, and G' and G'' are parallel, and largely frequency insensitive. A further discrimination can be made by adding excess solvent to the bulk "gel." Entanglement network systems will dissolve to form a more dilute polymer solution, whereas the gel will swell, but not dissolve.

Covalently cross-linked gels have been extensively investigated, and both the small and large deformation mechanical properties have been measured and compared, for example, with the expectation of the theory of rubber elasticity. Much of the recent experimental effort has been employed in making networks free from "defects," for example reducing cycles to give a tree-like network. The dynamics of entanglement networks have also been widely studied both experimentally and theoretically following the suggestion that their behavior may be described by "reptation" (2, 3). For solutions and melts of linear polymers the agreement between theory and experiment is usually very good, and efforts are now being made to extend the approach to polymers of more complex architecture, including the effects of introducing star and long chain branching.

Physical Gels

Although both classes of networks are still very active areas of interest, the behavior of both can be said to be largely understood, and any discrepancies between theory and experiment are largely at the refinable level. However, many systems fall between the above two categories, as they consist of chains which are "physically" cross-linked into networks, the cross-links themselves being of small but finite energy, and/or of finite lifetime; these are the so-called physical gels, which include both biological and synthetic polymers; for these it is fair to say that only very recently has much effort been employed in trying to elucidate structure-property relations (4).

The presence of non-covalent cross-links complicates any physical description of the network properties enormously, because their number and

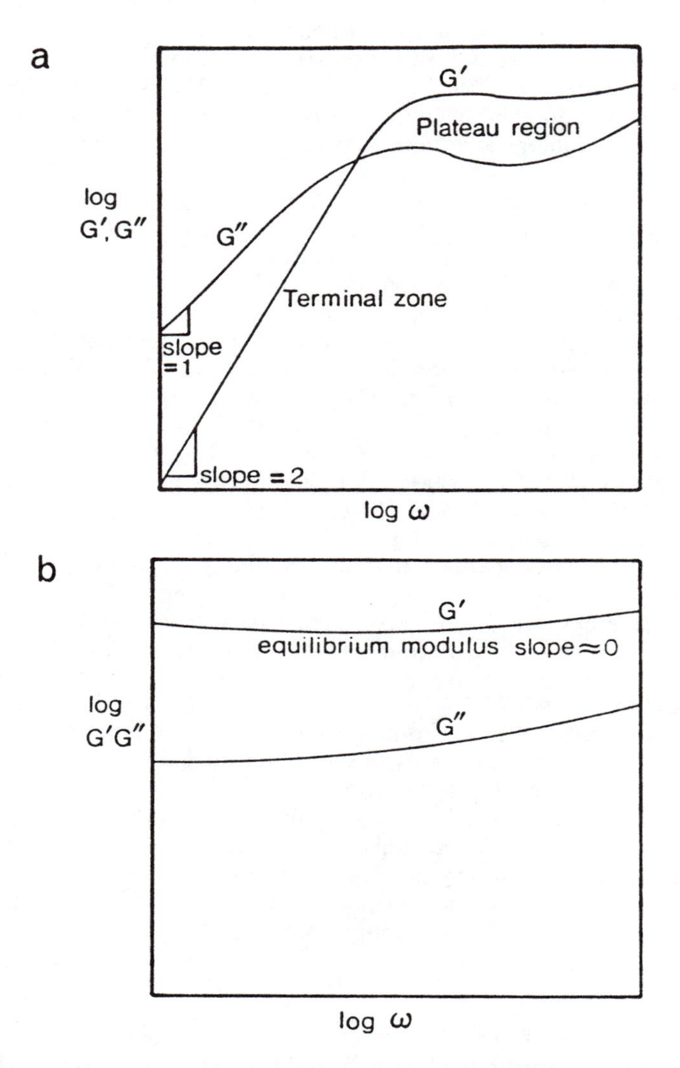

Figure 1. Dynamic mechanical spectra expected for the real (G′) and imaginary (G″) parts of the shear storage modulus for (a) an entanglement network system (pseudo-gel); (b) a covalently cross-linked network (CMr <∼ M_e), and $T > T_g$.

position can, and does, fluctuate with time and temperature. In many cases the nature of the cross-links themselves is not known unambiguously, often involving such disparate forces as, for example, Coulombic, dipole-dipole, van der Waals, charge transfer and hydrophobic and hydrogen bonding interactions (5). As far as biopolymer gels, in general, are concerned, non-covalent cross-links are formed by one or more of the mechanisms listed above, usually combined with more specific and complex mechanisms involving junction zones of known, ordered secondary structure, e.g., multiple helices, ion mediated "egg box" structures, etc. (6). Typically there is a specific, and often intricate, hierarchy of arrangements, a situation which is more familiar to molecular biologists than to polymer scientists.

A physical description of such networks is intrinsically rather difficult, because the *potential* number of junction zones per primary chain, the "functionality," f, and the extent (molecular weight) of the junction along the chain profile can be estimated only indirectly. In many cases there is a subsequent lateral aggregation of chains, after the initial contact. These factors must influence the actual number of physical cross-links and, consequently, the modulus of the final gel. Since this modulus will usually reflect both entropic (rubber-like elasticity) and enthalpic contributions, an *a priori* description of the modulus and mechanical response of these materials is bound to involve a number of approximations.

Amongst the biopolymer physical gels are, of course, gelatin, the seaweed and plant polysaccharides such as agarose, the carrageenans and pectin, starches and cellulose derivatives. More details of the structure and properties of these is given in our recent review (6). Physical gels can also be formed from synthetic polymers including isotactic polymers in certain solvents (i-polystyrene in decalin), ionomer systems in solvents of low dielectric constant, and a number of A-B-A type block copolymers—aggregation of heterostructural chains is often a common feature of such materials. In this case block A is compatible with a solvent and block B is incompatible. Under these circumstances, in solution the B-B interaction between units on adjacent chains forms the physical cross-link.

A further discrimination can be made between so-called "strong" and "weak" physical gels (7). Both respond as solids at small deformations, but whereas the strong gels, e.g., gelatin, are also solids at larger deformation, the weak gels are really structured fluids, so they flow almost as liquids at large deformations. Rheologically this means that whilst the mechanical spectrum within the linear (stress proportional to strain) viscoelastic regime is still as illustrated in Figure 1b, the maximum linear strain, γ, is very different. In the former case we usually say that $\gamma >\sim 0.2$, for weak gels it can be up to 1000 times smaller. Colloidal and particulate networks are often of this type.

More graphically this can be illustrated by the response of xanthan "gels." The typical behavior of these can be illustrated neatly by adding particles to a tube of a xanthan solution and shaking. When the shaking stops, the particles and the solution appear to "freeze" almost instantaneously, and in the absence of further perturbation, the particles are permanently suspended. The solution has a fast time recoverable, apparent "yield stress." The rheological characterization of such material properties is rather complex, but as well as xanthan, such properties are exhibited by other very "stiff" polysaccharides including gellan, rhamsan and wellan under appropriate conditions of concentration and counterion species and ionic strength (8). (Gellan can, of course, also form strong gels under other conditions.) The exact mechanism for this fast recovery has not been quantified, but certainly involves non-covalent interchain interactions, in some cases ion mediated (9), and an element of liquid crystallinity (10).

One view of physical gelation is that it implicitly involves a degree of heterogeneity. This is undoubtedly true at the level of detail of the chain backbone. For example, in the original Rees model for gels formed by carrageenan double helices (11), the existence of "rogue residues" was established chemically. In these, certain saccharide units are in the wrong conformation to allow the helix to propagate. This ensures that, rather than isolated helix pairs being formed, each chain can share portions of ordered helical structure with at least two other chains, an essential condition for branching and subsequent gel formation.

Rheological Characterization of the Gelation Process

As we have pointed out previously, where the small-deformation moduli and the gel time of biopolymer gels are of interest, the freedom to choose a range of polymer concentrations prior to the introduction of cross-links has a number of important consequences. If temperature alone is the driving force for gelation, both gel time and ultimate gel rigidity (the time independent modulus) obtained are related to the concentration of polymer present, the gel time diverging logarithmically to infinity at C_0 and the equilibrium modulus tending to zero.

For systems allowed to gel under a well-defined thermal regime, and allowed to achieve a limiting degree of cross-linking, there is a characteristic "cure" curve (a term taken form the curing of thermo-set resins) of log (G', G'') against time, Figure 2. This has an initial lag time, then both G'' and G' increase, but with G' increasing faster than G'' giving a cross-over time, (very often taken as the gelation time), and finally G' appearing to reach a plateau value. (Although for some gels, including gelatin, there is actually no final value of the modulus since the "shuffling" of helix partners allows log (G') to increase indefinitely when plotted against the log of time). G'' sometimes passes through a parabolic maximum, and then decreases to zero, and sometimes plateaus off. According to Valles and co-workers, in covalent networks the former effect is associated with the relaxation of "dangling chain ends" (12). We have observed and noted a qualitative correlation, with biopolymer gels formed from more flexible chains (e.g., gelatin) giving such a G'' parabola, whereas those from stiffer chains (e.g., agarose) give a slight maximum followed by a plateau (13).

From cure curves measured for different initial polymer concentrations, characteristic (long time) modulus-concentration and gelation time-concentration dependencies are found. For example, for the modulus G,

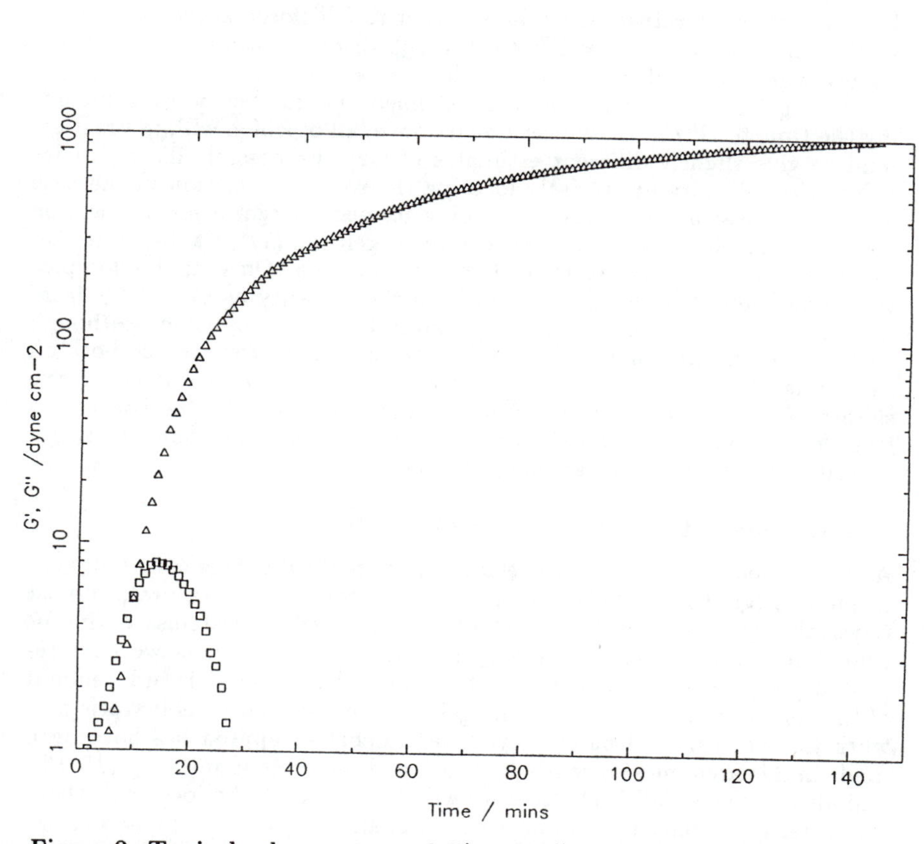

Figure 2. Typical gel cure trace of G′ and G″ *vs.* time; these data are for a 4% w/w gelatin gel so there is a clear maximum in G″ (cf. Figure 13 of Ref. 6).

near the critical gel concentration C_0, (the concentration below which no gel can be formed) a large and variable power law dependence of G on C is observed, whilst at much higher concentrations, a constant, and approximately C^2, limiting law emerges. Such behavior has been noted historically. In more recent literature, some data appear to fall closer to the asymptotic scaling prediction (3) $G \propto C^{2.25}$ (although discrimination between these values is quite difficult).

The gelation time (t_c) can be determined in a number of ways, including the time to G''/G' cross-over noted above, all of which will, in practice, tend to give slightly different estimates of t_c. The possible discrepancies reflect the difficulty in precisely locating the sol/gel transition in the case when non-covalent cross-links are being formed. Regardless of this, the concentration dependence of the reciprocal geltime ($1/t_c$) appears to follow a very similar curve to that of the gel modulus. On a double log plot the former function also has a logarithmic singularity as $C \longrightarrow C_0$ from above, and the curvature decreases steadily to a limiting value, (although with a different limit slope). In previous work we have commented both on modelling ($1/t_c$) vs. C data, and on the similarity between the two curve shapes. Here we discuss two applications of our model for the concentration dependence of G. Models for the concentration dependence of t_c are described in more detail elsewhere (14, 15).

The Concentration Dependence of Gel Modulus

As mentioned above, whilst the elasticity of ideal rubbers is essentially entropic in origin (modulus is proportional to absolute temperature), in most physical gels (and to a lesser extent in real covalent systems) enthalpic contributions are undoubtedly very important. Nevertheless we have repeatedly argued that the number of active network chains is independent of the contribution per chain to the gel modulus, so that combinatoric network theories can still be employed. A number of approaches have been made in this area, most recently by Clark and the present author (6, 15, 16), and also by Oakenfull (17). Here we will describe only the former method; the latter does differ in a number of ways, and a detailed comparison has recently been published (18).

In our work, we have used branching theory (19) to count the number of elastically active chains (EANCs) and so calculate the concentration dependence of modulus. To do this we abandoned the conventional (ideal entropy) assumption of a contribution RT to the modulus per mole of an EANC, and introduce a non-ideality parameter a such that aRT units would be the actual contribution. [Since a itself will generally be temperature dependent, the product aRT is a free energy, and a will in principle involve enthalpic as well as entropic contributions. In other words, the factorization into terms a and RT we have previously employed merely reflects a particular choice of energy units and (unlike the case of ideal rubbers) is *not* a statement about the expected temperature dependence of the gel modulus.]

Assuming that the making and breaking of cross-links occurs within the time window of our rheological experiment, and that the mechanical spectrum looks similar to Figure 1b, we can define an equilibrium constant for the reversible conversion of a cross-link into two free sites in the form of the Ostwald dilution law. Then if α is the proportion of cross-links formed from two free sites coming together (at this stage we attribute no molecular detail to either cross-links or sites) we can say

$$K'(= 1/K) = [(1-\alpha)^2/\alpha]N_0 f \quad (1)$$

where K' is an equilibrium constant, f is the functionality introduced above, and N_0 is the number of chains per unit volume. In 1965, Hermans solved a form of this equation valid for the case of "weak binding" *viz.* $K' \gg 1$ and $\alpha \ll 1$ (20). As we have shown previously, the more general case when K' is not $\gg 1$ is only a little more complex, and results in a solution valid for any f (16).

$$G = \frac{aRT(f-1)\alpha(1-v)^2(1-\beta)}{2K(f-2)^2} \times \frac{C}{C_0} \quad (2)$$

(Note that although α, v, and β are functions of f and C/C_0 only, the final result cannot be written explicitly, as v is determined by solution of a recursive equation. C_0 is *not* a parameter, as it is itself a function of f and K.) Best values of a, f and K can be determined by non-linear least squares, although if f (or a) can be estimated from knowledge of the gelling system and fixed, this is usually much better, since "independent" fits usually show these two parameters to be correlated. We have found in a number of different systems that the small deformation results do follow the concentration dependence of modulus predicted from this network theory above the critical gel concentration, and also that the enthalpic "front factor" a is related to chain stiffness in a predictable way. We will now illustrate the application of the method to some recent data for gelatin and amylose gels, and discuss the implications of the results.

Application of the Modulus/Concentration Model

Gelatin. In previous work (16) we have applied the cascade model to gelatin gels, but more recent data is now available (21). Shear creep experiments were performed using parallel plate geometry. The sample was a high molecular weight alkaline processed ossein gelatin supplied as a gift by Rousselot, and was identical to that used by Busnel and co-workers (22). It was a purified grade with $M_w \approx 1.55 \times 10^5$ and $M_w/M_n \approx 2.17$ (manufacturer's data), further details of the sample characteristics and experimental details are given elsewhere (21, 22). To ensure a reproducible time and cooling regime, the cooling bath thermostat was maintained at 15°C for exactly 19 hours (well into the cure plateaux) before beginning the creep phase of the experiment. The modulus G was obtained simply as $1/J_0$, with J_0 the "instantaneous" creep compliance—the error in this procedure is small as traces of log(J) vs. log(time) were very nearly horizontal.

Nine data points were obtained, over a concentration range from 2.4% to 15% w/w. Using Equation 2, the best global fit was obtained with a functionality of 58.6, and with $a = 0.35$, the corresponding C_0 as 0.45%. Although this fit is very good, it is always better if one can estimate one or more of the parameters independently using other available information. According to Peniche-Covas *et al.* (23) for an n-chain model, $f = (n-1)(x-1)+1$, where x is the number of junction zones per chain. To estimate the functionality we can use the fact that the minimum stable helix length is around 40 residues (22); since the molecular weight of a peptide residue in gelatin is ~ 105, this means that *per chain* the maximum number of helical sections should be around $M_n/(40 * 105) =\sim 17$. For the double chain hair-pin model (22) this gives f equal to 18, whilst for an independent triple helix model $f = 35$. The latter estimate is almost certainly too high, because of the need for the network to retain flexible regions, so that we feel a figure of $f \sim 10$ to 20 would be more reasonable.

We can also argue that a should not be less than 0.5 (the lowest value normally considered in the theory of rubber elasticity—a is often taken as unity). Using these "constraints" we find that very good fits can still be obtained with $f = 10$ (a is found to be 0.98), and $f = 20 (a = 0.65)$. The corresponding C_0's were 0.74% and 0.66%, respectively. The latter values agree quite well with independent light scattering measurements on the same gelatin sample by ter Meer, Lips and Busnel which have shown that $0.67\% < C_0 < 1.0\%$ (24). On the basis of this information we could actually eliminate the global best fit above, and the sum of squared deviations in all cases was quite similar (0.032 for the three parameter, and 0.042 and 0.036 for the two parameter models). A final piece of information is available from the work of Djabourov and Leblond (25) for a different sample. They found that at the gel point the amount of helix h_c *independent of temperature* was $\sim 7.0\%$. From the simple Flory gel point equation this would give a functionality of $h_c = 1/(f-1)$, i.e., $f \sim 15$, again in very good agreement with the above argument. Figure 3 illustrates the fit for $f = 15$; it also shows how important it is to make measurements in the regime $1 < C/C_0 < 3$, (i.e., below that in the present data) if a very critical comparison with theory is to be made.

Amylose. In amylose, the linear component of starch, it has long been established experimentally that the chain length at which precipitation replaces gelation is equal to 110 sugar residues. This is therefore an estimate of the distance between the start of consecutive junction zones and so in the fits the functionality, f, was assumed to be equal to $DP_n/110$. In the work by Gidley and co-workers, amylose samples were biosynthesized to have a range of molecular weights with a narrow MW distribution (26), and gels made for samples with a range of concentrations and MW's.

Figure 4 illustrates the application of the method to a range of amylose gels, which were measured using a series of oscillatory shear cure experiments (27). Under these conditions the fitted values of K and $a(\sim 10)$ are essentially independent of chain length. The apparent success of Equation 2 again requires us to examine the assumptions and conclusions very carefully.

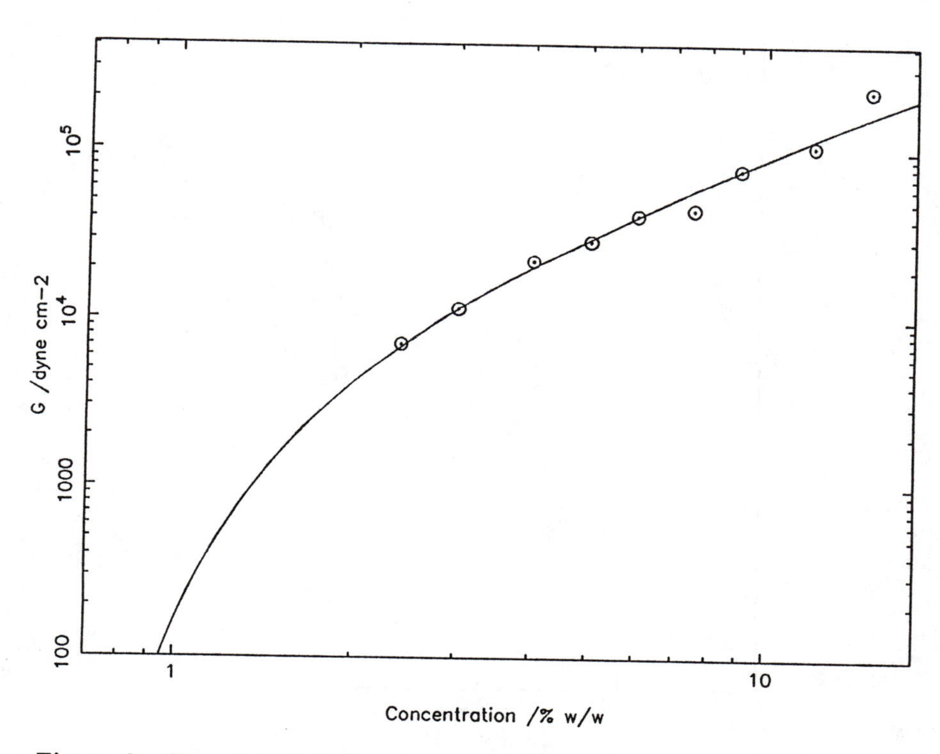

Figure 3. Experimental G vs. concentration (% w/w) data for gelatin gels from the creep compliance (21). The curve corresponds to Eqation 2 with $f = 15$; other parameters are $a = 0.77$ and $K' = 51.2$. C_0 is equal to 0.696%. On this scale the fits for $f = 10$ and 20 are almost indistinguishable.

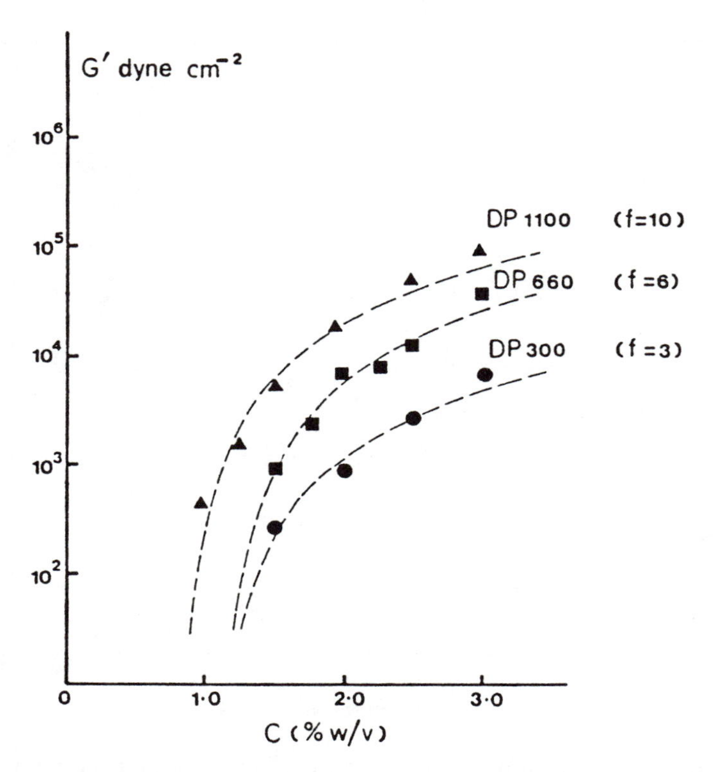

Figure 4. Experimental G′ *vs.* concentration (% w/v) data for amylose gels (•) DP = 300; (▪) DP = 660, (▲) DP = 1100. Theoretical fits obtained using the theory outlined above with f = 3, 6, and 10, respectively. (Reproduced from reference 26. Copyright 1989 American Chemical Society.)

The first assumption which may be challenged is that of a cross-linking equilibrium, especially since recent work on the de-swelling of agarose gels has suggested that cross-links in this system are effectively permanent (28), and that the degree of cross-linking in any system is largely under kinetic control. However, recently Clark has presented calculations in which cross-linking progresses by a pairwise irreversible association mechanism (15). Analysis of simulated modulus-concentration data obtained in this way suggests that, provided the gel curing process proceeds for long times, and for the same time for each system concentration, the same equations and fitting procedures can be adopted as in the equilibrium situation. Then K no longer has the significance of an equilibrium constant, but is a function of the forward rate constant and time.

Finally, we mentioned earlier how there is an apparent correlation between a and "chain stiffness." For other polysaccharide and some protein gels we have concluded in earlier work that a could be up to a factor of ten or more, although for gelatin, as above, it has always been found to be quite close to the classical value of unity (6,16). Whilst there is no unequivocal support for this deduction, recent theoretical progress has been made on a model of unzipped chains or "reel-chains" (29,30), and this establishes that, purely from entropy considerations, a parameter essentially equal to a, does indeed increase in approximately the expected manner as the chain becomes stiffer. A similar conclusion was reached from a quite different approach, *viz.* a computational investigation of the contribution of very short chains to network elasticity (31). Future progress requires a further combination of careful experiment on well defined systems, particularly including studies of synthetic analogues of biopolymer physical gels, and this is currently an area of considerable activity in a number of laboratories.

Acknowledgments

The author thanks co-chairs Prof. W. G. Glasser and Dr. H. Hatakeyama, and the American Chemical Society (Cellulose Division) for their hospitality and the opportunity to participate in this Symposium. He is grateful to the Royal Society, and to SERC for financial support during the course of his work.

Literature Cited

1. Ferry, J. D. *Viscoelastic Properties of Polymers, 3rd Ed.;* John Wiley: New York, 1980.
2. Doi, M.; Edwards, S. F. *The Theory of Polymer Dynamics;* Clarendon: Oxford, 1986.
3. de Gennes, P. G. *Scaling Concepts in Polymer Physics;* Cornell University Press: Ithaca, N.Y., 1979.
4. Burchard, W.; Ross-Murphy, S. B., Eds. *Physical Networks— Polymers and Gels;* Elsevier Applied Science: London, 1990.
5. Tsuchida, E.; Abe, K. *Adv. Polym. Sci.* 1982, **45**, 1.
6. Clark, A. H.; Ross-Murphy, S. B. *Adv. Polym. Sci.* 1987, **83**, 60.

7. Ross-Murphy, S. B. In *Biophysical Methods in Food Research—Critical Reports on Applied Chemistry, Vol. 5;* Chan. H. W.-S., Ed.; Blackwell: Oxford, U.K., 1984; p 138.
8. Robinson, G.; Manning C. A.; Morris, E. R. In *Food Polymers, Gels and Colloids;* Dickinson, E., Ed.; R.S.C.: London, 1991; p 22.
9. Ross-Murphy, S. B.; Morris, V. J.; Morris, E. R. *Faraday Discuss. Chem. Soc.* 1983, **18**, 115.
10. Allain, C.; LeCourtier, J.; Chauveteau, G. *Rheol. Acta* 1988, **27**, 225.
11. Rees, D. A. *Adv. Carbohydr. Chem. Biochem.* 1969, **24**, 267.
12. Bibbo, M. A.; Valles, E. M. *Macromolecules* 1984, **17**, 360.
13. Clark, A. H.; Richardson, R. K.; Ross-Murphy S. B.; Stubbs, J. M. *Macromolecules* 1983, **16**, 367.
14. Ross-Murphy, S. B. *Carbohydr. Polym.*, 1991, **14**, 281.
15. Clark, A. H. In *Food Polymers, Gels and Colloids;* Dickinson, E., Ed.; R.S.C.: London, 1991, p 322.
16. Clark, A. H.; Ross-Murphy, S. B. *Brit. Polymer J.* 1985, **17**, 164.
17. Oakenfull, D. G. *J. Food Sci.* 1984, **49**, 1103.
18. Clark, A. H.; Ross-Murphy, S. B.; Nishinari, K.; Watase, M. In *Physical Networks—Polymers and Gels;* Burchard, W.; Ross-Murphy, S. B., Eds.; Elsevier Applied Science: London, 1990; p 209.
19. Gordon, M.; Ross-Murphy, S. B. *Pure Applied Chem.* 1975, **43**, 1.
20. Hermans, J. R. *J. Polym. Sci.* 1965, **A3**, 1859.
21. Higgs, P. G.; Ross-Murphy, S. B. *Int. J. Biol. Macromol.* 1990, **12**, 233.
22. Busnel, J.-P.; Morris, E. R.; Ross-Murphy, S. B. *Int. J. Biol. Macromol.* 1989, **11**, 119.
23. Peniche-Covas, C. A. L.; Dev, S. B.; Gordon, M.; Judd, M.; Kajiwara, K. *Farad. Discuss. Chem. Soc.* 1974, **57**, 165.
24. ter Meer, H.-U.; Lips, A.; Busnel, J.-P. In *Physical Networks—Polymers and Gels;* Burchard, W.; Ross-Murphy, S. B., Eds.; Elsevier Applied Science: London, 1990; p 253.
25. Djabourov, M.; Leblond, J. In *Reversible Polymeric Gels and Related Systems;* Russo. P., Ed.); *ACS Symp. Ser.* 1987, **350**, p 211.
26. Gidley, M. J.; Bulpin, P. V. *Macromolecules* 1989, **22**, 341.
27. Clark, A. H.; Gidley, M. J.; Ross-Murphy, S. B.; Richardson, R. K. *Macromolecules* 1989, **22**, 346.
28. Lips, A.; Hart, P. M.; Clark, A. H. *Food Hydrocolloids* 1988 **2**, 141.
29. Nishinari, K.; Koide, S.; Ogino, K. *J. Phys. (Paris)* 1985, **49**, 763.
30. Higgs, P. G.; Ball, R. C. *Macromolecules* 1989, **22**, 2432.
31. Mark, J. E.; Curro, J. G. *J. Chem. Phys.* 1984, **81**, 6408.

RECEIVED December 16, 1991

Chapter 14

Effect of Water on the Main Chain Motion of Polysaccharide Hydrogels

H. Yoshida[1], T. Hatakeyama[2], and Hyoe Hatakeyama[3]

[1]Department of Industrial Chemistry, Tokyo Metropolitan University, Minami-ohsawa, Hachiohji-shi, Tokyo 192–03, Japan
[2]Research Institute for Polymers and Textiles and [3]Industrial Products Research Institute, Tsukuba, Ibaraki 305, Japan

The main chain motion of hyaluronic acid, xanthan and pullulan with various water contents ranging from 0 to 3 (grams of water per gram of dry polysaccharide) was investigated by differential scanning calorimetry (DSC). Water in the polysaccharide system was classified into three types: non-freezing, freezing bound, and free water. Hydrogel derived from polysaccharide, hyaluronic acid, and xanthan contained a large amount of freezing bound water. The water-polysaccharide systems formed the glassy state by cooling from 300°K to 150°K at the rate of 10°K/min. The glass transition temperature of the system was influenced by non-freezing water and freezing bound water. A part of the freezing bound water in hydrogels formed amorphous ice. As a result of cooperative motion of the polysaccharide, the glass transition of these systems absorbed non-freezing water and the amorphous ice.

Hydrogels have become increasingly important in industrial and medical fields, such as food processing, delivery systems, manufacture of artificial biological tissues, civil engineering, etc. The phase transition of gels induced by thermal and electrical stimulations, and also the gelation mechanism, have attracted much attention in both the theoretical and applied fields (1-4). Biological gels show better mechanical properties than synthetic gels, since biological gels are constituted of rigid molecules such as proteins which have helix coil conformation, or polysaccharides which consist of glucan or glucosamino glycan having β-glycoside linkages. Because these rigid biomolecules show high glass transition temperatures in the dry state, it is difficult to observe the main chain motions below those thermal decomposition temperatures. Not only for synthetic polymers but also

0097–6156/92/0489–0217$06.00/0

biopolymers, the main chain motion affects their mechanical and physical properties and also their functionality.

It is well known that the main chain motion of hydrophilic polymers depends on the interaction between water and polymers. In order to investigate the physical properties and functionality of hydrogels, it is important to consider the molecular motion of polysaccharide and water in hydrogels. We have reported the structural change of water in the water-polyelectrolyte system (5), the water-polysaccharide systems (6,7), and hydrogels (8-11) using differential scanning calorimetry (DSC) and nuclear magnetic resonance spectrometry (NMR). Most of the water-polyelectrolyte systems formed the liquid crystalline state at high concentrations due to the electrostatic interaction among the hydrated ions (5-7, 11). From DSC and NMR studies, water molecules interacting with polysaccharide molecules played an important role in forming the liquid crystalline state in these systems. Several kinds of water-polysaccharide systems form both a liquid crystalline phase and a hydrogel. It is expected that the higher structures, such as liquid crystals or junction zones of hydrogels, influence the main chain motion of polysaccharides.

In this study, three kinds of polysaccharides were used to investigate the effect of water on the main chain motion of polysaccharides. Hyaluronic acid and xanthan form hydrogel; xanthan also forms the liquid crystal phase. The water-pullulan system forms neither liquid crystal nor hydrogel. The effect of higher structures on the molecular motion of polysaccharides and water was also discussed.

Experimental

Materials. Three kinds of polysaccharides—hyaluronic acid, xanthan and pullulan—were used in this study. Hyaluronic acid and pullulan were obtained from Wako Pure Chemical Industries, Ltd., Tokyo, Japan. Xanthan used was Kelzan, supplied by Kelco Co., Ltd. Hyaluronic acid and xanthan were purified and neutralized as described elsewhere (9). The sodium content per tetraoligosaccharide unit of hyaluronic acid and xanthan was 2.5 and 2.6 mole/mole, respectively.

Apparatus. A Seiko differential scanning calorimeter Model DSC 200 connected to a Seiko Thermal Analysis System Model SSC 5000 was used. The scanning rate was 10°K/min. The sample weight used was 2-5 mg, depending on water content.

The sample was weighed and dissolved in distilled water in a DSC aluminum sample vessel. After the sample had completely dissolved, the water was slowly evaporated in order to obtain a sample with the predetermined water content (W_c). After sealing the DSC sample vessel, all samples were kept at room temperature for a few days in order to homogenize them. The W_c was defined as follows:

$$W_c(\mathrm{g/g}) = \frac{\text{amount of water/g}}{\text{dry sample weight/g}} \quad (1)$$

The dry sample weight was determined by the weight after heating to 460°K at which the endothermic peak due to the evaporation of water disappeared (12).

Results and Discussion

Type of Water in Hydrogels. The state of water in various polymers, including polymers in aqueous solutions, have been studied using thermal, dielectrical and spectroscopic analysis (13-16). From the value of their diffusion coefficients observed by NMR, water molecules in aqueous protein solutions were classified into three categories (15). Water molecules interacting with biomolecules are called trapped water, solid water, bound water, interfacial water, etc. In previous papers (5-12, 17-19), water molecules which showed first-order phase transition, such as crystallization and melting, were called freezing water, and those molecules which never crystallized even at 130°K were called non-freezing water. In the case of hydrophilic polymers insoluble in water, freezing water was classified into two types: free water and bound water (17). These two types of water were defined by the crystallization temperature on cooling at a given rate. At a cooling rate of 10°K/min, free water crystallized at 255°K and bound water at around 230°K. These values are independent of W_c (18, 19).

Three types of water in hydrogels of the water-polysaccharide system were found by DSC (20). Hydrogels and the water-polysaccharide system contain much freezing bound water, compared with other hydrophilic polymers. As the crystallization temperature of bound water in these systems changed continuously from 230°K to 250°K on cooling, it is difficult to classify freezing water into bound water and free water only by a DSC cooling curve (20). Figure 1 shows the schematic DSC curves for three types of water—non-freezing water, freezing bound water, and free water—undergoing cooling and heating at 10°K/min.

Free water showed a sharp exothermic peak due to crystallization on cooling and a sharp endothermic peak due to melting on heating. The starting temperature of crystallization on cooling (T_c) of free water in hydrogels was observed at temperatures between 254 and 256°K, depending on its thermal histories. The starting temperature of melting (T_{mi}) of free water was close to 273°K, and the melting peak temperature (T_m) was the same as that of pure water. The values of T_{mi} and T_m of free water scarcely depended on W_c and the thermal history of the system. On the other hand, freezing bound water shows a broad exothermic peak which can be attributed to the slow rate of crystallization and lower T_c on cooling. From the isothermal crystallization analysis of free and freezing bound water, the crystallization rate of freezing bound water was 10 times slower than that of free water (21). For freezing bound water, T_{mi} and T_m were much lower than those of free water. Both temperatures, T_{mi} and T_m of freezing bound water, depended on W_c and the thermal history of the system. Non-freezing water showed no first-order phase transition either on cooling or heating, but showed glass transition.

From the melting enthalpy of water, the freezing water content (W_f)

which contained both freezing bound and free water was calculated as shown in the following equation:

$$W_f(\text{g/g}) = \frac{\text{amount of freezing water/g}}{\text{dry sample weight/g}} \quad (2)$$

By the subtraction of W_f from W_c, the non-freezing water content (W_{nf}) was also calculated. The changes in W_{nf} for the water-polysaccharide systems are shown in Figure 2. Only non-freezing water existed in the W_c range below 0.5 for all polysaccharide systems. With increasing W_c, W_{nf} approached a constant value which depended on the type of polysaccharide. From the constant value of W_{nf} observed in the W_c range above 0.5, the number of moles of non-freezing water which interacted strongly with ionic groups or polar groups of polysaccharides was calculated. Sodium ions in the water-xanthan and the water-hyaluronic acid systems accompanied about 7 moles of non-freezing water in the hydration shell. In the W_c range above 0.5, the hyaluronic acid system showed the minimum value of W_{nf}. It may be appropriate to consider that the minimum value of W_{nf} was caused by the overlap of the hydration shell (22). In the water-pullunan system, each hydroxy group of pullulan bonded to one water molecule as non-freezing water.

Figure 3 shows the DSC heating curves of polysaccharides having the same W_c (ca. 0.6). All polysaccharide systems contained both non-freezing and bound water. The DSC curves of hyaluronic acid show glass transition temperature (T_g), an exothermic peak temperature due to cold-crystallization of water (T_{cc}), and an endothermic peak temperature (T_m) due to the melting of water. In addition to T_g, T_{cc} and T_m, the xanthan system shows the transition from the liquid crystalline state to the isotropic liquid state (11). T^* shows the peak temperature of this transition. The heat of liquid crystalline transition of the xanthan system was larger than that of the other lyotropic liquid crystals (23). The liquid crystalline state of the xanthan system differs from the cholesteric type liquid crystal which was observed in the case of the hydroxypropyl cellulose system (24, 25).

The DSC curve of the water-pullulan system shows T_g, T_{cc} and T_m as seen in Figure 3. The exotherm started immediately after T_g in the DSC heating curve of the pullulan system. Despite the same W_c, the values of T_g, T_{cc} and T_m are different in samples. The T_g values suggested that each polysaccharide system had different molecular mobility of main chain even at the same W_c. As can be seen in Figure 2, each polysaccharide system had different amounts of non-freezing water and freezing bound water. The lower T_m suggested that the water-hyaluronic acid and the water-xanthan system contained a large amount of freezing bound water even at the same W_c.

Glass Transition of the Water-Polysaccharide Systems. Figure 4 shows the W_c dependence on T_g of the water-hyaluronic acid system. The T_g's were observed for the sample which was prepared by cooling from 300°K to 150°K at 10°K/min. In the low W_c range, T_g decreased rapidly to the

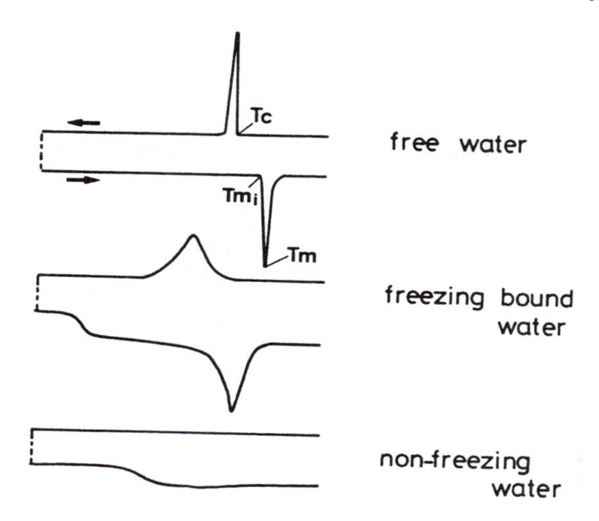

Figure 1. Schematic DSC cooling and heating curves of non-freezing, freezing bound and free water in hydrogel. T_c is the starting temperature of crystallization on cooling, T_{mi} and T_m are the starting temperature and the peak temperature of melting on heating.

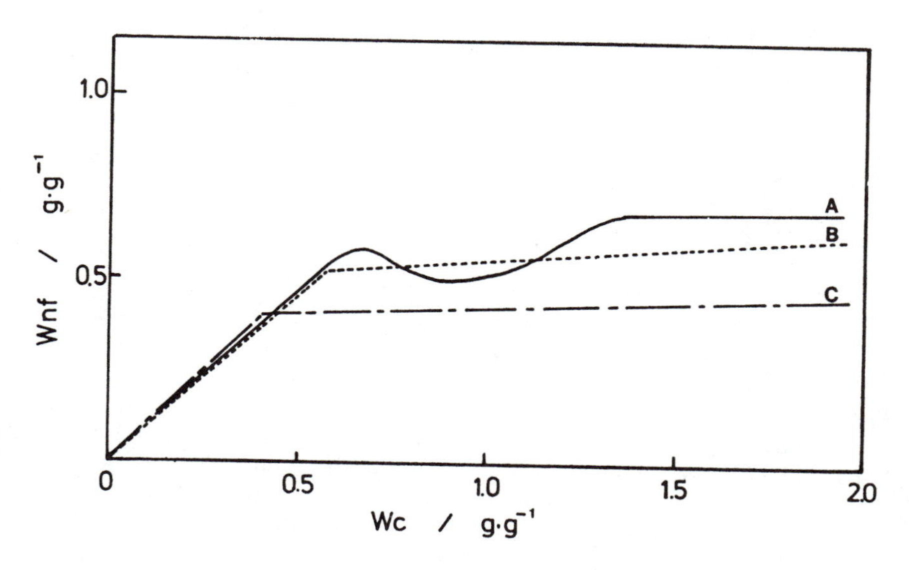

Figure 2. Relationship between non-freezing water content (W_{nf}) and water content (W_c) for the water-hyaluronic acid system (A), the water-xanthan system (B), and the water-pullulan system (C).

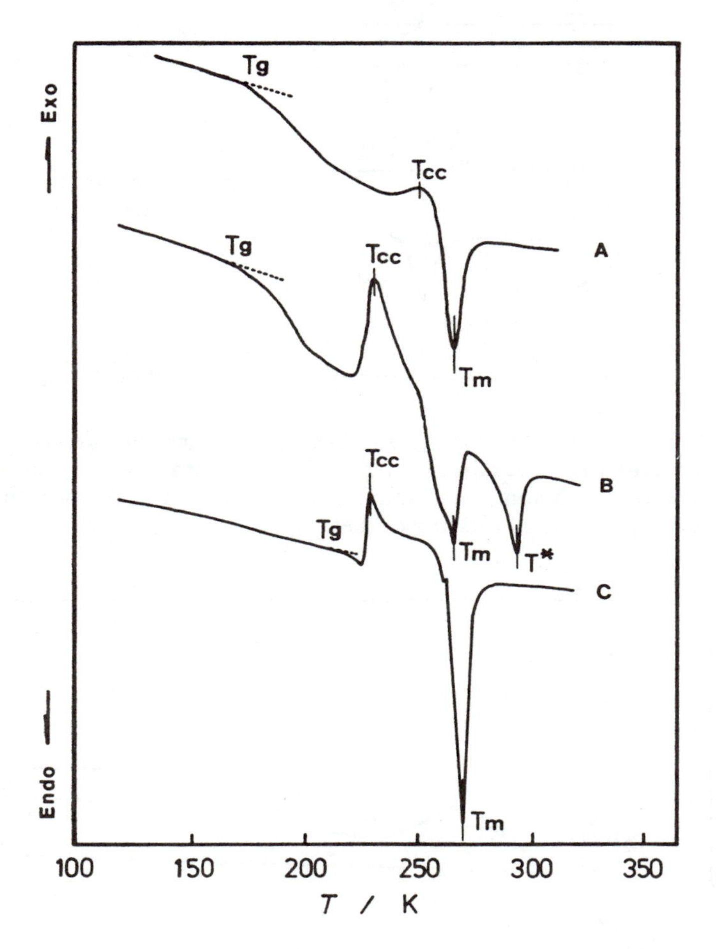

Figure 3. DSC heating curves of the water-hyaluronic acid system (A), the water-xanthan system (B), and the water-pullulan system (C) having the same W_c. W_c was about 0.6. T_g, T_{cc}, T_m, and T^* are the glass transition temperature, the cold crystallization temperature of amorphous ice, melting temperature of regular ice, and transition temperature from the liquid crystalline state to the isotropic liquid state, respectively.

minimum value with increasing W_c. After passing the minimum value, T_g increased gradually and approached the constant value with increasing W_c. Not only the water-hyaluronic acid system, but also other water-polysaccharide systems, showed a similar W_c dependency on T_g. Each polysaccharide showed a different value of W_c at which the minimum T_g was observed. In order to discuss the effect of water on T_g, the W_c range was divided into four regions as shown in Figure 4. Each region was defined in terms of type of water in the system.

Molecular Motion in Region I. In region I, the system which contained only non-freezing water showed glass transition only in the heating DSC curve. In the completely dry state, glass transition of all polysaccharides was not observed because T_g overlapped with their decomposition temperature. For all polysaccharide systems, T_g decreased markedly with the addition of water due to the plasticizing effect. The main chain motion of the polysaccharide was restricted by a large number of hydrogen bonds in the dry state. When the system absorbed a small amount of water, some water molecules broke the hydrogen bonds in the same way as the hydrogen bonds of hydrophilic polymers were broken (19).

Molecular Motion in Regions II and III. In regions II and III, the system contained non-freezing water and freezing bound water. When the system was cooled at 10°K/min, the crystallization of bound water was observed for the system in region III, but not for the system in region II. In region II, T_g still decreased gradually with increasing W_c. After reaching the minimum value, T_g increased and approached a constant value with increasing W_c in region III.

In regions II and III, the exothermic peak between T_g and T_m was observed in the heating DSC curve. The exothermic peak was influenced by annealing at temperatures between T_g and T_m. Figure 5 shows the schematic DSC curves of the water-polysaccharide system with W_c corresponding to regions II and III. The sample, which was prepared by cooling at 10°K/min from 300°K to 150°K, was used as the original sample. This original sample was heated to the end temperature of the exothermic peak, and then quenched to 130°K. The sample having this thermal history was used as the annealed sample. The heating DSC curve of the annealed sample does not show the exothermic peak. The value of T_g shifted to the higher temperature and the heat capacity difference between the glassy and liquid states at T_g (ΔCp) decreased by annealing. This fact suggested that the disordered part, which contributed to glass transition of the original sample, changed to the ordered part, which no longer contributed to glass transition.

The freezing bound water of the system in region II did not crystallize on cooling. However, the T_m of ice was observed in the heating DSC curve. This fact suggests that the freezing bound water crystallized on heating at around the exothermic peak and then melted at T_m. Therefore, we assume that the freezing bound water formed amorphous ice during cooling and the exothermic peak observed in the heating DSC curve was

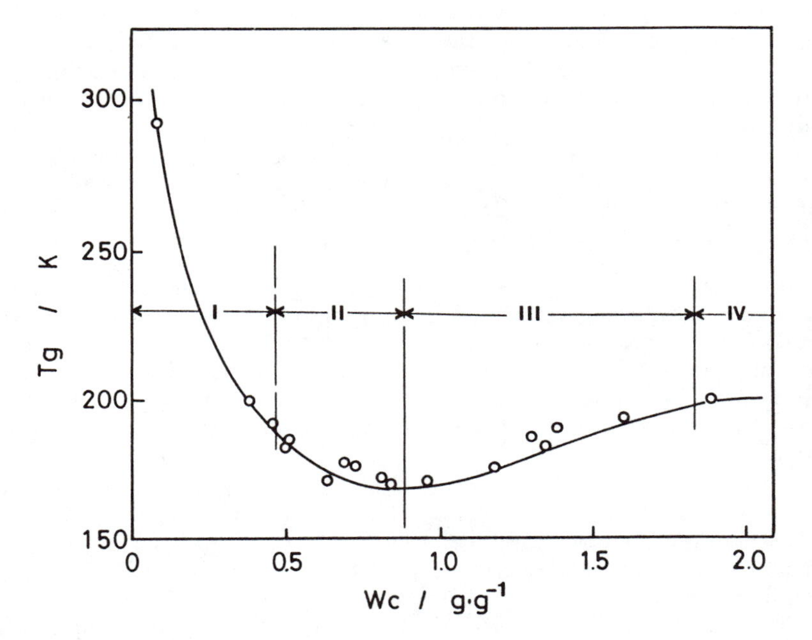

Figure 4. Relationship between glass transition temperature (T_g) and water content (W_c) for the water-hyaluronic acid system. Samples were prepared by cooling at 10°K/min from 300°K to 150°K. The W_c range is divided into four regions which were defined in terms of type of water in the system.

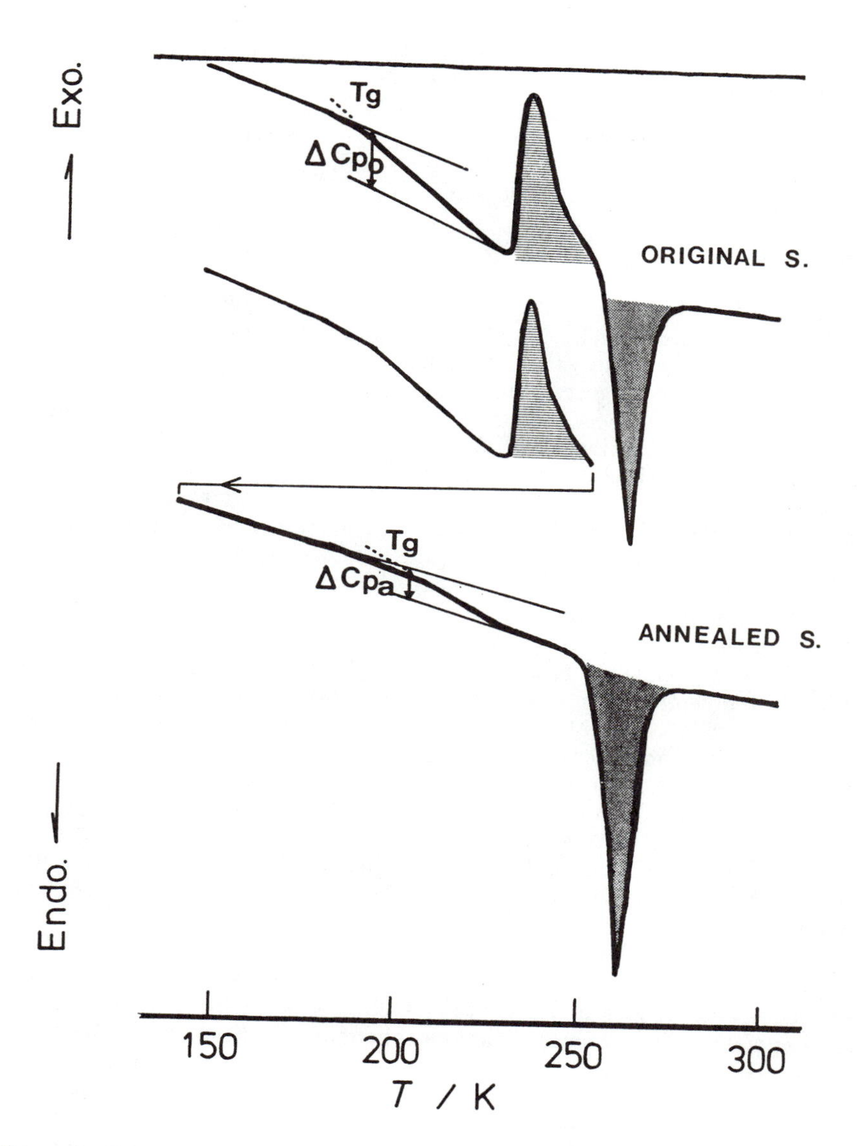

Figure 5. Schematic DSC curves of the water-polysaccharide system having different thermal history. The original sample was prepared by cooling at 10°K/min from 300°K to 150°K. T_g and ΔCp are glass transition temperature and heat capacity difference between glassy state and liquid state at T_g, respectively.

due to the crystallization of amorphous ice. From the melting enthalpy of ice and the area of the exothermic peak, the amount of amorphous ice which crystallized at T_{cc} (w_2) was estimated. The value of ΔCp of amorphous ice (ΔCp_2) was calculated using the following equation:

$$\Delta Cp_2 = \frac{(\Delta Cp_o - \Delta Cp_a)}{w_2} \quad (3)$$

where ΔCp_o and ΔCp_a were ΔCp of the original and the annealed sample, respectively.

The calculated values of ΔCp_2 are shown in Figure 6 as a function of W_c. For all polysaccharide systems, ΔCp_2 decreased with increasing W_c and approached the value of amorphous ice prepared from pure water (26). This fact suggests that freezing bound water formed amorphous ice when the system was cooled at 10°K/min and the molecular motion of amorphous ice contributed to the glass transition of the original sample in regions II and III. As can be seen in Figure 6, the formation of amorphous ice occurred in a wide W_c range in the hyaluronic acid and xanthan systems while the formation of amorphous ice occurred in a narrow W_c range in the pullulan system which did not form hydrogel readily. This fact indicated that the hydrogel contained a large amount of bound water and that part of the bound water changed easily to amorphous ice even if the system was cooled slowly. At a given W_c, the hyaluronic acid system showed the largest value of ΔCp_2, and the value of ΔCp_2 of the xanthan and pullulan systems decreased as shown in Figure 6. The large value of ΔCp_2 suggests that the structure of amorphous ice in the system adopts a more random arrangement than the system which showed the small ΔCp_2 values. Recently, it was reported that glass transition of amorphous ice in hydrogel prepared from poly(2-hydroxyethyl methacrylate) (PHEMA) was observed at 136°K by dynamic mechanical analysis (27). As the reported T_g is close to that of pure water which is obtained by calorimetric measurements (26, 28), amorphous ice in PHEMA hydrogel scarcely interacts with PHEMA in contrast to the water-polysaccharide systems.

In order to explain the molecular mechanism of W_c dependence on T_g in regions II and III, the following hypothesis was presented. As the system in regions II and III contained almost the same amount of non-freezing water, it was assumed that the change of T_g occurred as the result of the plasticizing effect of amorphous ice on the main chain motion of polysaccharide absorbed non-freezing water. Actually, the minimum value of T_g observed between regions II and III seemed to depend on amount of freezing bound water. The T_g change in the polymer-plasticizer system is well described by the Gordon-Taylor equation as shown below.

$$T_g = \frac{\phi_1 \Delta Cp_1 T_{g1} + \phi_2 \Delta Cp_2 T_{g2}}{\phi_1 \Delta Cp_1 + \phi_2 \Delta Cp_2} \quad (4)$$

where ϕ_1, ΔCp_1, and T_{g1} are weight fraction, ΔCp and T_g of i-th component, respectively. The amount of amorphous ice (w_2) which is calculated from the area of the exothermic peak gives the weight fraction of amorphous ice for each W_c in regions II and III. The values of T_{g1} and ΔCp_1 are obtained as the experimental values of the sample with W_c 0.5, which were 185.7°K and 0.321 J/gK for the hyaluronic acid system, 198°K and 0.345 J/gK for the xanthan system. The relationship between T_g and ϕ_2 of the original samples in regions II and III for hyaluronic acid and xanthan systems are shown in Figure 7. The solid lines show the calculated values using Equation 4 with 135°K and 1.94 J/gK as the values of T_{g2} and ΔCp_2, respectively (26). The experimental values correspond well with the calculated values. The result shows that the T_g of polysaccharide-absorbed non-freezing water decreased with increasing amounts of amorphous ice in region II. This fact suggests that the glass transition observed in the original sample was a result of the cooperative molecular motion of the polysaccharide and the amorphous ice in regions II and III.

Molecular Motion in Regions III and IV. It is well known that T_g depends not only on the amount of plasticizer but also the degree of crystallinity. In region III, a part of the freezing bound water was crystallized to the regular ice on cooling. The system in region IV contained free water in addition to non-freezing water and freezing bound water. All of the free water in the system formed regular ice at around 250°K on cooling at 10°K/min, since the crystallization rate of free water was much faster than the cooling rate (21). In order to explain the molecular motion of the system in regions III and IV, it is necessary to estimate the amount of regular ice in the system. The deviation of experimental value from the calculated value shown in Figure 7 was probably caused by the presence of regular ice which formed in the cooling process. From the heat of crystallization during cooling at 10°K/min, the regular ice content (W_{ice}) was obtained using the following equation:

$$W_{ice}(g/g) = \frac{\text{amount of ice formed during cooling/g}}{\text{dry sample weight/g}} \quad (5)$$

Figure 8 shows the relationship between T_g and W_{ice} of the original samples in regions III and IV for the water-hyaluronic acid system. The relation curve consists of two lines having different gradients corresponding to regions III and IV. This fact indicates that the increase of T_g in region III was due to the increase of the amount of regular ice which formed during cooling at 10°K/min. The system in region III was plasticized by amorphous ice. However, the main chain motion of the polysaccharide was restricted by the presence of the regular ice. When the system contained free water, the freezing bound water did not form the amorphous ice and the system showed a constant T_g. The constant T_g depended on the amount of non-freezing water irrespective of types of polysaccharide, because all of the freezing water crystallized on cooling in the system W_c corresponding to region IV. When the system contained a certain amount of regular ice at high W_c, the glass transition was difficult to detect. It was hard to determine the T_g of the water-hyaluronic acid system in the W_c range above 3.

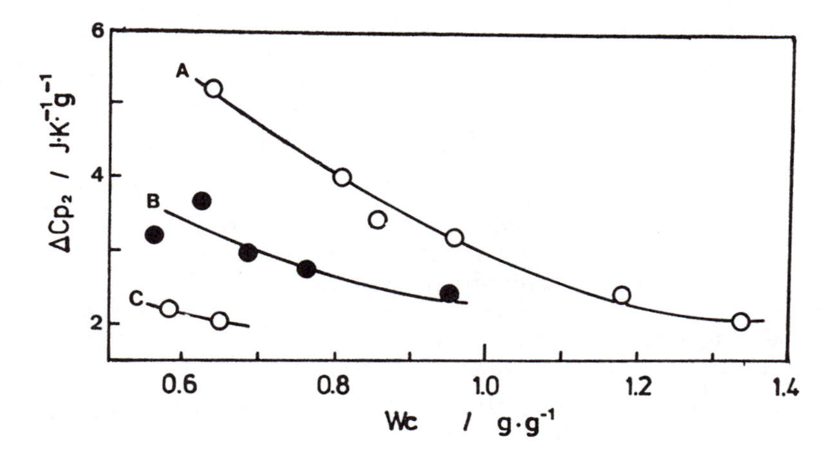

Figure 6. Relationship between heat capacity difference between glassy state and liquid state of amorphous ice (ΔCp_2) and W_c for the water-hyalurem (A), the water-xanthan system (B), and the water-pullulan system (C).

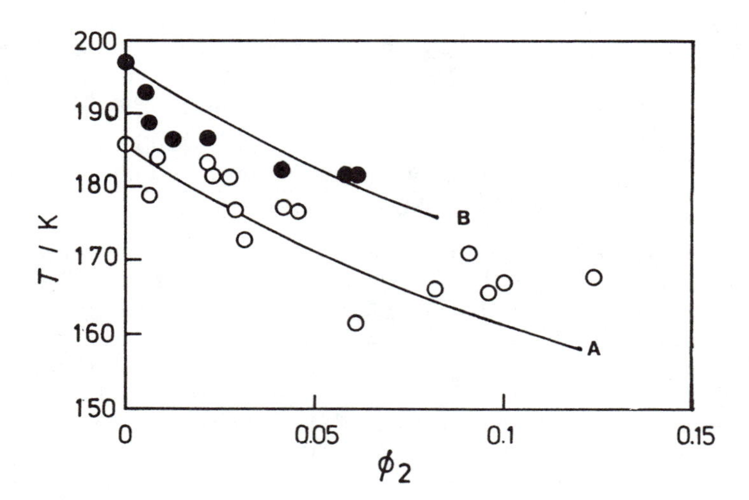

Figure 7. Relationship between T_g and amorphous ice content (ϕ_2) for the water-hyaluronic acid (A) and the water-xanthan (B) systems. The solid lines show the calculated value using Equation 4.

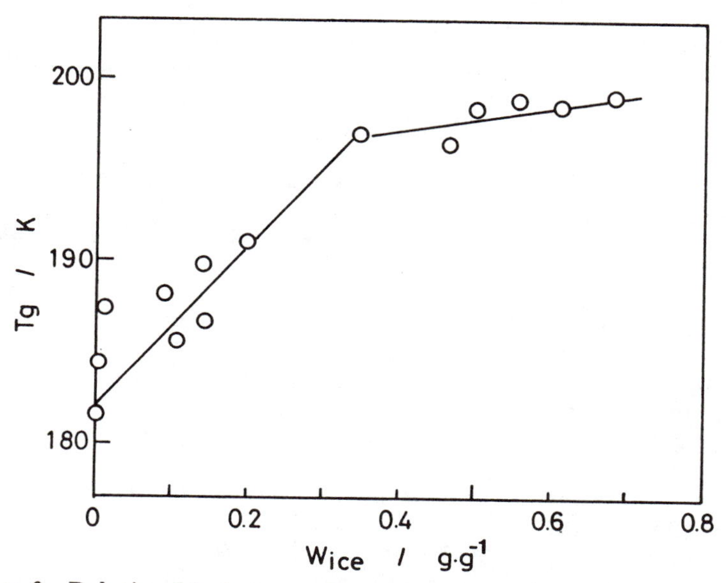

Figure 8. Relationship between T_g and ice content (W_{ice}) for the water-hyaluronic acid system.

Conclusions

The influence of water on the main chain motion of polysaccharides was discussed in four regions (I, II, III and IV) depending on W_c. In the completely dry state, the main chain motion of polysaccharides did not occur below the decomposition temperature. By sorption of water, glass transition of polysaccharides appeared and T_g decreased rapidly with increasing W_c. The sorbed water existed as non-freezing water in region I. The decrease of T_g of hydrogel was observed even if freezing water existed in the system in region II. In this region, freezing bound water formed amorphous ice by cooling at 10°K/min. The main chain motion of the polysaccharide-absorbed non-freezing water was plasticized by the amorphous ice in region II. With increasing W_c, T_g increased after passing the minimum value due to the presence of regular ice which formed on cooling in region III. In regions II and III, the glass transition of the system appeared as the result of the cooperative motion of polysaccharide, non-freezing water, and amorphous ice. In region IV where free water existed, the T_g of the system approached a constant value irrespective of the type of polysaccharide and was hard to observe with increasing W_c.

Literature Cited

1. Djabourov, M. *Contemp. Phys.* 1988, **29**, 273-297.
2. Amiya, T.; Tanaka, T. *Macromolecules* 1987, **20**, 1162-1164.
3. Andrade, J. D., Ed. *Hydrogels for Medical and Related Applications*; ACS Symp. Ser. 31; Washington, DC: American Chemical Society, 1976.

4. Nishinari, K. *J. Soc. Rheol. Jpn.* 1989, **17**, 100-109.
5. Hatakeyama, T.; Nakamura, K.; Yoshida, H.; Hatakeyama, H. *Thermochimica Acta* 1985, **88**, 223-228.
6. Hatakeyama, T.; Yoshida, H.; Hatakeyama, H. *Polymer* 1987, **28**, 1282-1286.
7. Hatakeyama, T.; Nakamura, K.; Yoshida, H.; Hatakeyama, H. *Food Hydrocolloids* 1989, **3**, 301-311.
8. Hatakeyama, T.; Yamauchi, A.; Hatakeyama, H. *Eur. Polym. J.* 1987, **23**, 361-365.
9. Yoshida, H.; Hatakeyama, T.; Hatakeyama, H. In *Cellulose: Structural and Functional Aspects*; Kennedy, J. F.; Phillips, G. O.; Williams, P. A., Eds.; Chichester: Ellis Horwood, 1989; p 305-311.
10. Yoshida, H.; Hatakeyama, T.; Hatakeyama, H. *Koubunshi Ronbunsyu* 1989, **46**, 597-602.
11. Yoshida, H.; Hatakeyama, T.; Hatakeyama, H. *Polymer* 1990, **31**, 693-698.
12. Hatakeyama, T.; Nakamura, K.; Hatakeyama, H. *Thermochimica Acta* 1988, **123**, 153-161.
13. Mauritz, K. A.; Fu, R. M. *Macromolecules* 1988, **21**, 1324-1333.
14. Wise, W. B.; Pfeffer, P. E. *Macromolecules* 1987, **20**, 1550-1554.
15. Uedaira, H. *Hyomen* 1975, **13**, 297-302.
16. Quinn, F.-X.; Kampff, E.; Smith, G.; McBrierty. *Macromolecules* 1988, **21**, 3192-3198.
17. Hatakeyama, T.; Nakamura, K.; Hatakeyama, H. *Netsusokutei* 1979, **6**, 50-55.
18. Nakamura, K.; Hatakeyama, T.; Hatakeyama, H. *Polymer* 1981, **22**, 473-477.
19. Nakamura, K.; Hatakeyama, T.; Hatakeyama, H. *Tex. Res. J.* 1981, **51**, 607-611.
20. Hatakeyama, T.; Yamauchi, A.; Hatakeyama, H. *Eur. Polym. J.* 1984, **20**, 61-64.
21. Yoshida, H.; Hatakeyama, T.; Hatakeyama, H. *Rept. Prog. Polym. Phys. Jpn.* 1987, **30**, 113-114.
22. Kjellander, R.; Florin, E. *J. Chem. Soc., Faraday Trans. 1* 1981, **77**, 2053-2058.
23. Navard, P.; Haudin, J. M.; Dayan, S.; Sixou, P. *J. Polym. Sci., Polym. Lett. Ed.* 1981, **19**, 379-385.
24. Gray, D. G. *J. Appl. Polym. Sci., Appl. Polym. Phys.* 1983, **37**, 179-192.
25. Navard, P.; Haudin, J. M.; Dayan, S.; Sixou, P. *J. Appl. Polym. Sci., Appl. Polym. Phys.* 1983, **37**, 211-221.
26. Sugisaki, M.; Suga, H.; Seki, S. *Bull. Chem. Soc. Jpn.* 1968, **41**, 2591.
27. Wilson, T. W.; Turner, D. T. *Macromolecules* 1988, **21**, 1184-1186.
28. Rasmussen, D. H.; Mackenzie, A. P. *J. Phys. Chem.* 1971, **75**, 967-973.

RECEIVED December 16, 1991

Chapter 15

Gelation and Subsequent Molecular Orientation of Silk Fibroin

Jun Magoshi[1] and Shigeo Nakamura[2]

[1]National Institute of Agrobiological Resources, Tsukuba, Ibaraki 305, Japan

[2]Department of Applied Chemistry, Kanagawa University, Kanagawa-ku, Yokohama 221, Japan

The mechanism of silk fiber formation of liquid crystal and gel state in cocoon spinning has been examined in detail. Nematic liquid crystal and gel state are found in the liquid silk flowing out from the anterior division of the silk gland.

Silk fiber is a fine, lustrous fiber produced by the silkworm and other insect larvae, generally to form their cocoons. These cocoons are constructed to protect themselves during resting stage or pupal stage. In many species, such as moths, bees, wasps, and butterflies, the cocoon is made entirely of silk. The silkworm, *Bombyx mori*, is cultivated by feeding it mulberry leaves until it is ready to spin the cocoon. The silk it spins is a fine and long double monofilament (1000-1500 m).

Fiber formation by the silkworm is carried out through the liquid crystal in the nematic state, which is transformed from a gel state to a sol state during spinning. Our work indicates that the thread spun by the silkworm is spun through the liquid crystal spinning process.

The mechanism of fiber formation from the liquid silk of the silkworm was first clarified by Hiratsuka (1) and Fóa (2). They found that the crystallization of the liquid silk to the silk filament is induced by the action of shear and elongation, in which the silkworm draws back its head. Since their work, many investigators have studied the spinning mechanism (3-5), but the spinning process of silk yarn is still not well understood.

The liquid silk of the silkworm is a highly viscous aqueous solution of two proteins, fibroin and sericin. The mechanism of silk thread formation from the liquid silk is the action of shear and elongational stresses acting on the silk fibroin, which causes the liquid silk to crystallize.

Investigation of the structure and properties of protein is very important for the utilization of biological functions. Therefore, we used fibroin as an example of a typical protein and studied its crystallization and conformational transition induced by various physical and thermal conditions.

0097–6156/92/0489–0231$06.00/0

Preparation of Samples. From a mature mulberry silkworm of *Bombyx mori,* the silk gland containing liquid silk was removed. Liquid silk used for experiments was obtained from the posterior part of the middle division of the silk gland unless otherwise specified.

Crystallization of Fibroin (6). Before the description of fiber formation and cocooning, the crystallization of fibroin is explained.

Fibroin has a relatively simple structure and is easily available from natural products. Three types of conformation—random coil; α-form (fibroin I, crank shaft pleated-sheet structure) (4, 5); and β-form (fibroin II, antiparallel-chain pleated-sheet structure) (4, 5)—have been observed by x-ray diffraction and infrared spectroscopy. These forms of fibroin film are obtained by varying the preparation conditions, such as the concentration of fibroin and the casting or quenching temperature of the aqueous solution.

The α- and β-form conformations are very stable to heating and treatment with organic solvents. However, conformational changes occur easily from the random coil to the α-, β- and well-oriented β-forms by mechanical shearing, treatment with organic solvents, heat treatment and under several other conditions.

The crystallization of fibroin from dilute solution and from amorphous random-coil film was studied in detail. The effects of various physical and thermal conditions on the conformation of fibroin are summarized in Figure 1.

The conformation of fibroin in the film cast from an aqueous solution largely depends on the casting conditions. The α-form and β-form crystalline and amorphous random coil films can be obtained by varying the fibroin concentration and temperature. Crystalline films in the α- and β-form conformations can be obtained by casting a concentrated solution at 0 to 40°C and above 50°C, respectively, whereas casting of a dilute solution at 0 to 40°C results in amorphous films with a random coil conformation. Crystallization of fibroin from a 25% aqueous solution starts with a weight loss of the solution of about 45% during the evaporation of water when the casting temperature ranges from 20 to 120°C. This weight loss corresponds to about 65% of the fibroin concentration. In conditions above 120°C, the critical crystallization concentration of fibroin increases with increasing temperature.

When the aqueous solution of fibroin is frozen by quenching prior to drying, β-form crystals are obtained at quenching temperatures in the range of −2 to −20°C irrespective of the concentration of fibroin. On the other hand, when the solution is quenched below −20°C, the α-form and random-coil conformations are obtained, depending on the concentration of fibroin. Fibroin molecules in dilute aqueous solution are in the random-coil conformation. Therefore, extremely rapid freezing of a very dilute solution does not permit the change to the β-form to occur and the random coil conformation in a dilute solution is maintained in fibroin films. These results indicate the relationship between the conformation, the casting and quenching temperature, and the fibroin concentration. The phase diagram was constructed as shown in Figure 2.

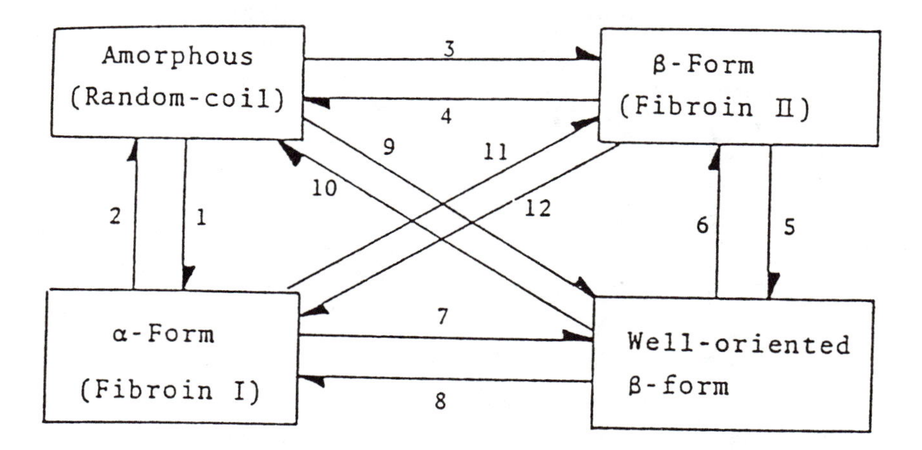

Crystallization induced under various conditions.

1) Casting from aqueous solution (>5%) at 0 to 45°C
   Quenhcing aqueous solution (>5%) below −20°C prior to drying
   Crystallization from aqueous solution at pH above 6.0
   Epitaxial growth on nylon 66
   Treatment of film with water at 0 to 50°C

2) Application of high pressure

3) Casting from aqueous soltution (>5%) above 45°C
   Quenching aqueous solution (>5%) at 0 to −20°C prior to drying
   Mechanical shearing
   Application of an electric field to aqueous solution
   Treatment with polar organic solvents
   Crystallization from aqueous solution at pH below 4.5
   Heat treatment of film at 190 to 200°C
   Treatment of film with hot water at above 60°C
   Treatment of aqueous solution with enzyme (Chymotrypsin)

4) Application of high pressure

5) Drawing of liquid fibroin or film

6) Application of high pressure
   Treatment with salt

7) Heating to 270°C
   Mechanical shearing

8) Treatment with salt

9) Mechanical shearing
   Epitaxial growth
   Drawing of liquid fibroin or film

10) Application of high pressure

11) Mechanical shearing

12) Heat treatment by wet process

Figure 1. Conformational change of silk fibroin.

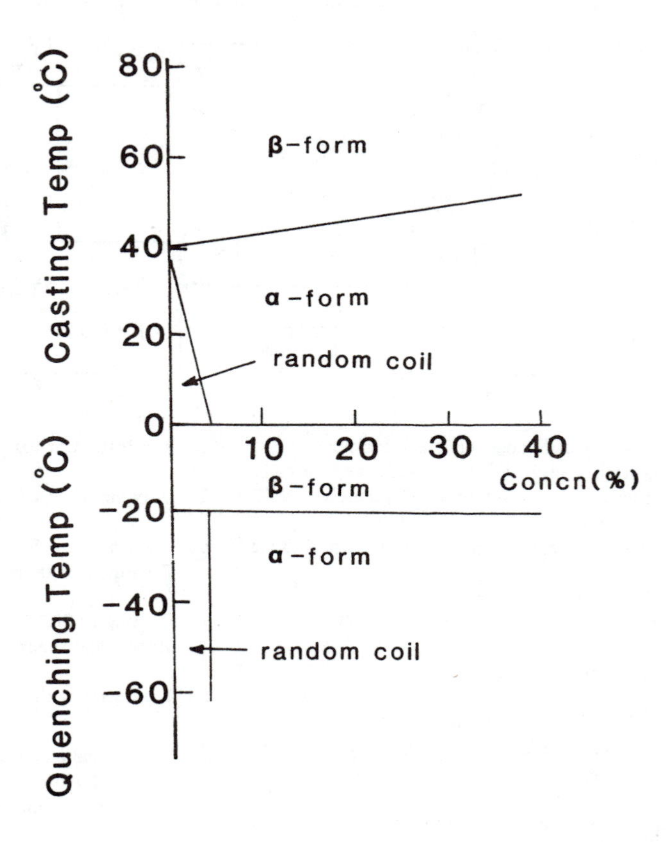

Figure 2. Phase diagram of the conformation—casting or quenching temperature—fibroin concentration.

The conformation of the fibroin film is also affected by the substrate onto which the fibroin film is cast. Amorphous fibroin film is obtained from a dilute aqueous solution at 20°C cast onto the substrates including polyethylene, polypropylene, polycarbonate, polystyrene, polyvinyl chloride, polyacrylonitrile, polymethyl methacrylate and polytetrafluoroethylene films and a glass plate. On the other hand, α-form crystals are obtained by casting onto nylon 66 film and a mercury surface.

In the spherulite formation of fibroin from aqueous solution, the optical sign depends on the drying temperature, drying rate and quenching temperature. Negative birefringent spherulites with the α-form are obtained in the film cast at temperatures in the range of 0 to 40°C and with a high drying rate at 20°C, whereas positive β-form spherulites appear at higher temperatures up to 80°C and with a low drying rate at 20°C. Positive β-form spherulites are also obtained by freezing the fibroin solution at −2°C to −20°C and the drying at 20°C. Positive β-form spherulites are formed on the surface of well-oriented β-form fibroin immersed in fibroin solution at 20°C.

β-form crystals are produced when liquid fibroin is dried after immersion in polar hydrophilic organic solvents such as methanol. The α-form crystals are obtained when liquid fibroin is treated with hydrophilic organic solvents. However, no conformational changes take place in higher alcohols such as n-propanol and in hydrophilic solvents. The random coil to β-form transition occurs very rapidly by immersion in methanol. X-ray diffraction patterns of a specimen wet with methanol indicate that the crystallization of amorphous random-coil fibroin to β-form crystals is induced immediately after immersion in methanol.

Upon treatment with water at below 60°C, the random-coil fibroin in the film is converted to the β-form, and above 70°C to a mixture of α- and β-conformations, and the β-form content increases with increasing immersion temperature. The β-form conformation is not affected by immersion in water.

Well-oriented fibrils with the β-form conformation can be obtained by drawing the liquid fibroin at extension rates higher than 500 mm/min. No conformational change of the random-coil amorphous fibroin film is induced by drawing up to a draw ratio of 4.0. Beyond a draw ratio of 4.0 the random-coil conformation is converted to the β-form crystals. Crystallization to the β-form crystals is almost completed at a draw ratio of 6.0.

Crystallization occurs by heating amorphous random-coil film. Intra- and intermolecular hydrogen bonds are broken between 150 and 180°C after evaporation of water from the amorphous fibroin film. The glass transition occurs at 173°C. The random coil to β-form transition accompanied by reformation of hydrogen bonds takes place above 180°C. Thermally induced crystallization to the β-form transition occurs at 173°C. Thermally induced crystallization to the β-form crystals starts above 190°C.

Silk Gland. The structure of the silk gland has already been described in detail (7). The silk glands are a pair of tubes lying one on each side of the larvae as shown in Figure 3.

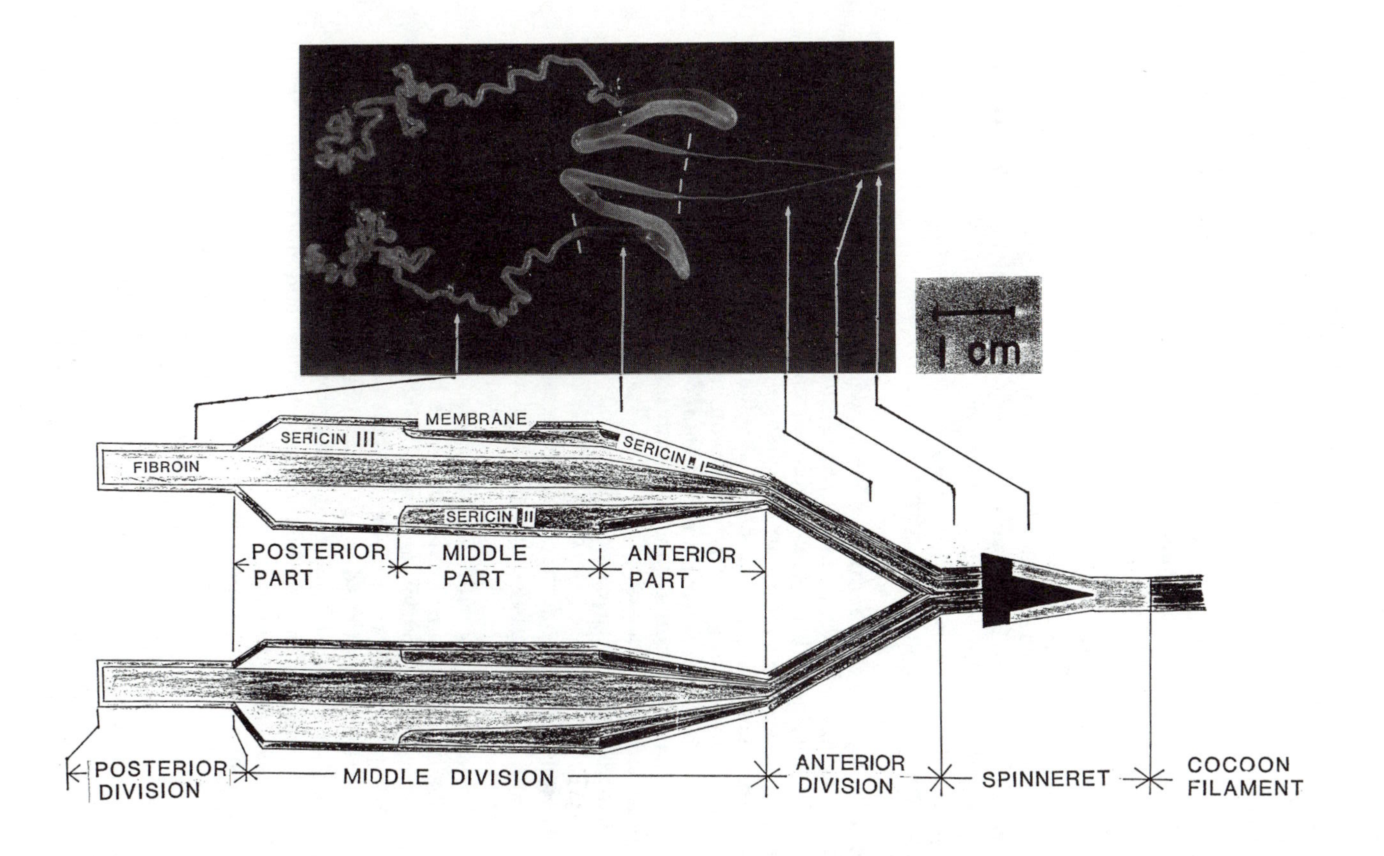

Figure 3. Silk gland of the larva of the silkworm *Bombyx mori.*

The schematic diagram and photograph are the silk gland of the silkworm, *Bombyx mori*. Each gland can be divided into three sections: posterior, middle, and anterior. The epithelial wall of the gland consists of large hexagonal cells, where the silk fibroin and silk sericin are synthesized. The silk proteins are stored inside the gland.

The posterior division, which is closed at one end, is narrow and very convoluted, where the main silk protein, fibroin, is synthesized. The fibroin moves forward into the wider middle division that serves as a reservoir. The second silk protein, sericin, is synthesized in this section and accumulates as a separate layer surrounding the fibroin core. The two proteins move forward together without mixing into the anterior division which becomes narrower as it approaches the spinneret. The two glands join together just before the spinneret. At the spinneret twin cores of fibroin are surrounded by a layer of sericin (Figure 4).

The spinneret of the silkworm consists of the silk press and the orifice as seen from the micrograph in Figure 4. The common gland which contains two separate fibroin solutions and an outer layer sericin solution goes to the silk press without mixing. The silk press acts as a nozzle in the spinning of synthetic fibers. The cross-section of the silk press is in the form of two crescent shapes. On extrusion through the spinneret into the air, the proteins are converted from a concentrated viscous solution into a thin, glossy, strong, filament which cannot easily be reconverted into solution. A remarkable feature of this cocoon spinning is the stretching, brought about by the drawing back of the silkworm's head. This stretching causes orientation of the long fibrous molecules in the direction of drawing. The water does not evaporate in the spinneret but it is lost in the air.

Figure 5 shows the scanning electron micrographs of the silkworm extruding silk filament through the spinneret into the air. Before the picture was taken, the specimen in Figure 5a was vacuum-dried. The specimen in Figure 5b was quenched to −190°C and then freeze-dried. Two fibroin filaments are observed in Figure 5a. No shrinkage occurs in the freeze-dried specimen, even with loss of water during drying.

Superdrawing. At the end of the fifth instar, the larva of the silkworm produces a strong and glossy cocoon consisting of long filaments of fibroin covered with sericin. When the anterior division of the spinning silkworm is cut off at this time, and the liquid silk flowing out is drawn at a rate corresponding to that of the cocoon spinning, the liquid silk flows very smoothly in the direction of drawing (Figure 6). This is the so-called elongational flow. The resulting filament is in the well-oriented β-form conformation. The diameter of the undrawn liquid silk is about 12.8 times the diameter of the drawn silk. Therefore, the silkworm draws the liquid silk at about 150 times at the undrawn length. This phenomenon for synthetic fibers is called superdrawing.

Liquid Crystal. We have found that silk fibroin molecules flowing from the anterior division of the silk gland of *Bombyx mori* assume a liquid crystalline order. The anterior region of the silk gland is removed from a

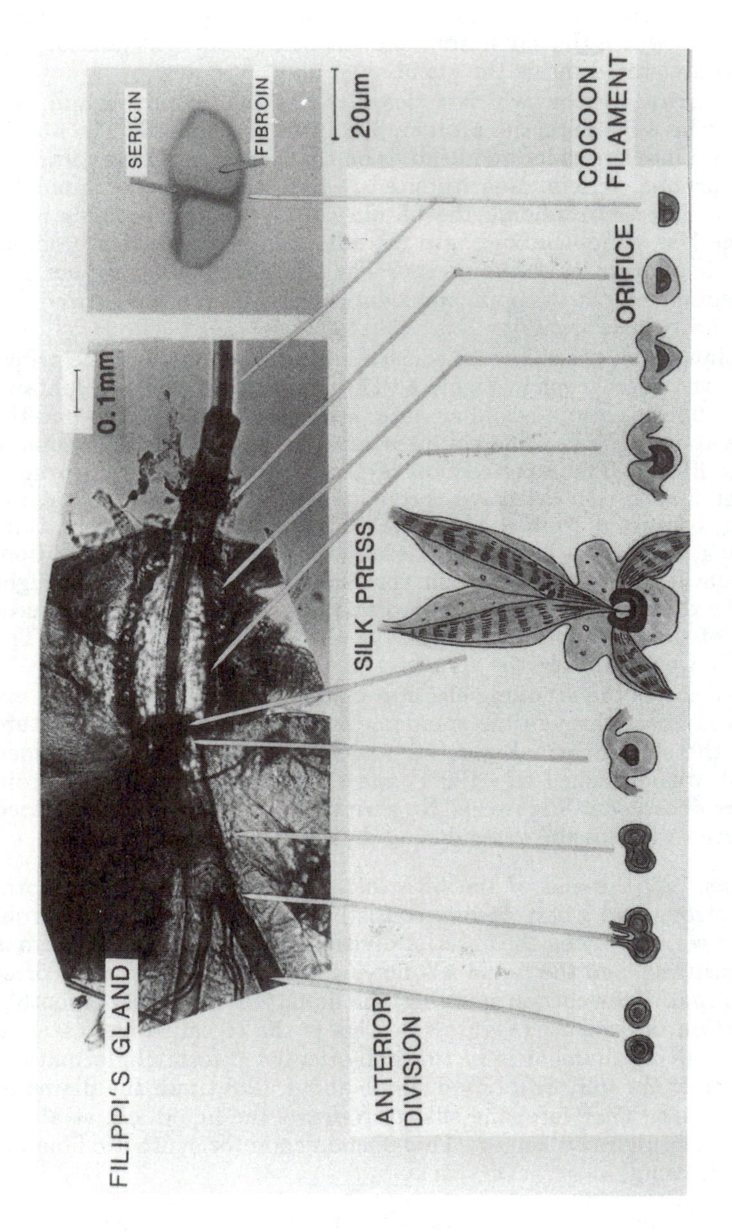

Figure 4. Micrograph and schematic cross-sections of the spinneret of the silkworm.

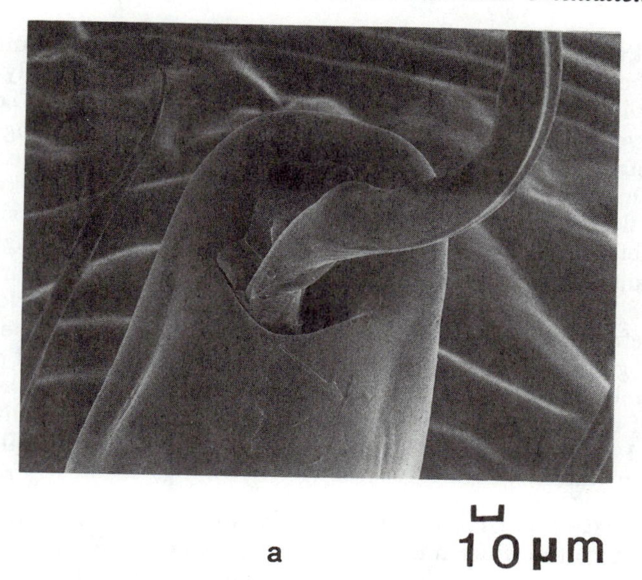

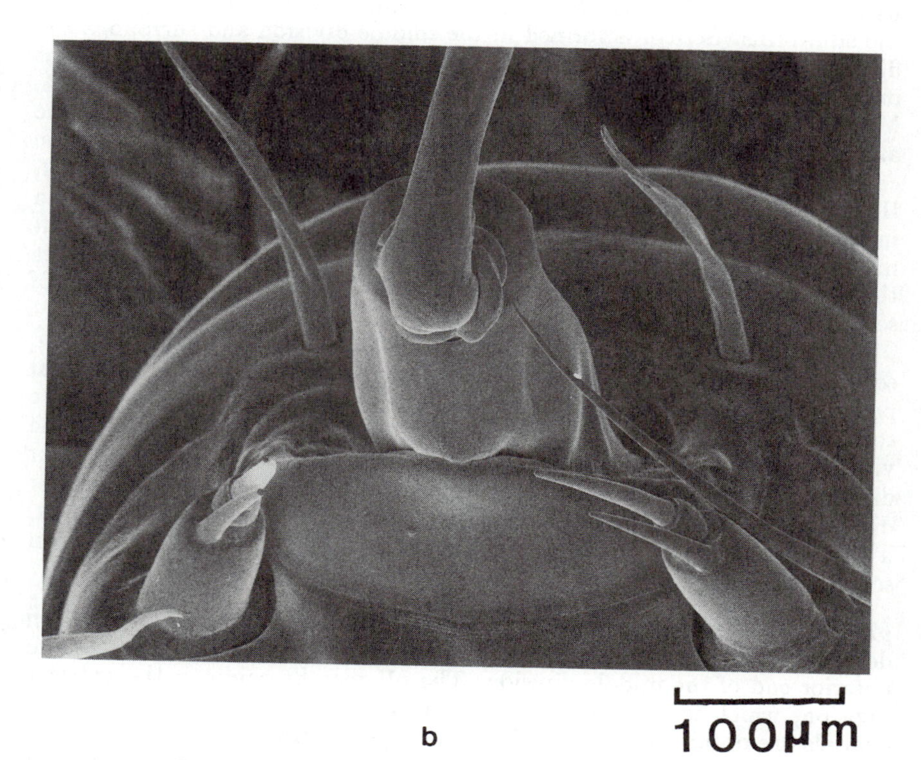

Figure 5. Scanning electron micrographs of the silkworm extruding filaments into the air. (a) Dried in vacuum; (b) quenched to −190°C and then freeze-dried.

mature silkworm (one day before spinning) and then its end is immediately cut off. The liquid silk flows out spontaneously and continuously and does not show birefringence by examination under polarized light (Figure 7).

However, the anterior division of the silk gland of a spinning silkworm shows birefringence. When the end of the anterior division of the silk gland is cut off immediately after being taken out of the cocooning silkworm, the liquid silk flows out spontaneously and continuously. The liquid silk flowing out is slightly birefringent under polarized light, as shown in Figure 8.

The anterior division of the silk gland is also birefringent. In polarized light under sensitive color, the liquid silk flowing out of the anterior division is not birefringent, whereas the liquid silk is birefringent in the direction of the silk gland. These results indicate that the nematic liquid crystalline phase can be produced in the liquid silk coming out from the anterior division of the silk gland. The liquid crystal of silk fibroin is dispersed in water for 1 hr as shown in Figure 9. The gel of liquid crystal silk is easily dissolved in water.

Gel State in the Silk Gland. Fibroin is synthesized in the posterior division and moves into the wider middle division as about a 12% aqueous solution with Ca, Mg, and K ions which came from the leaves of mulberry eaten by the silkworms. Sericin is formed in the middle division and surrounds the fibroin core as a separate layer. The water in the fibroin solution goes out into the sericin layer and the fibroin solution is concentrated to about 23%. Moreover, the water in the sericin solution goes out into the membrane and the ions go into the sericin and fibroin solutions. The middle division consists of three parts—anterior, middle, and posterior—where sericin I, II, and III are synthesized, respectively. The total concentration of sericin in liquids is about 7%. The four proteins move forward together without mixing into the anterior division, which becomes narrower as it approaches the spinneret, and the twin cores of fibroin are surrounded by a matrix of sericin.

During cocooning of the silkworm and after the epithelial wall and sericin are removed, the viscosity of the liquid silk is determined at several parts of the middle division by measuring the stress required for a probe to penetrate into the specimen at a predetermined rate (Figure 10). The viscosity of the gel in the liquid silk decreases as it approaches the anterior division. Therefore, the viscosity of the liquid silk may be low in the anterior division even though the liquid silk is too thin to carry out a measurement. The decrease in viscosity suggests the formation of ordered structure in solution.

pH of Silk Gland. The pH of the fibroin solution in the middle division decreases as it approaches the anterior division and becomes 4.9 at the anterior end of the middle division. The pH may be acidic in the anterior division, which is favorable for the β conformation, because the β-form crystals are obtained at pH 4.5 from aqueous solution.

The yield stress (strength of gel) of the aqueous solution of regenerated silk fibroin changes with pH. In the pH range 5.0-6.5, the yield stress is

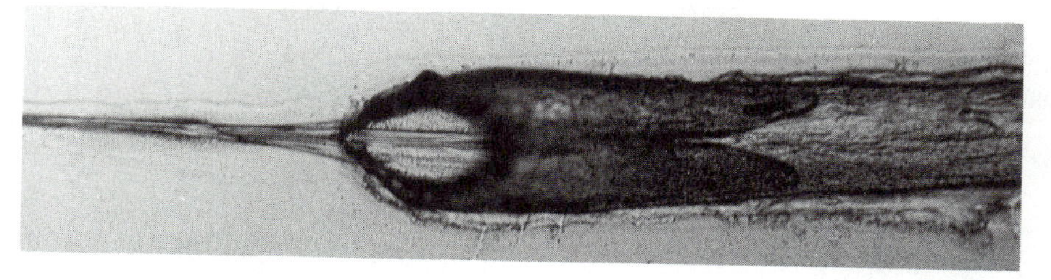

Figure 6. Drawing of the liquid silk from the anterior division.

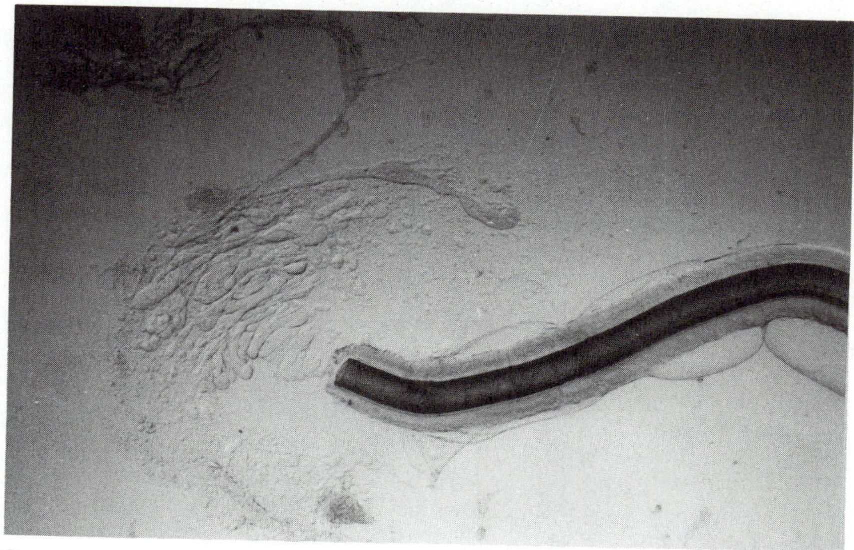

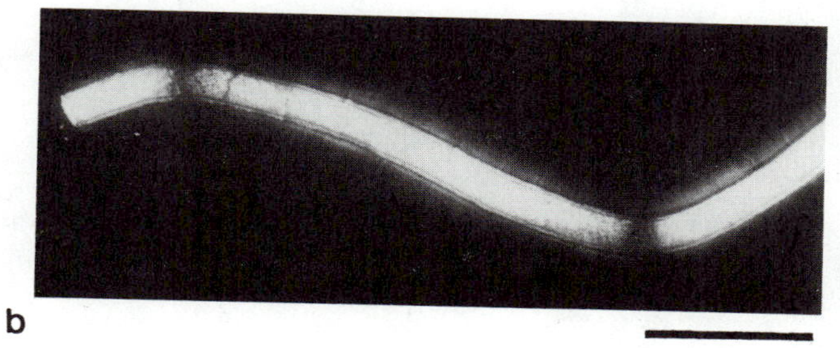

Figure 7. Micrographs of the liquid silk from the anterior division of the silk gland. (a) In unpolarized light; (b) in polarized light.

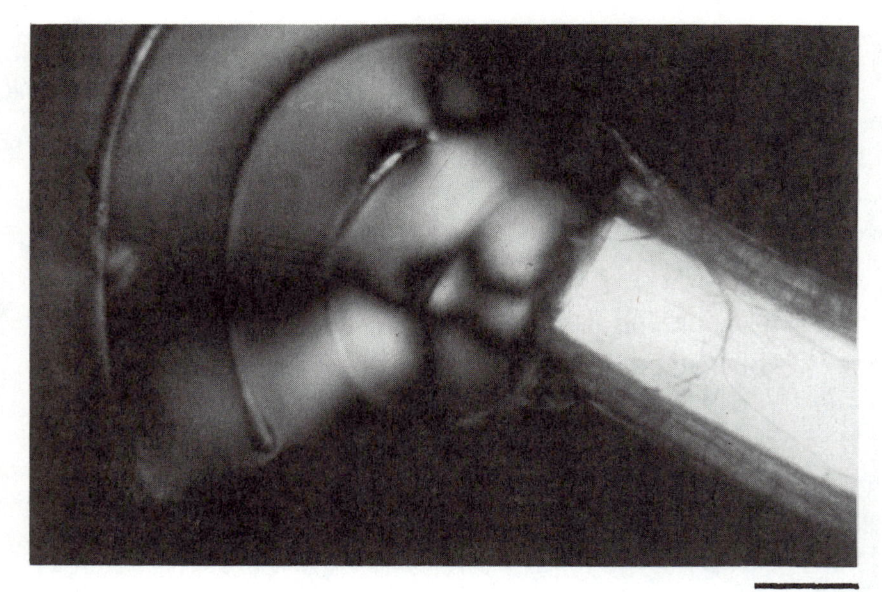

Figure 8. Liquid crystal of liquid silk from the anterior division in polarized light.

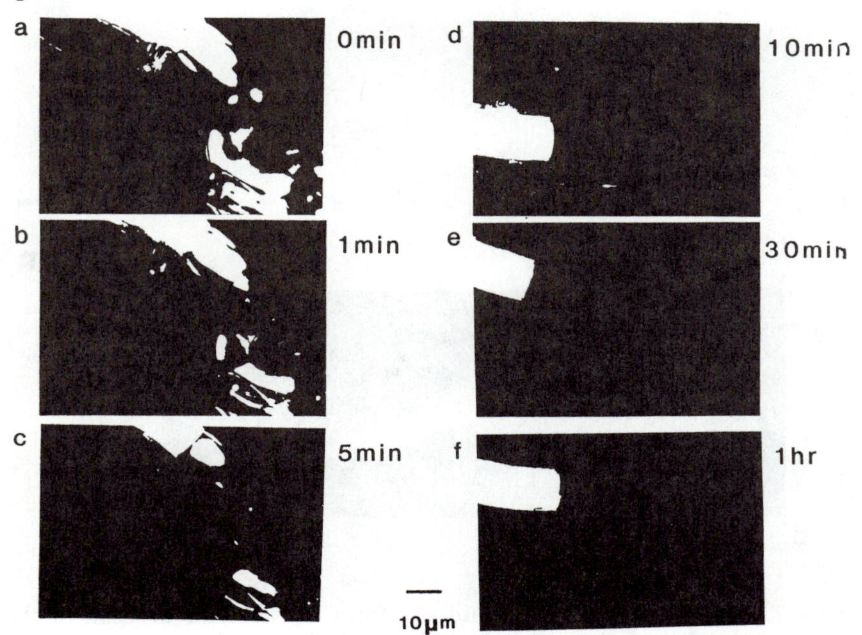

Figure 9. Liquid crystal of liquid silk from the anterior division immersed in water.

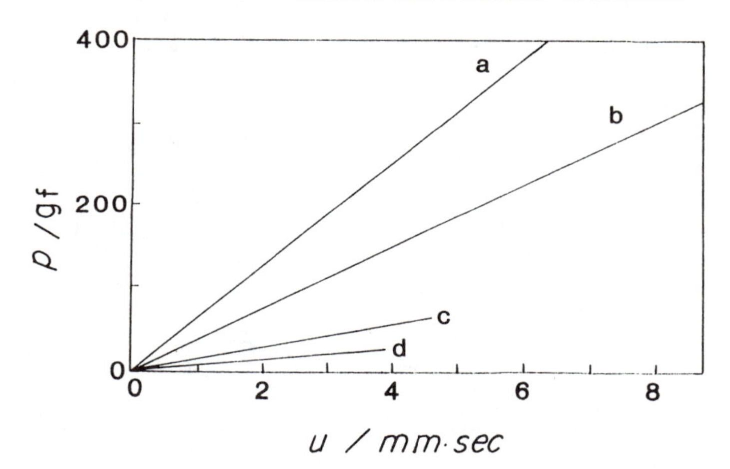

Figure 10. Stress required for penetration vs. rate of penetration at different parts of the middle division of liquid silk. (a) Middle portion; (b) posterior part of anterior portion; (c) middle part of anterior portion; and (d) anterior part of anterior portion of the middle division of liquid silk.

high compared to other solutions, as shown in Figure 11. As the pH value decreases, the yield stress becomes lower and the viscosity also becomes lower. Therefore, in the cocooning of the silkworm, the liquid silk in the anterior division has a low viscosity due to the low pH and the formation of liquid crystals.

Figure 12 shows the change in yield stress of the gel state with pH at various temperatures. The yield stress of fibroin is shifted to the alkaline side with increasing temperature.

Gel-Sol Transition. The metal ions in the liquid silk increase from the posterior division to the anterior division. The strength of the gel decreases with increasing concentration of Ca^{2+} ions in the silk fibroin solution, as shown in Figure 13. Therefore, the gel-sol transition occurs in the anterior division by increasing Ca^{2+} ion concentration.

High Speed Spinning. The liquid silk in the middle division of the silk gland is changed to fiber by drawing at a high rate. When the liquid silk was drawn on a Tensilon tensile tester at 20°C, the liquid silk behaved as a non-Newtonian fluid in the extension rate range of 10-1,000 mm/min (Figure 14).

As the extension rate was increased, the stress required to stretch liquid silk increased and the yield points became remarkable. At slower rates of 10-75 mm/min, neither conformational change nor orientation of fibroin molecules was observed by x-ray diffraction and birefringence measurements (Figure 15). At rates higher than 500 mm/min, the stress-strain curves became irregular beyond the yield points and the specimen turned opaque.

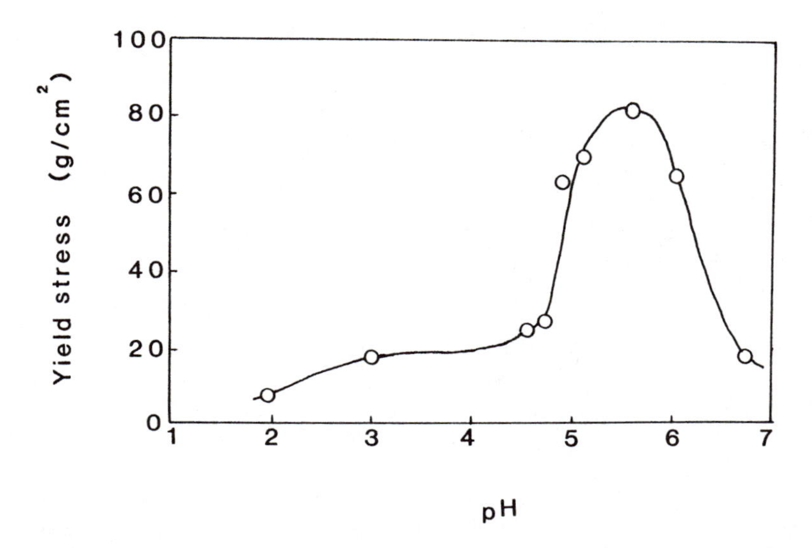

Figure 11. Change of yield stress (gel strength) of silk fibroin solution with pH at 20°C.

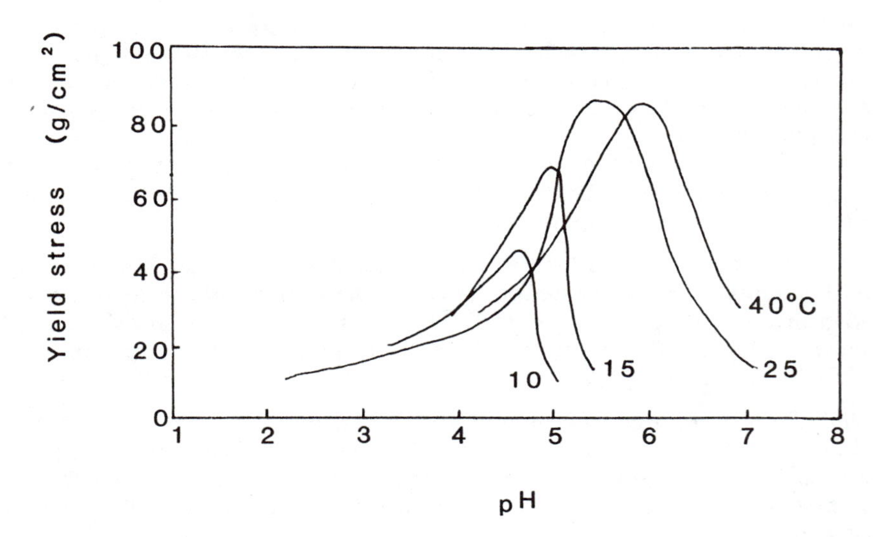

Figure 12. Change of yield stress of silk fibroin solution with pH at various temperatures.

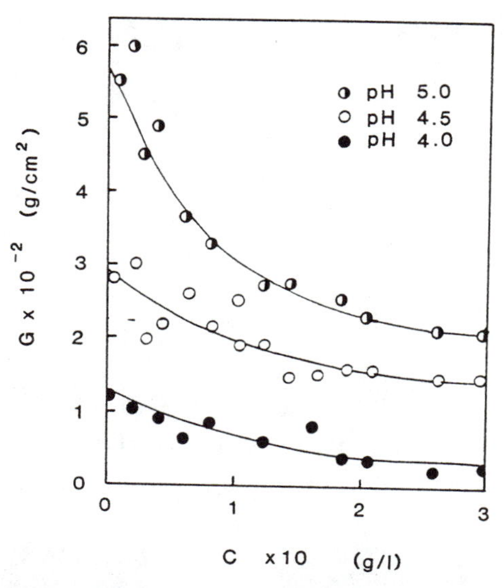

Figure 13. Change of yield stress of silk fibroin solution with Ca^{2+} concentration at various pH values.

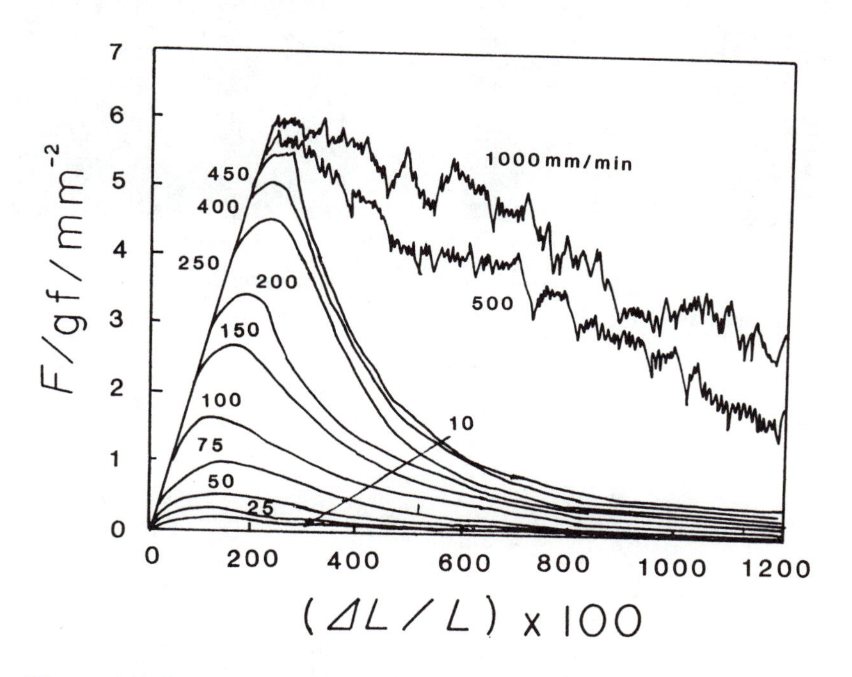

Figure 14. Stress-strain curves of liquid silk of the silkworm at 20°C.

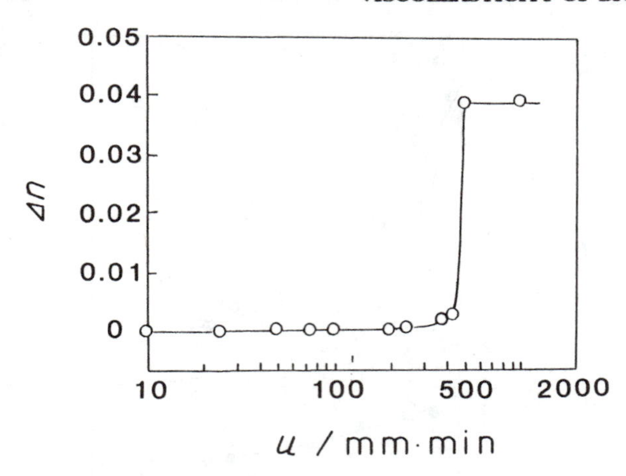

Figure 15. Birefringence vs. extension rate of liquid silk.

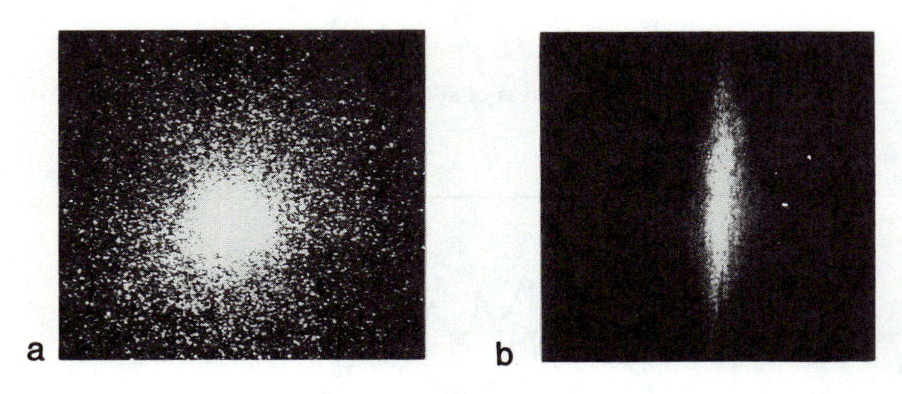

Figure 16. Laser light scattering patterns of undrawn liquid silk (a) and drawn liquid silk (b).

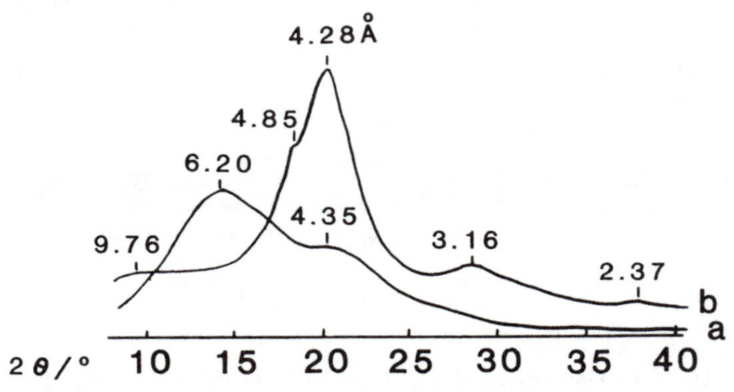

Figure 17. X-ray diffraction patterns of undrawn liquid silk (a) and drawn liquid silk (b) in the equatorial direction.

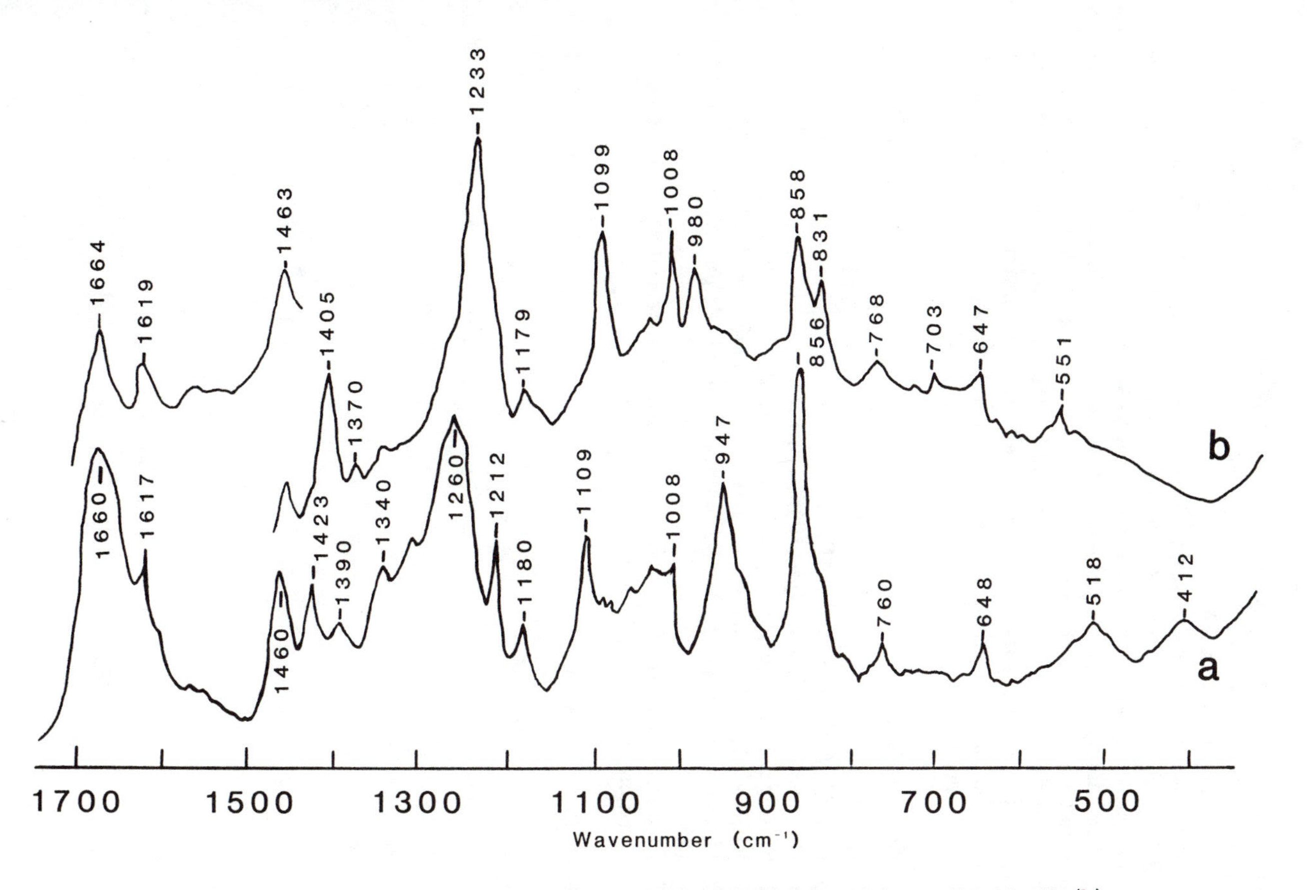

Figure 18. Laser Raman spectra of undrawn liquid silk (a) and drawn liquid silk (b).

Most of the water initially present is squeezed out by drawing. The yield points shift to higher elongation with increasing extension rate. At lower extension rate the yield force is very low, and at rates of 50-450 mm/min this value is almost constant.

From laser light scattering patterns (Figure 16), x-ray diffraction patterns (Figure 17), and laser Raman spectra (Figure 18), we know the conformation was transformed to the β-form by drawing.

Accordingly, until the extension rate of 500 mm/min is attained, the conformational change does not occur. At rates higher than 500 mm/min, the conformation of fibroin in liquid silk transformed from the random coil to the β-form, crystallization to the well-oriented fibrils took place, and the silk filament was produced by drawing liquid silk at rates higher than 500 mm/min at 20°C.

Literature Cited

1. Hiratuska, E. *Bull. Sericult. Expt. Sta.* 1916, **1**, 181.
2. Fóa, C. *Kolloid Z.* 1912, **10**, 7.
3. Oka, S. *J. Phys. Soc. Jpn.* 1946, **19**, 1481.
4. Shimizu, M. *J. Sericult. Sci. Jpn.* 1951, **20**, 155.
5. Kratky, O.; Schauenstein, E. *Discuss. Faraday Soc.* 1951, **11**, 171.
6. Magoshi, J.; Magoshi, Y.; Nakamura, S. *Appl. Polym. Symp.* 1985, **41**, 188.
7. Lucas, F.; Rudall, K. M. *Comprehensive Biochemistry*; Elsevier: Amsterdam, 1977; Vol. 26B, p 477.

RECEIVED December 16, 1991

Chapter 16

Magnetic Rheometry of Bronchial Mucus

Peter A. Edwards and Donovan B. Yeates

Departments of Medicine and Bioengineering, University of Illinois, Chicago, IL 60680

To ascertain the viscous, elastic and yield properties of mucus, a magnetic rheometer was developed. Ferromagnetic particles, suspended in viscous fluids or canine tracheal pouch mucus and initially aligned by a magnetic field, were subjected to pulsed magnetic fields. The remanent magnetism of the particles was recorded with a fluxgate gradiometer after each pulse. The time courses of the remanent magnetism from particles suspended in viscosity standards were used to derive indices of viscosity and to estimate the yield stress. The decrease in remanent magnetism due to the recoverable shear strain (elastic recoil) of the mucus was recorded when the field was turned off. For fresh canine tracheal pouch mucus, the viscosity decreased from 1798 to 4.6 poise (p) at strain rates increasing from 0.01 to 7.3 s^{-1}. The recoverable shear strain increased from 0.21 to 0.92 with strains increasing from 0.23 to 1.0; but did not vary with the strain rate. The yield stress ranged from 5.6 to 13 dynes/cm^2. Application of N-acetylcysteine (NAC), 0.034 M, resulted in a greater decrease in yield (74% of control) than in viscosity (39%) and recoverable strain (22%). Five human sputum samples exhibited lower viscosity values 68 to 2.5 p at 0.5 to 12 s^{-1} and yield stress values 0.5 to 2.6 dynes/cm^2. NAC, 0.034 and 0.68 M did not alter the rheology of the sputum samples. This magnetic system can utilize small opaque mucus samples (down to 200 μl) while applying minute strains that are unlikely to cause shear degradation of the mucus. The rheological behavior of canine mucus under the conditions of magnetic rheometry does not support the use of a Maxwell model for mucus. We suggest that the prop-

0097–6156/92/0489–0249$06.00/0

erties of mucus when the applied stress is less than the yield stress are important for cilia-mucus interactions. The yield stress may prevent retrograde mucus flow while the viscosity and recoverable shear strain are important to bulk mucus flow within the airways.

Mucociliary transport in the bronchial airways is responsible for the removal of inhaled deposited particles and organisms as well as intrapulmonary cellular debris. The effectiveness of this transport system is dependent on the viscoelastic properties of the mucus and its interaction with the beating cilia (1). The maximum velocity of the mucociliary transport process appears to require mucus of an optimum viscosity and elasticity (2,3). The lack of retrograde mucus flow under non-pathological conditions suggests the presence of a yield stress.

The rheological properties of mucus and their relation to mucus transport have been recently reviewed (22). Pulmonary mucus, the viscoelastic gel phase of the pulmonary airway secretions, is shear thinning (4) and the elasticity is linear only for small strains (5). Also, mucus is thixotropic (6) and has measurable spinnability (7). These non-Newtonian rheological properties are primarily caused by the presence and interaction of glycoproteins and proteoglycans in an aqueous medium (8). Viscous and elastic properties of mucus have been evaluated *in vitro* using a number of types of rheometers (9-12). None of these are adaptable to the measurement of mucus viscosity and elasticity *in vivo*. We demonstrate, *in vitro*, the feasibility of a magnetic rheometry system whose principle of operation is applicable to the measurement of mucus rheology *in vivo*. In this system, micron-sized ferromagnetic particles were suspended in the test medium and rotated under the influence of pulsed magnetic fields. The time courses of rotation were sensed remotely by fluxgate magnetometry. Viscosity, elasticity and yield were derived from these. This magnetic rheometer measured viscosity, elasticity and yield stress of microliter samples of mucus utilizing minute strains. We measured the viscosity, recoverable shear strain (elastic recoil) and yield stress of fresh canine tracheal pouch mucus samples as small as 200 μl and on opaque mucus samples over a range of shear rates and strains.

Magnetic Rheometer

Magnetic rheometry relies on the ability of ferromagnetic particles, suspended in a fluid, to align their magnetic moments with external magnetic field lines; the resultant magnitude of the field generated by these moments represents the orientation of the particles with respect to the sensing probe.

The magnetic rheometer is shown schematically in Figure 1. The specimen and the magnetizing and sensing elements of the magnetic rheometer were contained within two concentric mumetal cylinders. These shields reduced steady and alternating magnetic fields of environmental origin. A copper wire torus around these shields was used for their periodic degaussing. The test specimen, a 5 ml suspension of 100 mg of ferromagnetic

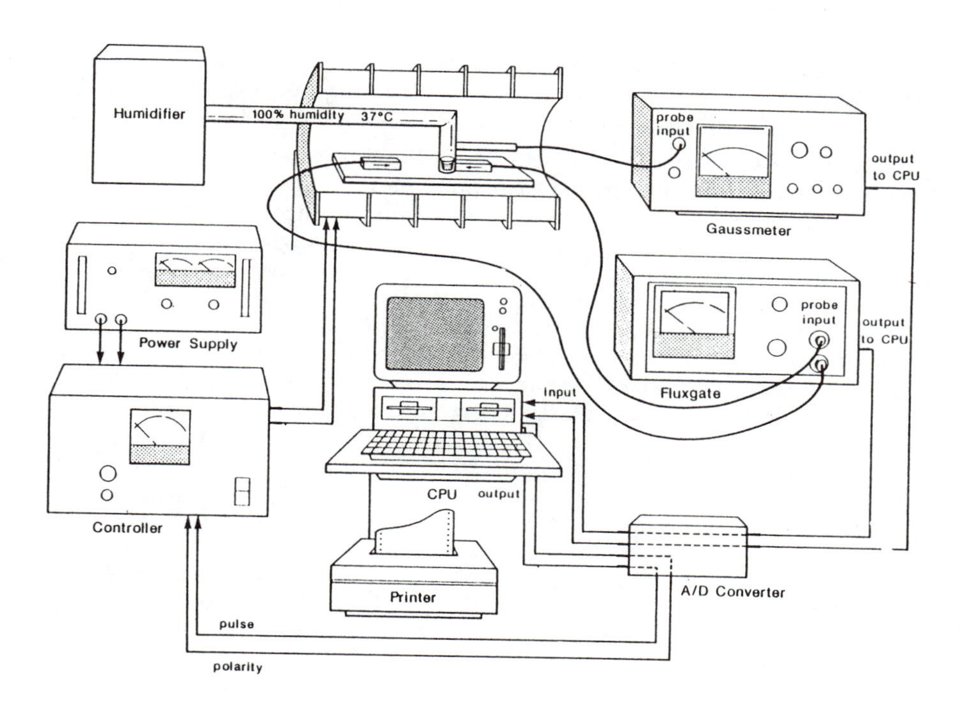

Figure 1. Schematic diagram of the magnetic rheometer. The magnetizing coil (outer diameter = 20.3 cm, inner diameter = 10.3 cm, and length = 38.5 cm) was powered by three separate DC power supplies: (1) Lambda LES-F-02-0V (0–18 v) for magnetic fields from 0 to 20 G; (2) Newark Electronics-Brute II (5–55 v) for secondary fields of 10 to 200 G; and (3) Displex, Inc. (50–110 v) for primary magnetization and strong secondary fields of 200 to 500 G. The output of the power supplies was controlled by the Apple IIe microcomputer A/D converter via the controller. The output of the fluxgate gradiometer and the Gaussmeter were interfaced with the Apple IIe microcomputer for data collection, storage, and hardcopy production. The temperature and humidity of the sample was maintained at 37°C and 100% humidity by a temperature controlled humidifier. The magnetic shielding is not shown.

particles, was contained in a plastic microbeaker (Fisherbrand 5 ml disposable). This suspension was placed between two antiparallel fluxgate Brand gradiometer probes, 7 cm apart, that were aligned coaxially with the magnetizing coil. As any resultant magnetic field generated by the particles decreases rapidly with distance, the placement of the specimen immediately adjacent to only one of the probes enabled the measurement of the field generated by the particles while subtracting out any remaining low frequency fields of environmental origin that were common to both probes.

To measure viscosity and yield stress, the time course of the summation of the remanent fields from the suspended particles was recorded. Initially, the particles were randomly oriented and thus produced no resultant magnetic field. The particles' magnetic moments were aligned with the axes of the probes by applying a strong magnetic field, usually 200–400 G. This alignment process, termed "primary magnetization," utilized the magnetizing coil, a 5500-turn copper wire solenoid positioned such that its axis was coaxial with that of the shields. Since the magnetic moments of the particles were aligned parallel to the axes of the gradiometer probes, a maximum remanent field was measured. To ascertain the viscous and yield properties of the fluid, the time course of the reorientation of the particles, caused by a magnetic field of opposite direction, "secondary magnetization," was measured. As the remanent magnetic field of the particles could not be measured by the gradiometer while the magnetizing coil was energized, the secondary magnetism was generated by applying, to the magnetizing coils, a sequence of square wave current pulses of constant magnitude. The pulses were typically 5–10 s long with 5–15 s pauses between them. The remanent fields due to the reorientation of the particles were measured and recorded between the pulses and after the last pulse of each sequence. The angular velocity with which the particles rotated was a function of both the strength of the applied magnetizing field and the resistance to rotation by the viscous fluid in which they were suspended. The strain and rate of strain for the magnetic rheometer were dependent upon the magnitude and duration of each pulse. The strength of the magnetic fields produced by the magnetizing coil were measured by an incremental Gaussmeter (Bell 640). The magnetic rheometer was automated by interfacing the data acquisition and experimental control functions with a microcomputer via an A/D converter. This automated rheometer was used to characterize the rheological behavior of mucus at a variety of strains and strain rates. A hard copy of the stored data was obtained using a digital plotter (Gould DS10).

To measure elasticity, the relaxation of the remanent field immediately following each pulse of secondary magnetism was measured. This relaxation was a measure of the elastic recoil of the specimen. The recoverable shear strain was the normalized difference between the magnitude of the remanent field immediately following primary magnetization and the magnitude of the remanent field immediately after secondary magnetization. The normalized elastic recoil was the recoverable shear strain. The difference between the shear strain and recoverable shear strain is the non-recoverable shear strain (flow).

To measure the yield stress of the test specimen, the maximum shear stress that would not produce non-recoverable shear strain was estimated. First, the minimum secondary field strength that would result in non-recoverable shear strain was determined by repeated measurements using successively greater secondary field strengths (beginning with a secondary field of zero strength). The secondary field strengths that were less than the minimum resulted in only recoverable shear strain. Second, the viscosity of the specimen was determined using successively weaker secondary fields, beginning with secondary field strengths greater than the minimum for non-recoverable shear strain and approaching the minimum secondary field for non-recoverable shear strain. Third, comparison of the viscosity and shear rate during each of the viscosity measurements mentioned above allowed the calculation of the average shear stress during each measurement. Finally, from a plot of the calculated shear stresses, the shear stress at the minimum secondary field strength was extrapolated. The extrapolated shear stress was the estimate of the yield stress.

Theory

The theory of operation of the magnetic rheometer was derived by modeling the suspended particles as a collection of fluid-mechanically independent spheres, which (after prealignment of their magnetic moments) rotate in unison. The differential equation governing the rotation of a suspended sphere during magnetic rheometry was coupled to the differential equation for the fluid motion by the boundary condition at the surface of the particle.

The equations of motion for the sphere were developed by performing a force balance. The law of conservation of angular momentum dictates that the rate of change of angular momentum must equal the sum of any applied torques. The two torques for this case were the magnetic torque (T_M) due to the applied magnetic field and the viscous torque (T_V) due to the shear stress at the surface of the particle:

$$I(dW/dt) = T_M + T_V \quad (1)$$

where I = moment of inertia
and W = angular velocity

A uniform magnetic field exerts a force on a ferromagnetic particle immersed in a medium that is equal to the cross product of the magnitude of the applied field (H) and the magnetic moment of the particle (M) multiplied by the permeability of the medium (μ_M):

$$T_M = \mu_M(M \times H) \quad (2)$$

In this analysis the magnetic properties of the particles were assumed to be isotropic.

For simplicity of notation, consider a spherical coordinate system (r, θ, F). For the case where the applied field was uniform, in the z direction (where F is the angle a vector makes with the z axis), and the magnetic vector is in the same plane as the field vector (the plane defined by r and Φ). The resulting torque causes rotation (no translation) of the particle in that same plane (provided that the two vectors are not parallel) and is a function of only r. The magnitude of magnetic torque was

$$T_M = \mu_M M_o H_Z \sin\Phi \tag{3}$$

The magnetic torque was opposed by a viscous torque resulting from drag at the surface of the particle. Insertion of Equation 3 into Equation 1 gave:

$$T_V = \mu_M M_o H_Z \sin\Phi - (2ma^2/5)[d^2\Phi/dt^2] \tag{4}$$

This differential equation could not be solved in closed form, even for a simple viscous fluid, unless the inertial term was neglected. However, neglecting inertia resulted in a solution that could not satisfy the initial conditions of a particle at rest. Consequently, we utilized a semi-empirical method. The magnetic torque was calculated from the appropriate physical constants and was subtracted from the rate of change of angular momentum which was calculated from the experimental data. The result of this subtraction was the viscous torque on the particle which was used to calculate the stress on the surface of the particle while the strain and rate of strain at the particle surface were determined from the data (*i.e.*, the time course of the rotation).

The resultant of the magnetic fields of all suspended particles was used to determine the time course of movement of the particles. The data interpretation was simplified by assuming that the signal measured could be taken as the sum of the signals from the individual particles. Thus, the remanent particle field $B_r(t)$ was related to the rotation angle, Φ, by the following relation (13):

$$B_r(t) = B_{ro}\cos[\Phi_0 - \Phi(t)] \tag{5}$$

where $\Phi_0 =$ the initial angle ~ 0 and $B_{ro} = B_r$ when $\Phi = \Phi_0$.

The viscous torque on the sphere was the cross product of the force and the moment arm. For the case of a magnetic sphere under the influence of an external magnetic field the moment arm was simply the radius (a) and the unit area was the surface area ($4\Pi a^2$). Thus the stress, τ, was defined as follows:

$$\tau = T_V/4\Pi a^3 \tag{6}$$

A value for the strain at the surface of the particle was also obtained from the gradiometer signal ($B_r(t)$). The 90° rotation was used as a characteristic strain since it approximates the angle swept out by cilia in mucus during the "power stroke" phase of ciliary motion. Thus, all measures of deformation angle were divided by 90 for nondimensionalization.

$$\gamma(t) = \text{the strain} = \Phi(t)/90 = \cos^{-1}[B_r(t)/B_{ro}]/90 \tag{7}$$

(*Note:* The deformation of the fluid at the surface of the particle is actually equal to the rotation angle multiplied by the radius of the particle; however, the radius in the numerator and denominator cancel, leaving the result the same.)

The average rate of strain over any given time period was equal to the strain divided by the elapsed time. The apparent viscosity was equal to the quotient of the stress and the rate of strain. Thus, equations 6 and 7 were used to calculate the viscosity of a test fluid provided the particles were spherical and the appropriate physical constants were known. A more practical approach was to empirically calibrate the instrument using viscosity standards. The latter approach was used for the experiments described in subsequent sections. Nickel spheres, however, were used to validate the theoretical equations.

The nickel spheres were suspended in a viscoelastic polymer solution Polyox WSR–301 (1%). The manufacturer's low shear rate viscosity data gave a value of 250–270 poise for this solution. The average value calculated from the theoretical equations for magnetic rheometry and experimental data was 217 p. Using empirical calibration (see below) the values obtained were 250 and 248 p. Magnetic rheometry shows the good agreement between theory and experiment for nickel spheres, as well as agreement with the manufacturer's data. Unfortunately, the magnetic characteristics of the nickel spheres made them unsuitable for use with the secondary field strengths necessary to rotate them in materials as viscous as mucus. Thus, the magnetite particles were used and necessitated the empirical calibration method.

Calibration and Methods

The magnetic rheometer was empirically calibrated using viscosity standards (Brookfield) of 9.80, 48.80, 130, 307.2, 620.0, and 960 poise. Magnetite particles (Pfizer), which ranged from 0.2 to 3.0 μm in diameter, were suspended in 5 ml samples of each viscosity standard. The suspensions also contained a small number of aggregates, which ranged from 5.0 to 30.0 μm in diameter. For calibration purposes, 100 mg of the magnetic particles were suspended in each sample. The time course of remanent magnetism of particles in the 620.0 p viscosity standard after primary magnetization and during secondary magnetization is shown in Figure 2a. The remanent magnetism changed sequentially from its maximum "positive" value. As the particles rotated, the remanent field measured by the gradiometer decreased, reaching zero when the net magnetic moment of the particles was perpendicular to the axes of the probes and approaching a maximum negative value as their magnetic moments approached alignment in the opposite direction. In the case presented in Figure 2b, the average value corresponding to each of the first six plateaus of remanent field, expressed as a percent of the maximum initial field, was plotted against the time for which the secondary field was applied. The time required for the remanent field to decrease to zero "T_{B0}" was used as an index of viscosity. Measurements of T_{B0} were independent of the concentration of the suspended particles over

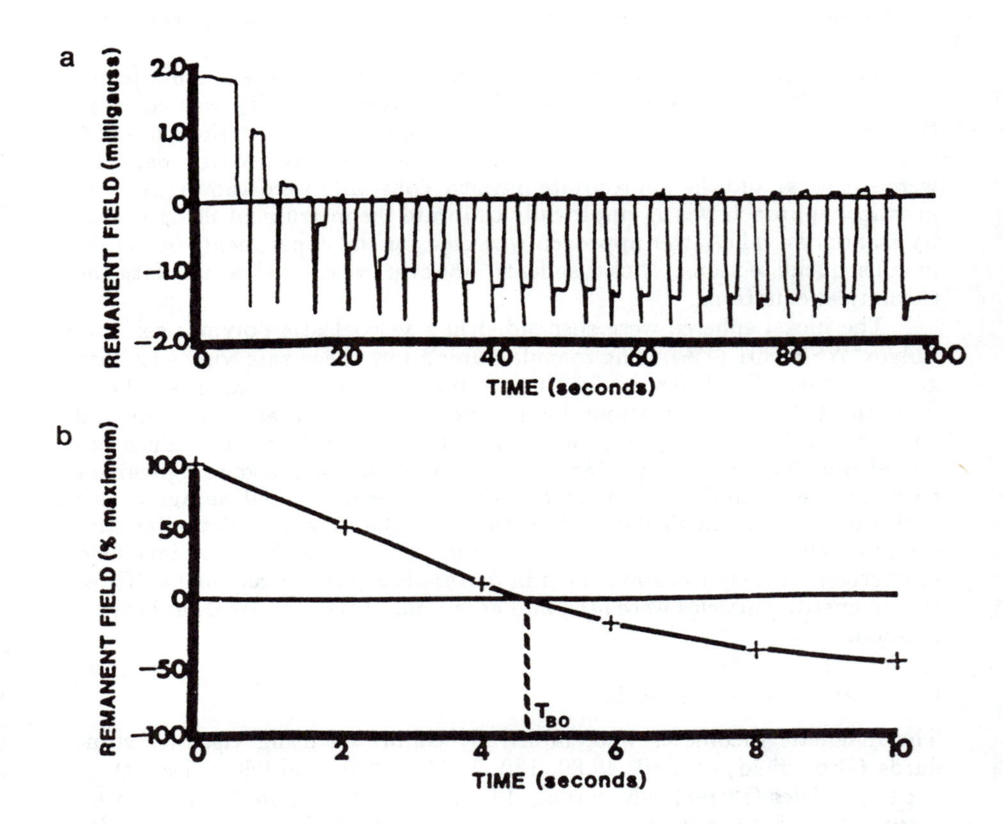

Figure 2. Measurement of T_{B0} for Fe_3O_4 particles in a viscous fluid. (a) After primary magnetization (320 G), the remanent magnetic field of ferromagnetic particles suspended in the 620 p viscosity standard decreases in response to the 1.5 s pulses of secondary magnetization (200 G) which were delivered at a pulse rate of 0.25 Hz. During each secondary pulse the gradiometer output was approximately zero, resulting in a series of plateaus along the zero line. The spikes of remanent magnetism after each pulse were due to a transient response of the gradiometer probes to the strong magnetic field of the magnetizing coil. (b) The normalized value of each plateau in Figure 2 a is shown as a function of time that the secondary field was applied (*i.e.*, the time periods for recording the remanent plateaus were eliminated).

the range of 0.1–20 mg/ml as demonstrated by the coincidence of the T_{B0} (Figure 3) for the three suspensions.

The T_{B0} was derived for each viscosity standard using secondary field strengths of 60–340 G (40 G increments). The magnitude of T_{B0} increased with increasing viscosity and decreased with increasing field strength. The viscosity is plotted against the T_{B0} for most of the secondary field strengths and is shown in Figure 4. The family of such curves forms a series of calibration curves from which the viscosity of a test sample may be determined from its T_{B0} at the appropriate secondary field strength.

To evaluate the rheological properties of mucus obtained from a subcutaneous tracheal pouch of an adult male beagle dog (14), the magnetite particles were blown onto the surface of the mucus sample using a Wright dust feeder. To avoid shear degradation due to mechanical mixing, the particles were drawn below the surface by a permanent magnet.

Results

The time course of remanent magnetism from particles suspended in mucus during the magnetic rheometry process is shown in Figure 5a. In contrast to the viscous oils, the remanent plateaus were actually relaxation curves. These relaxation curves were due to the elastic nature of mucus. The average values of the relaxed remanent fields after each pulse of secondary magnetism as a percent of the initial remanent field are shown as a function of the time the secondary field was applied in Figure 5b.

An example of the complete relaxation of the remanent field from particles suspended in mucus is shown in Figure 6. The elasticity of the mucus sample was measured by applying one 4.5 s pulse of secondary magnetization (200 G) and recording the complete relaxation of the remanent field. The magnitude of the relaxation was the elastic response. An example of complete relaxation of the remanent field from particles suspended in mucus is shown in Figure 6. The difference between the initial plateau magnitude and the magnitude after complete relaxation was the viscous response. The recoil time constant for mucus, *i.e.*, the time required for the relaxation to reach 63.2% of its final value, was 2.1 s.

An example of the remanent magnetic fields recorded during a determination of yield stress is shown in Figure 7. Figure 7a shows the determination of the minimum secondary field strength for non-recoverable shear strain (H_m), while Figure 7b shows the results of the successive viscosity measurements that lead to an estimation of the yield stress for a mucus sample.

Ten canine tracheal pouch mucus samples exhibited viscosity values ranging from 1798 to 4.6 p at 0.1 to 7.3 sec^{-1} rates of strain, respectively. The normalized strain increased from 0.23 to 1.0 (see Figure 8), while the recoverable shear strain at fixed strain did not increase with the rate of strain (see Figure 9). The yield stress for these same samples ranged from 5.6 to 13.5 $dynes/cm^2$.

Five human sputum samples exhibited lower viscosities (68 to 2.5 p at 0.5 to 12 sec^{-1}) and yield stress values of 0.5 to 2.6 $dynes/cm^2$. The elastic

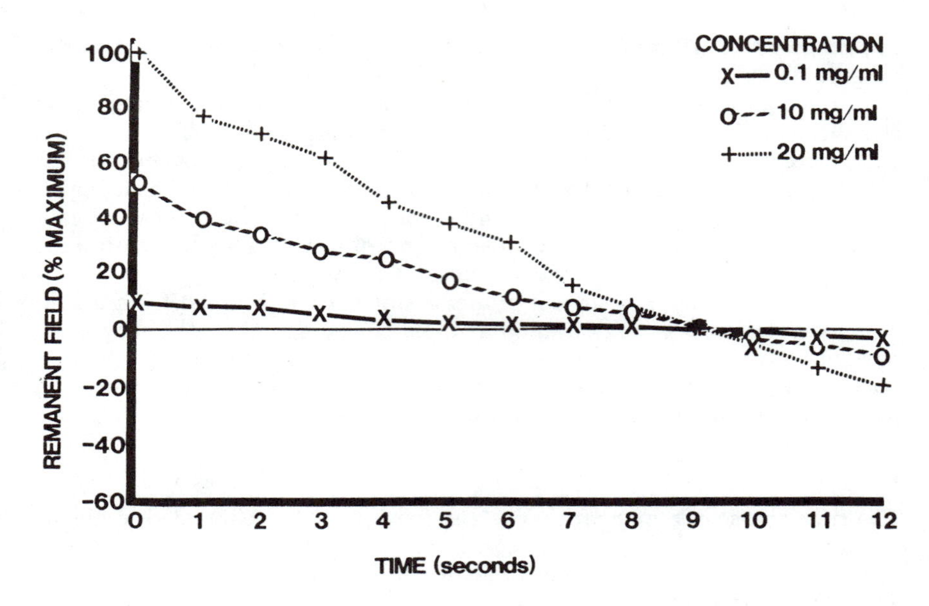

Figure 3. The effect of particle concentration on magnetic rheometry. The T_{B0} was the same over the concentration range 0.1–20 mg/ml.

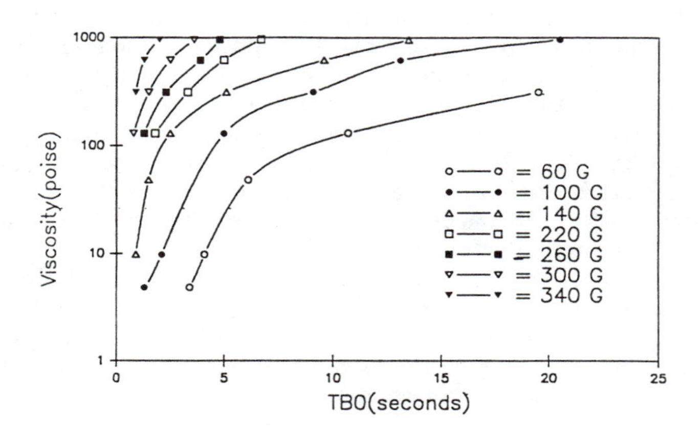

Figure 4. Magnetic rheometer calibration curves. The viscosity is plotted as a function of the measured T_{B0} for secondary fields varying from 60 to 340 G (40 G increments) resulting in calibration curves (not all shown here) for the rheometer.

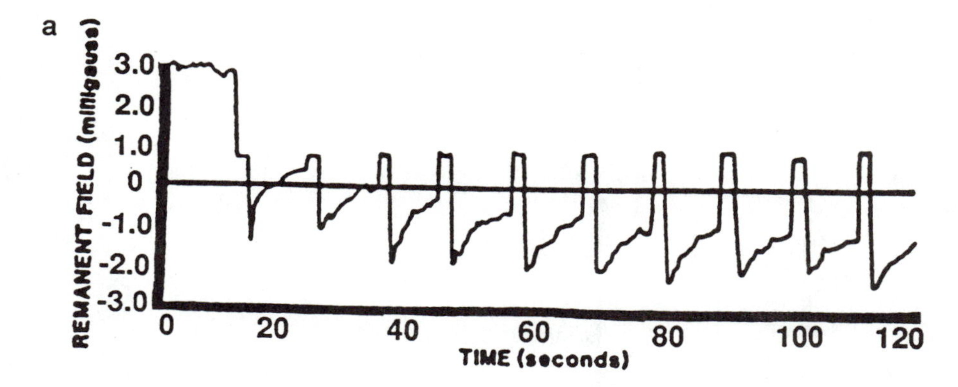

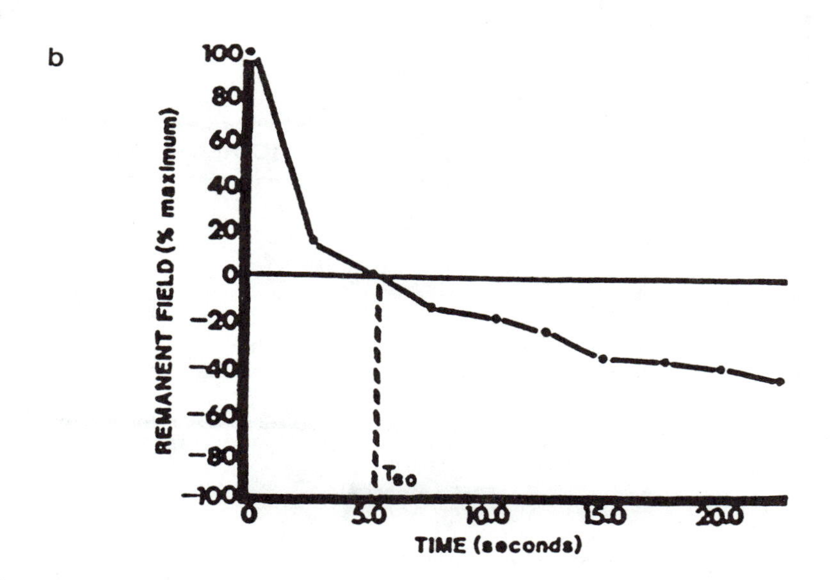

Figure 5. Measurement of T_{B0} for Fe_3O_4 particles in mucus. (a) After primary magnetism (320 G), the remanent magnetic field of ferromagnetic particles, suspended in mucus, changes due to particle rotation which was caused by 5.0 s secondary magnetic pulses (200 G), delivered at a pulse rate of 0.1 Hz. After each pulse the elastic recoil of the mucus caused particles to rotate and resulted in the relaxation of the remanent field. (b) The normalized value of each relaxation curve in Figure 5a is shown as a function of the time that the secondary field was applied. The T_{B0} was the time required for the curve fitting the normalized values to reach zero.

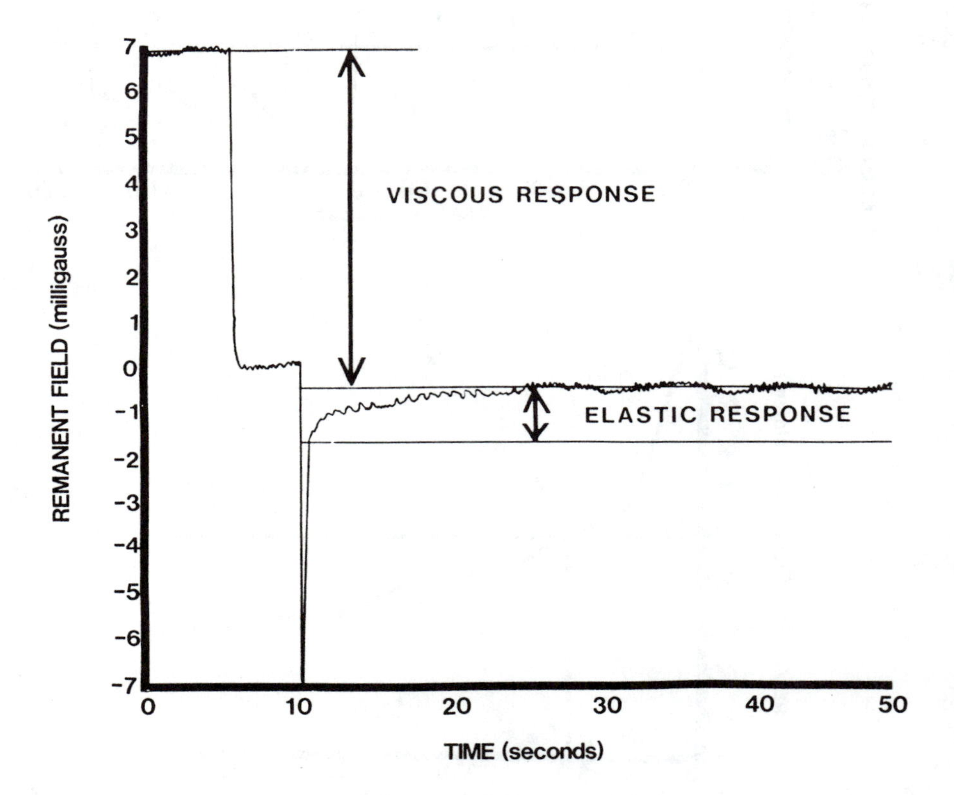

Figure 6. The elastic recoil of mucus. After primary magnetization (320 G), ferromagnetic particles in mucus were subjected to a single 4.5 s 200 G secondary pulse and the particles were allowed to recoil completely. The magnitude of the recoil was the elastic response while the difference of magnitude between the initial and completely relaxed conditions was the viscous response.

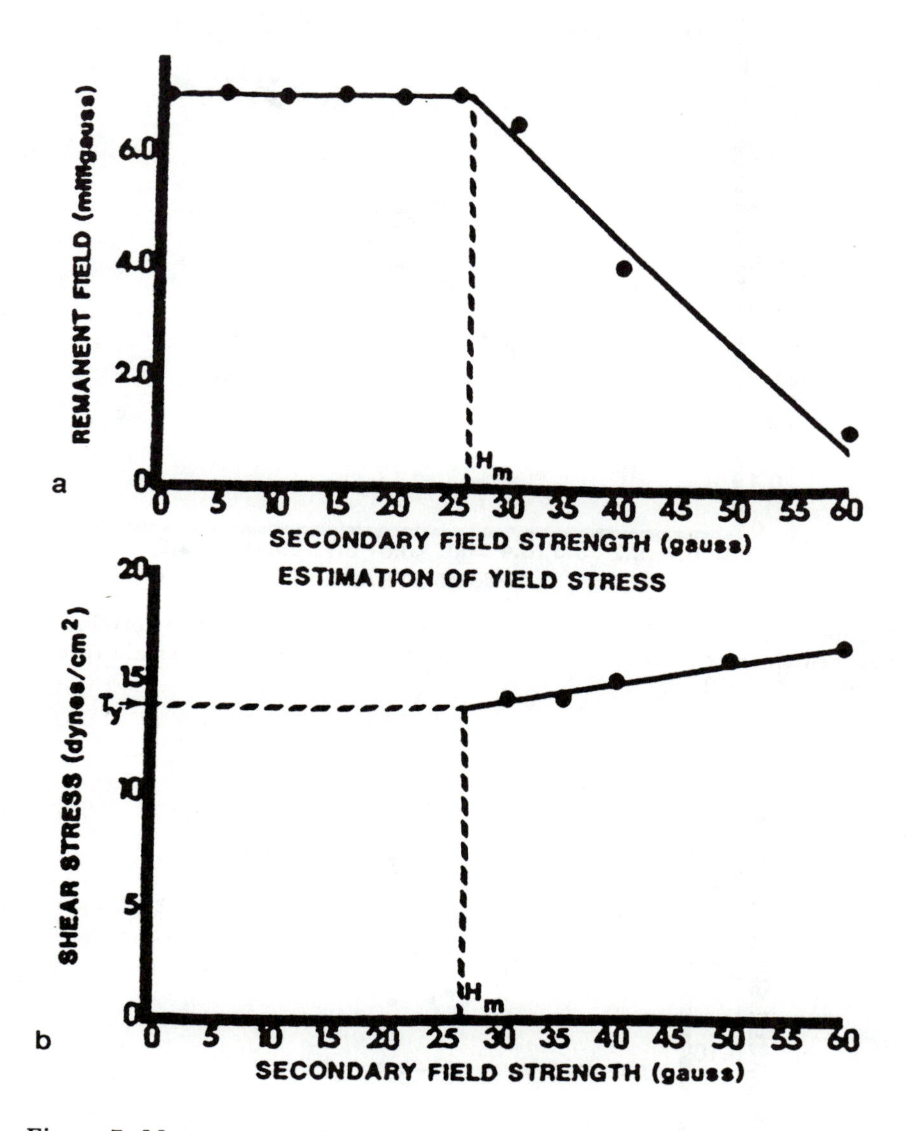

Figure 7. Measurement of the yield stress of mucus. (a) The magnitude of the particle remanent field (30 seconds after a 30-second pulse of secondary magnetism) vs. the corresponding secondary field strengths. H_m is the weakest secondary field strength which results in a particle remanent field that does not return to its original value even after mucus relaxation. (b) Estimation of the yield stress by extrapolation of shear stress data for secondary fields approaching H_m.

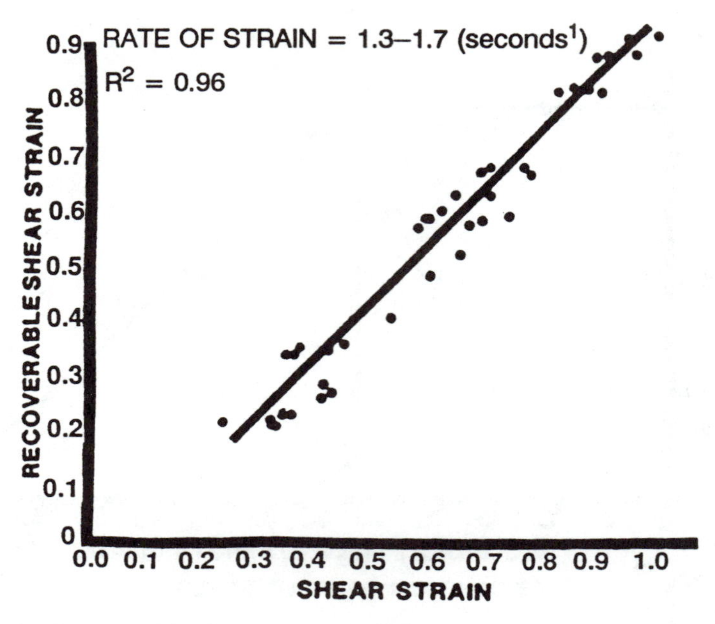

Figure 8. Recoverable shear strain increases with strain. The correlation between the recoverable shear strain and the strain when the strain rate is fixed. The r = 0.98.

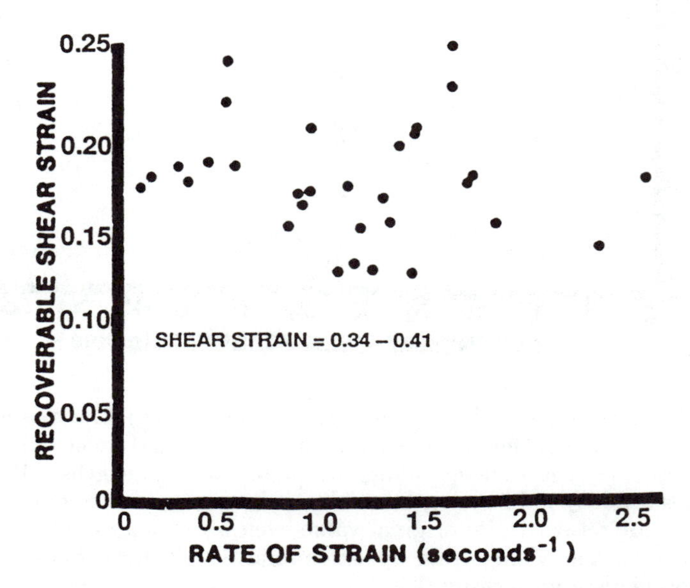

Figure 9. Lack of correlation between the recoverable shear strain and the strain rate at constant strain.

response was similar to that for the canine mucus. The recoverable shear strain at fixed strain rate increased from 0.4 to 1.0 as the strain increased from 0.3 to 0.8.

N-acetylcysteine (NAC), when administered in 1 ml aliquots of a 0.034 M solution to fresh canine tracheal pouch mucus, resulted in a greater decrease in yield stress (74% of control) than in viscosity (39%) and recoverable shear strain (22%). Figure 10 plots the viscosity of mucus vs. the strain rate before and after administration of NAC. In contrast to the response of canine mucus to NAC, five human sputum samples from patients with chronic bronchitis were not altered by NAC at either 0.034 or 0.68 M concentrations. The viscosity vs. strain rate plot for sputum is shown in Figure 11.

Discussion

These studies demonstrate the feasibility of measuring both the viscous and elastic properties of small volumes of dilute magnetic suspensions by the application and measurement of external magnetic fields.

The majority of reported rheological data for canine tracheal pouch mucus has been obtained from dynamic (oscillatory) measurements systems, which report the shear loss modulus ("viscous response") and the shear storage modulus ("elastic response") as functions of frequency. The dynamic storage moduli and loss moduli increase from 1 to 500 dynes/cm^2 over the range 0.05–500 rad/sec (9,10,15). The measurements of King *et al.* (10), made under *quasi* steady state conditions and Shake *et al.* (16), on canine tracheal pouch mucus demonstrated that the viscosity decreased from 1250 to 5 poise between 0.01–100 sec^{-1}. The measurements by the magnetic rheometer were within this range and exhibited the characteristic decrease with increasing rate of strain. Davis (9) and Sturgess *et al.* (17) reported similar values for human sputum viscosity. The values of recoverable shear strain measured by the magnetic rheometer were also comparable to those reported (9,16).

The strain imposed on the fluid by the magnetic rheometer was two to three orders of magnitude less than either the capillary or the cone and plate rheometers. For instance, the deformation was approximately 0.8 μm compared to 10,000 μm for a capillary rheometer (16), at least 1100 μm for a cone and plate (17), and 2.5–7.5 μm for the oscillating sphere magnetic microrheometer (10).

The viscosity of mucus decreased with increasing rate of strain; a phenomenon well documented in the literature. The recoverable shear strain increased with increasing shear strain when the strain rate was fixed. However, the recoverable shear strain did not increase with increasing strain rate when the shear strain was fixed. Thus, the elastic behavior of mucus under the conditions of magnetic rheometry was inconsistent with a Maxwell model for mucus.

The yield stress values of canine mucus were higher but of the same order of magnitude as the estimate of the force of gravity on the mucus layer *in vivo*. For comparison, the stress due to gravity in a vertical trachea

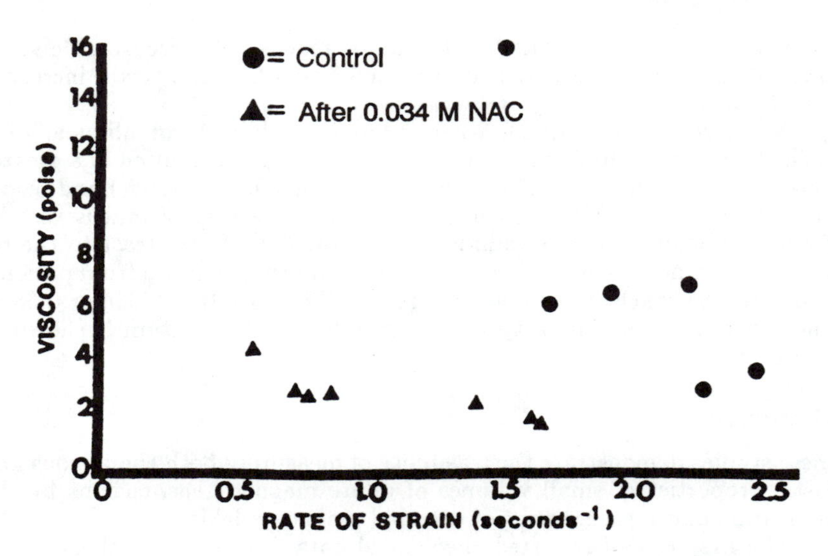

Figure 10. The decrease of mucus viscosity after N-acetylcysteine treatment. The viscosity of canine tracheal pouch mucus is altered by the application of 1 ml of 0.034 M NAC directly to the top of the sample. The decrease in viscosity was evident at both low and high shear rates.

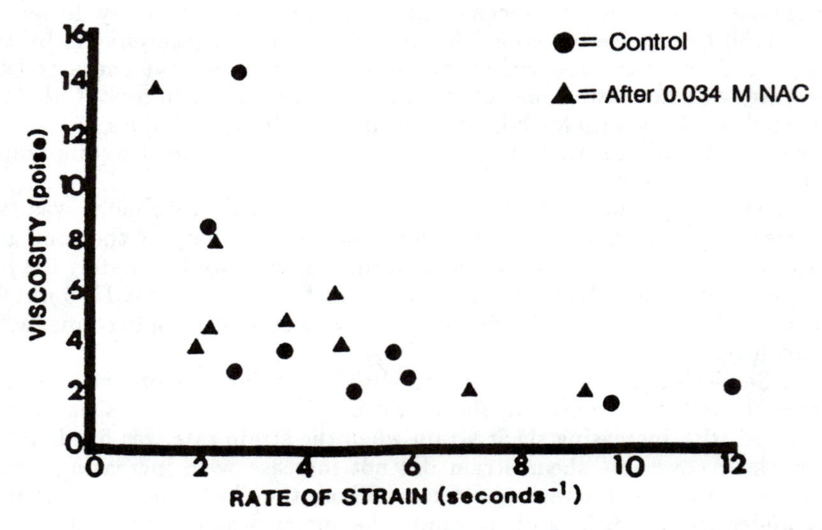

Figure 11. N-acetylcysteine fails to alter the viscosity of human sputum. The viscosity of human sputum is plotted as a function of strain rate. The administration of NAC failed to decrease the viscous response as it did for the canine mucus.

with a 6μm deep coating of mucus is estimated to be approximately 1 dyne/cm^2. Therefore, the yield stress may be responsible for preventing retrograde flow in the lung. Conversely, a diminished yield stress may allow retrograde flow and thereby inhibit mucociliary clearance in disease. It is interesting to note that some of the yield stress values for the sputum from chronic bronchitics were less than the estimates for gravity-induced flow. In the case of thick sputum, the yield stress may need to be modified for effective physiotherapy. Mucolytics are widely used for this purpose.

The administration of 0.034 M N-acetylcysteine (NAC) significantly reduced the viscosity, elasticity and yield stress of canine tracheal pouch mucus. The results of these studies agreed both qualitatively and quantitatively with other literature values. However, the rheological properties of human sputum from patients with chronic bronchitis and aspiration pneumonias were unaffected by the application of 0.034 and 0.68 M NAC. This discrepancy in the mucolytic action of NAC may be indicative of differences in structure between the two fluids or may reveal that the pathology of the disease states of the patients causes mucolysis and therefore further mucolytic action could not be detected. Concurrent biochemical analysis and rheological investigation would be necessary to confirm these possible explanations.

The magnetic rheometer has some limitations. The rheological properties of fluids can only be evaluated when they are sufficiently viscous to maintain a suspension long enough to make the measurements. The least viscous viscosity standard successfully used had a viscosity of only 0.505 p, which is well below the reported values for mucus. The ferromagnetic particles used in the test require a high coercivity such that the applied magnetic fields cause particle rotation rather than a reorientation of the magnetic domains within the particles. The nickel spheres mentioned above were inadequate in this respect and could only be used for low secondary field tests on less viscous materials. The shape of the particles used in this study was not spherical, as was assumed for strain estimation purposes. However, the particles did serve to demonstrate the feasibility of this method. In addition, the concentration of particles in the suspension must be low enough to avoid significant particle-particle interaction and to not materially affect the viscosity of the test fluid.

The measurement principle utilized by the magnetic rheometer relies upon the detection of the resultant remanent magnetism of the particles in a particular direction. Ideally, the particles rotate in only one plane. Anisotropy of the mucus sample could result in direction-specific viscosity measurements. This effect could be minimized by varying the position of the sample cup in the sample holder. No attempt was made to determine or correct for sample anisotropy in this study.

When using the magnetic rheometer, the rate of strain increased with increasing secondary field strength. As the particles were rotated faster by the application of stronger fields, it was necessary to shorten the pulse width so that the particles would not rotate past 90 degrees during the first pulse, resulting in insufficient data for the viscosity measurement. The time constant of the magnetizing coil limited the brevity of the secondary pulses and thus there was an upper limit of 15 sec^{-1} for the rate of strain. Similarly, when the secondary field strength was reduced to lower the rate of strain, there was a limitation due to the yield stress of the test sample. Secondary field strengths of less than 50 G did not always provide enough force to the particles to overcome the yield stress.

Individual cilia within the mucociliary transport system must apply microscopic strains due to their dimensions. The stresses associated with these would be expected to be equally minute. Also, the current understanding of pulmonary ciliary movements suggests that a three-phase pattern of ciliary movement results in a transient stress-strain history for mucus *in vivo* (19,20). The mucus is subjected to shear in the effective stroke only. The mucus is allowed to relax for the remainder of the cycle, *i.e.,* rest period and recovery stroke, with essentially no reverse stress. After this, the cycle repeats but the stress is unidirectional. This stress history and the mucus rheology combine to dictate the rate of mucociliary transport. The transient strains applied during the secondary magnetic pulses were much more similar to those applied *in vivo* than are the strain histories associated with capillary, rotational or dynamic rheometers. The time course of the strains applied during magnetic rheometry would more closely mimic cilia if the time between secondary pulses could be shortened to approach the rest and recovery period of cilia. The long mucus relaxation times afforded by the present system are convenient for simultaneous measurements of viscosity and elasticity, but the prestressed condition that may occur *in vivo* is eliminated by this technique. More sophisticated automation will be necessary to address this problem and to incorporate the calibration curves and shear stress calculations into the automation software.

The use of magnetic pulses to turn particles obviated the need for rotating the specimen as described by Valberg *et al.* (18). Thus, this system should potentially be adaptable to measuring the viscoelastic properties of mucus *in vivo* after the administration of a ferromagnetic aerosol. Such a procedure would be not unlike that attempted by Nemoto (21). Nemoto's measurements were inherently difficult and probably inaccurate because they relied on the ability to accurately estimate the forces that caused spontaneous reorientation of magnetized particles in the lung. Also, Nemoto did not measure elastic behavior. Valberg and Albertini, using a technique which was similar to that for the magnetic rheometer, measured cytoplasmic viscosity of macrophages *in situ* and much more directly than Nemoto. For the magnetic rheometer, an *in vivo* measurement and collection procedure would utilize the same pulse generation and detection scheme as the *in vitro* system. Preliminary studies have demonstrated the feasibility of deposition of the iron oxide particles. Trial efforts to measure *in vivo* mucus rheology by magnetic rheometry in anesthetized dogs uncovered the need for additional signal processing and magnetic shielding to improve the signal to noise ratio. Nevertheless, the potential for *in vivo* magnetic rheometry clearly exists and would add an important new tool for the study of mucus rheology and mucociliary transport as a whole.

Acknowledgments

This work was supported by the Medical Service of the Veterans Administration, Illinois Minority Graduate Incentive Program, and NIEHS ES04317. These laboratories were also supported by HLBI HL33461.

Literature Cited

1. Gelman, R. A.; Meyer, F. A. *Amer. Rev. Respir. Dis.* 1979, **129**, 553-557.
2. Shih, C. K.; Litt, M.; Khan, M. A.; Wolf, D. P. *Amer. Rev. Respir. Dis.* 1977, **115**, 989-995.
3. Chen, T. M.; Dulfano, M. J. *J. Lab. Clin. Med.* 1978, **91**, 423-431.
4. Sturgess, J. M.; Palfrey, A. J.; Reid, L. *Clinical Sci.* 1970, **38**, 145-156.
5. Lutz, R. J.; Litt, M.; Chakrin, L. In *Rheology of Biological Systems*; Gabelnick, H. L.; Litt, M., Eds.; Springfield, IL: Charles C. Thomas Publ., 1973; pp 119-195.
6. Puchelle, E.; Zahm, J. M.; Duvivier, C.; Dideleon, J.; Jacquot, J.; Quemada, D. *Biorheology* 1985, **22**, 415-423.
7. Puchelle, E.; Zahm, J. M.; Petit, A. In *Advances in Experimental Medicine and Biology 144: Mucus in Health and Disease—II*; Chantler, E. N.; Elder, J. B.; Elstein, M., Eds.; New York: Plenum Press, 1982; pp 397-398.
8. Litt, M. *Chest Suppl.* 1981, **80**, 846-849.
9. Davis, S. S. In *Rheology of Biological Systems*; Gabelnick, H. L.; Litt, M. Eds.; Springfield, IL: Charles C. Thomas Publ., 1973; pp 158-193.
10. King, M.; Macklem, P. T. *J. Appl. Physiol.: Respirat. Environ. Exercise Physiol.* 1977, **42**, 797-802.
11. Crick, F. H. C.; Hughes, F. W. *Exp. Cell. Res.* 1950, **1**, 37-80.
12. Crick, F. H. C. *Exp. Cell. Res.* 1950, **1**, 505-533.
13. Valberg, P. A.; Butler, J. P. *J. Biophys.* 1987, **52**, 537-550.
14. Wardell, J. R.; Chakrin, L. W.; Payne, B. S. *Am. Rev. Resp. Dis.* 1970, **101**, 741-754.
15. Litt, M.; Khan, M. A.; Charkin, L. W.; Wardell, J. R.; Christian, P. *Biorheology* 1974, **11**, 111-117.
16. Shake, M. P.; Dresdner, R.; Gruenauer, L.; Yeates, D. B.; Miller, I. F. *Biorheology* 1987, **24**, 231-235.
17. Sturgess, J. M.; Palfrey, A. J.; Reid, L. *Clinical Sci.* 1970, **38**, 145-156.
18. Valberg, P. A. *J. Cell. Biol.* 1985, **101**, 130-140.
19. Sleigh, M. A. *Cell Motility Suppl.* 1982, **1**, 19-24.
20. Sleigh, A.; Blake, J. R.; Liron, N. *Am. Rev. Respir. Dis.* 1988, **137**, 726-741.
21. Nemoto, I. *Inst. Electr. Electron. Eng. Trans. Biomed. Eng. BME* 1982, **29**, 745-752.
22. Yates, D. B. In The Lung: Scientific Foundations; Crystal, R. G.; West, J. B., Eds.; New York: Rover Press Ltd.; 1991. pp 197-205.

RECEIVED December 16, 1991

Chapter 17

Molecular Origin for Rheological Characteristics of Xanthan Gum

Masakuni Tako

Laboratory for Chemistry of Sugar Technology, Department of Agricultural Chemistry, University of the Ryukyus, Nishihara, Okinawa 903–01, Japan

The molecular basis for the rheological characteristics of xanthan, a polysaccharide produced by the plant pathogen *Xanthomonas campestris*, was studied. Possible modes of intramolecular associations in xanthan between an alternate hydroxyl group at C-3 and the adjacent hemiacetal oxygen atom of the D-glucosyl residues, and between the methyl group of the acetyl residue and the adjacent hemiacetal oxygen atom of the D-glucosyl residue, were proposed. Possible binding sites for D-mannose-specific interaction between xanthan and galactomannan have also been proposed. The interaction may take place between the hemiacetal oxygen atom of the inner D-mannose side-chain of xanthan and the hydroxyl group at C-2 of the D-mannose residue of the galactomannan molecule with hydrogen bonding. Such analogous counterparts may play a dominant role in the interaction. The univalent cation (K^+) on the carboxyl group of the intermediate D-glucuronic acid in the side-chain of xanthan may also take part in the interaction by electrostatic force of attraction. The interaction between the extracellular bacterial polysaccharide and typical galactomannan components of the plant cell wall may play a part in the host-pathogen relationship, and may provide the existence of D-mannose-specific binding sites used for cell recognition by the pathogen.

The bacterial polysaccharide xanthan, produced by *Xanthomonas campestris*, is of interest not only for its unique rheological properties (1,2,3), but also for its formation of a mixed gel with plant galactomannan (4,5). The primary structure of xanthan is a $(1 \rightarrow 4)$-linked β-D-glucan backbone (cel-

0097–6156/92/0489–0268$06.00/0

lulose) substituted through 0-3 on alternate glucosyl residues with a charged trisaccharide side-chain (6, 7). The internal D-mannose of the side-chain is substituted at 0-6 with acetyl groups. About one-half to two-thirds of the terminal D-mannose residues bear a pyruvic acid, depending on the culture conditions (8). We have proposed that xanthan molecules are associated in quaternary structure through charged trisaccharide side-chains (9, 10), on which the methyl groups of pyruvic acid residues may contribute, while the methyl groups of the acetyl residues may contribute to its intramolecular association. The unusual viscosity and dynamic viscoelasticity of xanthan, showing a sigmoid curve with increasing temperature, might be attributed to the formation of the intra- and intermolecular associations.

Galactomannans are reserve carbohydrates found in the endosperms of some legume seeds. They consist of a (1 → 4)-linked β-D-mannan main-chain to which are attached various amounts of single α-D-galactosyl residues (11, 12). Such galactomannans have attracted attention in relation to their role in polysaccharide interaction, with κ-carrageenan (13, 14), agarose (15, 16), and xanthan (4, 17).

The strength of the xanthan-galactomannan gelling interaction increased with decreasing content of D-galactose in the galactomannan (17, 18). Early workers attributed gelation to an intermolecular binding between the xanthan helix and the unsubstituted region of the galactomannan backbone (18, 19). Recently, Cairns *et al.* (20, 21) have proposed an alternate association mechanism between the cellulosic backbone of xanthan and the mannan backbone of the galactomannan. The established models have ruled out the function of the side chains of the xanthan molecule.

On the other hand, we demonstrated a new mode of an intermolecular interaction involving the side chains of the xanthan and backbone of the galactomannan (17, 22, 23).

We report herein the non-Newtonian behavior and dynamic viscoelasticity of denatured xanthan solution and of that solution mixed with galactomannan. The rheological properties of xanthan are analyzed with respect to their association characteristics in more detail in order to propose a structure of intramolecular associations within the xanthan molecule and binding sites with the galactomannan molecule in aqueous solution.

Results

By spectrophotometrical estimate as the 2,4-dinitrophenyl derivative, and saponification and back titration, native xanthan had pyruvic acid and acetic acid contents of 5.7 and 5.4% respectively (W/W), which meant that 67 and 98% of the terminal and inner D-mannosyl groups bore pyruvic and acetic acid moieties, respectively. The pyruvate groups were removed by heating a solution of native xanthan (1 g/l in 1 mM oxalic acid) at 95°C for 2 h. Under these conditions, about 84% of the pyruvate groups were removed, but acetyl groups were unaltered. The acetyl-free xanthan was prepared by dissolving the native, or depyruvated, xanthan in distilled water and altering the solvent to 10 mM KOH, then letting it stand at room temperature for 10 h. This product was labelled as a deacetylated or deacylated xanthan.

The flow curves at concentrations below 0.5% for the deacylated xanthan resembled classical shear-thinning behavior, and at concentrations above 0.8%, approached plastic behavior. The yield values were estimated to be 20 and 35 dyn/cm^2 at concentrations of 0.8 and 1.0%, respectively. Despite showing lower yields for deacylated xanthan than for depyruvated material, the flow curves of the former shifted over higher shear-stress than those of the latter at high concentrations. This may be due to the free form intramolecular association, to which the acetyl group contributes; therefore, the deacetylation may lead to a more flexible and extended conformation than the depyruvation for xanthan in aqueous solution.

The apparent viscosity of native xanthan in a 0.5% solution showed a sigmoid curve (Figure 1) where it decreased a little with the increase of temperature up to 20°C, then increased gradually and showed a maximum at 45°C; after that it decreased a little again and stayed high even at 80°C. The deacetylated xanthan, however, showed a high value at a low temperature (0°C) and stayed constant up to 35°C, then decreased gradually, but remained high even at 80°C, as in native xanthan. Such a phenomenon, showing high viscosity at temperatures $< 35°C$, may be caused by the free form intramolecular association by deacetylation. The viscosity of depyruvated xanthan showed the lowest value at a low temperature and stayed constant until 55°C, then it decreased rapidly pointing to 55°C as a transition temperature. Though the viscosity of the deacylated xanthan was higher than that of any of the other polymers except the deacetylated material at low temperature, after the temperature reached 25°C, it decreased rapidly, indicating 25°C was a transition temperature. The apparent viscosities of native, deacetylated, depyruvated, and deacylated xanthan were lower than that of the polymer alone in the presence of urea (4 M). A transition temperature was observed at 65°C in a solution of the native and deacetylated xanthan in the presence of urea. This indicates that an intermolecular association of the native and deacetylated xanthan, to which the pyruvate groups may contribute at high temperatures, breaks down above the transition temperature—a phenomenon which is not observed in the polymer alone.

Though the dynamic modulus showed a sigmoid curve in a solution of native xanthan even at a concentration of 0.5% during increase in temperature, for the deacylated xanthan it decreased at a concentration below 0.5%; but for 0.8 and 1.0%, it increased a little up to 25 and 30°C, which was estimated to be a transition temperature; then it decreased rapidly (Figure 2). The $\tan\delta$ value of the deacylated xanthan decreased from 2.20 to 0.53 with increases in the concentration from 0.3 to 1.0% at low temperature. The values at both concentrations were higher than those of native, deacetylated, and depyruvated xanthan (3, 9, 10, 24).

Because the intermolecular interaction between xanthan and galactomannan molecules is closely correlated with the degree of substitution of the mannan chain, the degree of substitution of locust-bean gum and

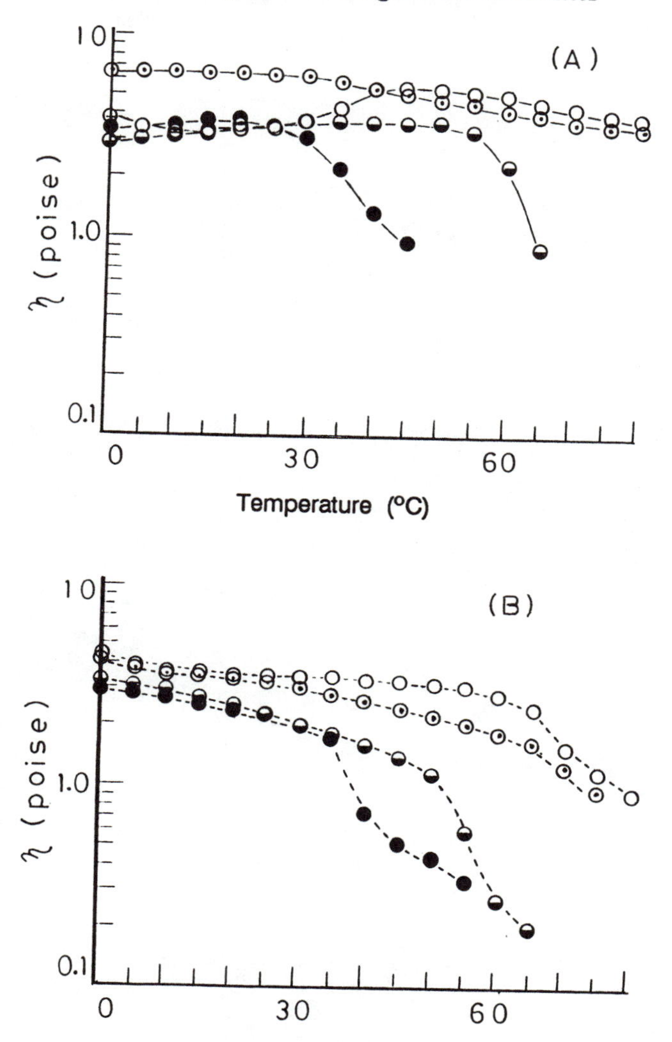

Figure 1. Effect of temperature on viscosity of various types of xanthan at 0.5% and 9.50 sec^{-1} with addition of 4.0 M urea. The solid lines refer to the polymer alone and the broken lines to the addition of urea. ●, native; ⊖, deacetylated; ○, depyruvated; ⊙, deacylated xanthan.
(Reproduced with permission from reference 34. Copyright 1989 Japan Society for Bioscience, Biotechnology, and Agrochemistry.)

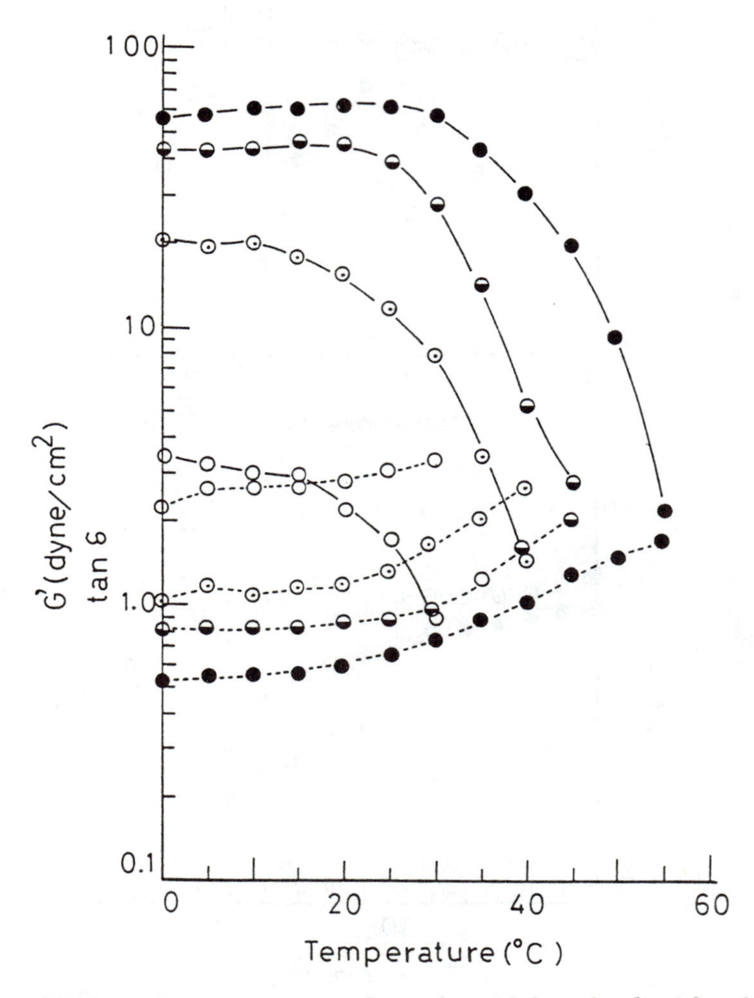

Figure 2. Effect of temperature on dynamic modulus of a deacylated xanthan at various concentrations and 3.77 rad/sec. The solid lines refer to the dynamic modulus and the broken lines to the tan δ. Concentrations: ○, 0.3; ⊙, 0.5; ◒, 0.8; ●, 1.0%.

(Reproduced with permission from reference 34. Copyright 1989 Japan Society for Bioscience, Biotechnology, and Agrochemistry.)

guar gum was determined by liquid chromatography and calculated to be D-mannose to D-galactose, 4.1 : 1.0 and 2.0 : 1.0, respectively.

Although neither xanthan nor locust-bean gum gelled alone, a mixture of the gums gave a gel at 0.2% total gums. Figure 3 shows the effect of the ratio of xanthan (native, deacetylated, depyruvated, and deacylated) to locust-bean gum in solution on the dynamic modulus at 25°C. Maximum dynamic modulus was achieved when the mixing ratio of xanthan to locust-bean gum was 1:2. In the case of deacylated xanthan and deacetylated xanthan, a much stronger gel was observed, about twice as strong as the mixture with native or depyruvated xanthan, indicating that much more intense intermolecular interaction was produced by deacetylation. In spite of an increase in dynamic modulus of native, deacetylated, depyruvated, and deacylated xanthan alone by addition of salt, a very small dynamic modulus was observed in a mixture with locust-bean gum on addition of $CaCl_2$ (6.8 mM), indicating that carboxyl groups of the D-glucuronic acid residues of the intermediate side-chains of xanthan may take part in the interaction with locust-bean gum molecules.

On the other hand, for mixtures of guar gum with native and depyruvated xanthan, little synergistic increase in dynamic modulus was observed at room temperature (25°C), as shown in Figure 4. This may be due to the presence of side-chains, at every other unit (11), at two to four units (25), or pairs and triplets (12) of the guar gum molecule. However, the synergistic interaction was enhanced in the mixture with deacetylated and deacylated xanthan, indicating that the xanthan molecules had become more flexible and could associate with guar gum molecules more easily, probably because they were free from the intramolecular association to which the acetyl groups contributed. The maximum dynamic modulus was achieved when the mixing ratio of xanthan to guar gum was 2:1; this ratio was reversed for the mixture with locust-bean gum (Figure 3).

The effect of temperature on the dynamic modulus of a mixed solution of locust-bean gum with native, deacetylated, depyruvated, and deacylated xanthan in a ratio of 1:2 at 0.2% total gums was measured with a rheogoniometer. Although the dynamic modulus of a mixture with deacylated xanthan showed a very large value at 20°C, as it did in a mixture with deacetylated xanthan, a mixture with depyruvated material stayed at a low value as was the case in a mixture with native xanthan. The dynamic modulus of the mixed solutions decreased rapidly with increasing temperature. The dynamic modulus of a mixture with deacylated xanthan stayed at very low values in the presence of urea (4.0 M) even at a temperature of 20°C. This suggests that hydrogen bonding may have a dominant role in the interaction, because urea is known as a hydrogen bond breaker.

Discussion

A transition temperature at which viscosity decreased rapidly was observed at 25 and 55°C at a concentration of 0.5% for the deacylated and depyruvated xanthan, respectively. This suggests that the xanthan molecule is involved in an intramolecular association even after deacylation.

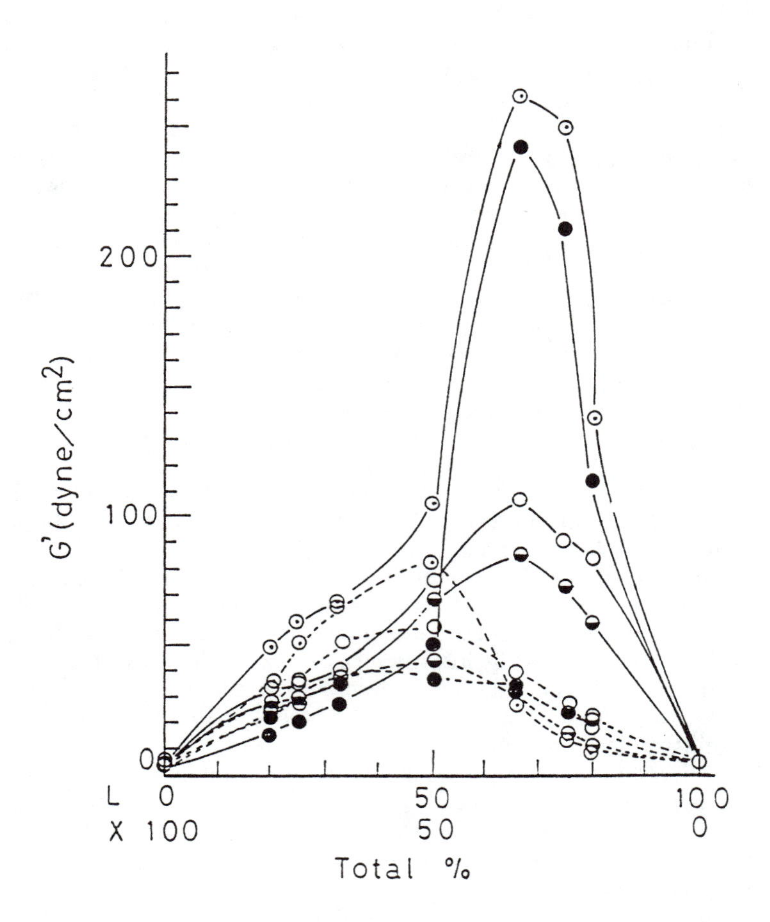

Figure 3. Dynamic modulus, at 3.77 rad/sec and 25°C, of a 0.2% xanthan-locust-bean gum solution as a function of the ratio of components: (○), native; (⊙), deacetylated; (◒), depyruvated; (●), deacylated xanthan. The full lines refer to the mixture of polysaccharide alone and the broken lines to the addition of $CaCl_2$ (6.8 mM). (L), locust-bean gum; (X), xanthan.
(Reproduced with permission from reference 35. Copyright 1991 Marcel Dekker.)

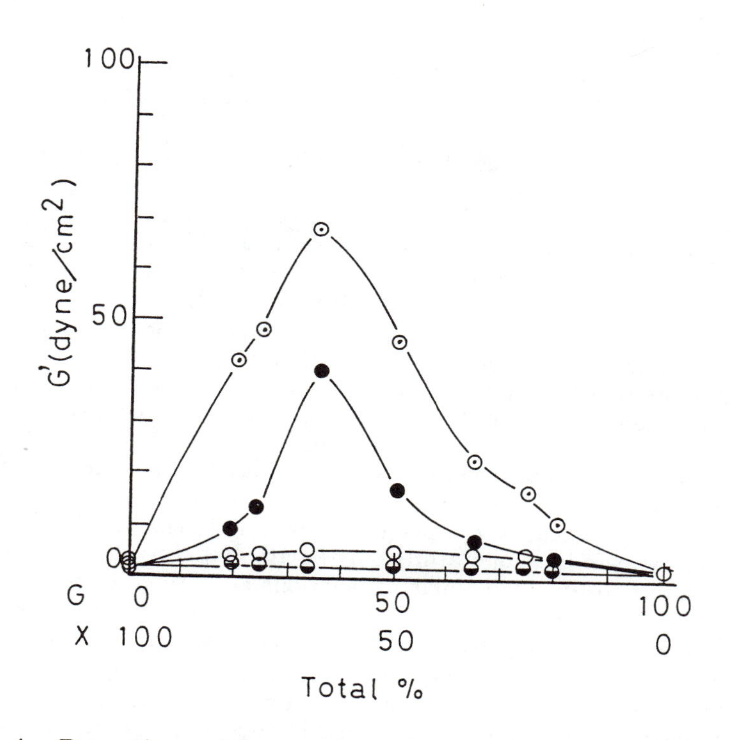

Figure 4. Dynamic modulus, at 3.77 rad/sec, of a 0.2% xanthan-guar gum solution as a function of the ratio of components: (○), native; (⊙), deacetylated; (◒), depyruvated; (●), deacylated xanthan. (G), guar gum; (X), xanthan.

Thus, as illustrated in Scheme 1, we conclude that the xanthan molecule is involved in an intramolecular association between an alternate hydroxyl group at C-3 and adjacent hemiacetal oxygen atom of the D-glucosyl residues with hydrogen bonding, as in the solid state, which makes a rigid conformation of the cellulose backbone. The alternate intramolecular hydrogen bonding of the xanthan molecule, after deacylation, may dissociate rapidly above the transition temperature (25°C). Though a molecule of the deacylated xanthan is involved in the alternate intramolecular hydrogen bonding, it seems to adopt a more flexible and extended conformation than that of the native and depyruvated xanthan. This may be due to the absence of another alternate intramolecular association, to which the acetyl groups may contribute (9, 24).

Thus, we also conclude that the methyl group of the acetyl residue may associate with the adjacent hemiacetal oxygen atom of the D-glucosyl residue with hydrogen bonding, as illustrated in Scheme 1. The hydrogen bonding may exist below transition temperatures of 55, 60, and 65°C, which resulted from the depyruvated xanthan at concentrations of 0.5, 0.8, and 1.0% as shown in Figure 1 and a previous study (24).

Though order-disorder transition on heating did not occur in the solution of the native and deacetylated xanthan (Figure 1) owing to the formation of an intermolecular association in which the methyl groups of the pyruvate residues might contribute as in methylcellulose molecules, it was observed in the presence of urea at a temperature of 65°C in a 0.5% solution. This indicates that the intermolecular association becomes more liable to dissociate above the transition temperature in the presence of urea. Accordingly, a sigmoid curve of apparent viscosity and dynamic viscoelasticity, during an increase in temperature, may be attributed to the breakdown of the two alternate intramolecular associations at a range of temperatures from 25-35°C or 55-65°C, and to the formation of the intermolecular association between side chains on different molecules at temperatures > 45°C.

The xanthan molecule may keep an ordered, rod-like, rigid, and less extended conformation owing to the formation of the intramolecular associations (Scheme 1) at room temperature in aqueous solutions. We suggest that the conformational transition of the xanthan molecules in aqueous solutions may consist of a three-step process on heating which may cause a sigmoid curve. Whether the xanthan molecules exist as a single- or double-stranded helix remains a topic of current debate (26-28), but the mechanism of the intramolecular associations corresponds to a single-stranded helix.

The synergistic interaction between xanthan and locust-bean gum has been confirmed from the results of a mixed solution of deacylated xanthan and locust-bean gum. Namely, in spite of 1/10 and free form substitution of pyruvic and acetic acid groups on the terminal and inner D-mannose side-chains of xanthan molecules, very strong gelation was observed, as in a mixture with deacetylated xanthan, at room temperature. This indicates that the pyruvate groups of xanthan molecules have no major role in the interaction with locust-bean gum molecules. The deacetylation of xanthan molecules marginally improved its gelation behavior even after depyruva-

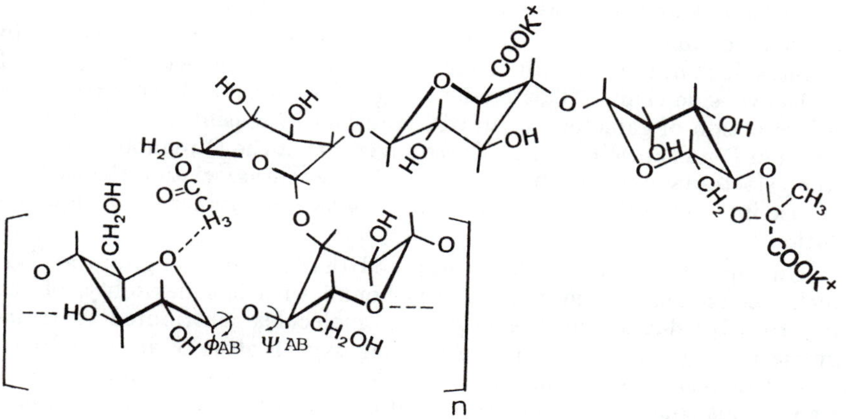

Scheme 1. Possible mode of intramolecular associations in xanthan molecule in aqueous solution. The dotted lines refer to the hydrogen bonding. The conformational changes in the interaction between alternate neighboring sugar residues are expressed in the terms of the two angles of rotation, ϕ_{AB} and ψ_{AB}. The tertiary structure of xanthan in aqueous solution may adopt five-fold, single-stranded helix as in the solid state. The xanthan molecule may keep an ordered, rod-like, rigid, and less extended conformation owing to the formation of the intramolecular associations at room temperature in aqueous solution.

(Reproduced with permission from reference 36. Copyright 1991 Marcel Dekker.)

tion with locust-bean gum. This indicates that the molecules of xanthan become more flexible and can associate with locust-bean gum more easily, probably because they were free from the intramolecular association. The breakdown of the gel by addition of $CaCl_2$ suggests that the univalent cations (K^+) on carboxyl groups of the D-glucuronic acid residues of the xanthan molecules seem to take part in the interaction with electrostatic force of attraction as in κ-carrageenan (29). Furthermore, the very small value of the dynamic modulus in the mixture of deacylated xanthan and locust-bean gum in the presence of urea (4.0 M) suggests that another association site seems to exist with hydrogen bonding.

Galactomannans have a characteristic of self-association themselves in aqueous solution, the strength of which increases with decreasing content of D-galactose side-chains. This tendency agrees with that of the strength of the interaction of galactomannan with xanthan increasing with decreasing content of D-galactose residues. These suggest that the D-mannose residues of the galactomannan molecules take part in the self-association themselves, and also interact with D-mannose residues of xanthan molecules in aqueous solution.

On the other hand, we have proposed that a hydroxyl group at C-2, which has an axial orientation and is bound with less flexibility, of the anhydro-α-L-galactose residue in agarose molecules contributed to an intermolecular hydrogen bonding on the ring oxygen atom of an anhydro-α-L-galactose residue with different molecules in aqueous solution (30). Such an analogous counter-association may also exist in the interaction between xanthan and galactomannan molecules.

Thus, we propose a possible binding site between xanthan and locust-bean gum in aqueous solution. The interaction may take place between the hemiacetal oxygen atom of the inner D-mannose side-chain of xanthan, which may be an acceptor, and the hydroxyl group at C-2, which may be a donor, of the D-mannose residue of mannan backbone of locust-bean gum with hydrogen bonding, as illustrated in Scheme 2. This bonding is likely owing to less flexibility of the side-chains of xanthan, which is substituted at C-3 of the D-glucose backbone, and of the hydroxyl group at C-2 at the axis of the mannan backbone of locust-bean gum molecules. Furthermore, the inner D-mannose residue of the side-chain of the xanthan molecule adopts an alternate pyranose-ring conformation due to bearing at axial C-2 the D-glucuronic acid residue, whose hemicacetal oxygen atom may attract the adjacent hydroxyl proton at C-2 of the mannan backbone of the locust-bean gum molecule by a cage effect (30, 31).

The univalent cation (K^+) which associates with the carboxyl oxygen atom with ionic bonding on D-glucuronic acid residue of the xanthan side-chain may also take part in the interaction with the adjacent hemiacetal oxygen atom of the mannan backbone of the locust-bean gum molecule with electrostatic force of attraction, as illustrated in Scheme 2. As the mannan backbone of the locust-bean gum molecule has a rigidity owing to an intramolecular hydrogen bonding, O(5)—O(3′) (32, 33), the side-chains of the xanthan molecule, the tertiary structure of which may keep a five-

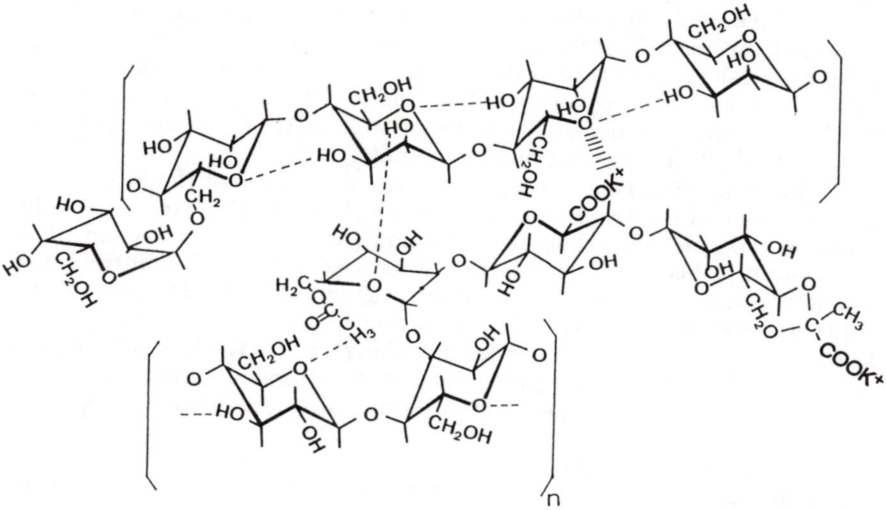

Scheme 2. Possible binding sites for D-mannose-specific interaction between xanthan and galactomannan in aqueous solution. The dotted and broken lines refer to hydrogen bonding and electrostatic force of attraction. As the tertiary structure of the xanthan molecule may keep a five-fold, single-stranded helix, its side-chains are inserted into the adjacent, unsubstituted segments of the backbone of the galactomannan molecule. A molecule of xanthan may combine with two or more molecules of galactomannan (locust-bean gum), the ratio depending on the favored conformation in aqueous solution. As the side-chains of the native and depyruvated xanthan molecules are somewhat rigid because of the intramolecular association contributed by acetyl groups, an incomplete interaction may exist in part and greater interaction may result from deacetylation.
(Reproduced with permission from reference 36. Copyright 1991 Elsevier.)

fold, single-stranded helix, are inserted into unsubstituted segments of the mannan backbone which is extended into a two-fold, ribbon-like structure in aqueous solution.

Such a binding mechanism provides an explanation for less interaction in a mixture with guar gum at room temperature. The side-chains of guar gum molecules may prevent the insertion of the charged trisaccharide side-chains of the xanthan molecule into the backbone of guar gum. However, synergistic interaction occurred with deacylated xanthan, as with deacetylated material, at room temperature, indicating that the xanthan molecules had become more flexible and could associate with the guar gum backbone more easily, probably because they were free from the intramolecular association to which the acetyl groups contribute.

Although it has been reported that the side-chains of the locust-bean gum molecule are distributed in uniform blocks along the backbone of the mannan molecule, the mode of interaction (Scheme 2) is independent of the structure, because each junction may take place within three sugar residues including the acetyl and pyruvate groups of the side-chains of the xanthan molecule. The stronger interaction in the mixture with deacetylated xanthan suggests some regular branching units in the locust-bean gum molecules. The analogous counterpart between D-mannose residues of the xanthan and locust-bean gum molecules may play a dominant role in the interaction.

Finally, the interaction between the extracellular bacterial polysaccharide, xanthan, and typical galactomannan components of the plant cell wall may suggest a part in the host-pathogen relationship, because *Xanthomonas campestris* is one of the plant-pathogen bacteria. Furthermore, the mode of interaction may provide the existence of D-mannose-specific binding sites in several cell recognition processes.

Experimental

The xanthan, locust-bean gum, and guar gum were identical with those used in our preceding studies (9, 10, 14, 16, 17, 22-24) and purified as previously described (22, 23, 34). Viscosity and dynamic viscoelasticity were measured with a rheogoniometer (IR 103, Iwamoto Seisakusho Co., Ltd.), and apparent viscosity (η), shear rate (D), shear stress (S), dynamic viscosity (η'), and dynamic modulus (G′) were calculated as described in a previous paper (34).

Literature Cited

1. Jeanes, A.; Pittsley, J. E.; Senti, F. R. *J. Appl. Polym. Sci.* 1961, **5**, 519-526.
2. Holzwarth, G. *Biochem.* 1976, **15**, 4333-4339.
3. Tako, M.; Nagahama, T.; Nomura, D. *Nippon Nogei-Kagaku Kaishi* 1977, **51**, 513-518.
4. Rees, D. A. *Biochem. J.* 1972, **126**, 257-273.
5. Kovacs, P. *Food Technol.* 1973, **27**, 26-30.

6. Jansson, P.-E.; Kenne, L.; Lindberg, B. *Carbohydr. Res.* 1975, **45**, 275-282.
7. Melton, L. D.; Mindt, L.; Rees, D. A.; Sanderson, G. R. *Carbohydr. Res.* 1976, **46**, 245-257.
8. Cadmus, M. C.; Rogovin, S. P.; Burton, K. A.; Pittsley, J. E.; Knutson, C. A.; Jeanes, A. *Can. J. Microbiol.* 1976, **22**, 942-948.
9. Tako, M.; Nakamura, S. *Agric. Biol. Chem.* 1984, **48**, 2987-2993.
10. Tako, M.; Nakamura, S. *Agric. Biol. Chem.* 1987, **51**, 2919-2923.
11. Baker, C. W.; Whistler, L. *Carbohydr. Res.* 1975, **45**, 237-243.
12. McCleary, B. V. *Carbohydr. Res.* 1979, **71**, 205-230.
13. Dea, I. C. M.; McKinnon, A. A.; Rees, D. A. *J. Mol. Biol.* 1972, **68**, 153-172.
14. Tako, M.; Nakamura, S. *Agric. Biol. Chem.* 1986, **50**, 2817-2822.
15. Dea, I. C. M.; Clark, A. H.; McCleary, B. V. *Carbohydr. Res.* 1986, **147**, 275-294.
16. Tako, M.; Nakamura, S. *Agric. Biol. Chem.* 1988, **52**, 1071-1072.
17. Tako, M.; Nakamura, S. *FEBS Lett.* 1986, **204**, 33-36.
18. Dea, I. C. M.; Morris, E. R.; Rees, D. A.; Welsh, E. J.; Barnes, H. A.; Price, J. *Carbohydr. Res.* 1977, **57**, 249-272.
19. Morris, E. R.; Rees, D. A.; Young, G.; Walkinshaw, M. D.; Darke, A. *J. Mol. Biol.* 1977, **110**, 1-16.
20. Cairns, P.; Miles, M. J.; Morris, V. J. *Nature* 1986, **322**, 89-90.
21. Cairns, P.; Miles, M. J.; Morris, V. J.; Brownsey, G. J. *Carbohydr. Res.* 1987, **160**, 411-423.
22. Tako, M.; Asato, A.; Nakamura, S. *Agric. Biol. Chem.* 1984, **48**, 2995-3000.
23. Tako, M.; Nakamura, S. *Carbohydr. Res.* 1985, **138**, 207-213.
24. Tako, M.; Nakamura, S. *Agric. Biol. Chem.* 1988, **52**, 1585-1586.
25. Hoffman, J.; Svensson, S. *Carbohydr. Res.* 1978, **65**, 65-71.
26. Moorhous, R.; Walkinshaw, M. D.; Arnott, S. *ACS Symp. Ser.* 1977, **45**, 90-102.
27. Norton, I. T.; Goodall, D. M.; Frangou, S. A.; Morris, E. R.; Rees, D. A. *J. Mol. Biol.* 1984, **175**, 371-394.
28. Sato, M.; Norisue, T.; Fujita, H. *Polym. J.* 1984, **16**, 341-350.
29. Tako, M.; Nakamura, S. *Carbohydr. Res.* 1986, **155**, 200-205.
30. Tako, M.; Nakamura, S. *Carbohydr. Res.* 1988, **180**, 277-284.
31. Tako, M.; Sakae, Ayano; Nakamura, S. *Agric. Biol. Chem.* 1989, **53**, 771-776.
32. Zugenmaier, P. *Biopolym.* 1974, **13**, 1127-1139.
33. Atkins, E. D. T.; Farnell, S.; Mackie, W.; Sheldrick, B. *Biopolym.* 1988, **27**, 1097-1105.
34. Tako, M.; Nakamura, S. *Agric. Biol. Chem.* 1989, **53**, 1941-1946.
35. Tako, M. *J. Carbohydr. Chem.* 1991, **10**, in press.
36. Tako, M. *J. Carbohydr. Chem.* 1991, **16**, 239-252

RECEIVED December 16, 1991

Chapter 18

Cure Analysis of Phenol–Formaldehyde Resins by Microdielectric Spectroscopy

Timothy G. Rials

Southern Forest Experiment Station, U.S. Department of Agriculture Forest Service, 2500 Shreveport Highway, Pineville, LA 71360

The influence of prepolymer functionality and pH conditions on the curing behavior of phenol-formaldehyde resins was studied by differential scanning calorimetry and microdielectric spectroscopy. Excellent agreement was found for these two analytical methods, readily distinguishing between the acid and alkaline cured systems. In terms of the rheology of the cure process, observed by changes in the resin conductivity, the greatest disparity was found between the two early stage resins when cured at acid pH. A much more rapid development of crosslinks and an apparently more extensive network structure observed for the earliest stage resin (PF-I) were attributed to the availability of more unsubstituted para positions on the aromatic ring for reaction with hydroxymethyl phenol groups. This effect was readily observable under both dynamic and isothermal curing regimes.

The curing reaction of thermosetting polymers is of fundamental significance in determining the structure and properties of the polymeric network. It also presents unique problems in terms of characterization, primarily as a result of the phase transformation from liquid prepolymer to an insoluble network. This conversion to an "infinite" molecular weight polymer effectively prohibits the use of many conventional analytical techniques to study the critical stage of the polymerization reaction occurring beyond gelation. In addition, materials that cure by condensation reactions further complicate analysis of the curing reaction by the inherent elimination of low molecular weight molecules. As a direct consequence of these difficulties, the formation of polymeric networks is a process that is adequately understood for only a few ideal systems.

Phenol-formaldehyde polymers, as a group, epitomize the unique problems associated with thermoset cure analysis. Despite their widespread use,

Published 1992 American Chemical Society

they remain one of the least understood systems in terms of their curing reactions and the rheology of the curing process. This limited understanding mainly stems from the complexity of the apparently simple curing reaction, and the array of synthesis variables which influence polymerization including formaldehyde to phenol ratio, catalyst content, pH, and cure temperature (1). Not surprisingly, a similar array of analytical tools have been used to address this question; however, only a few have been applied to that region beyond the gel point. Of these, differential scanning calorimetry remains the most routinely utilized method for characterization of phenolic resin cure (2,3). Over the years, however, several relaxation spectroscopy techniques have demonstrated considerable utility in elucidating the process of network formation in phenol-formaldehyde systems.

Perhaps the greatest contribution in terms of chemical structure definition has been provided by nuclear magnetic resonance spectroscopy (^{1}H and ^{13}C-NMR). Research in this area (4–6) has expanded with the increased capabilities of instrumentation and will, no doubt, continue to enhance the level of understanding of phenol-formaldehyde resin cure chemistry. The gap between chemical structure and material properties was first filled by Gillham and Lewis (7) with the development of the torsional braid analyzer, providing the means to generate the now famous time-temperature-transformation diagram for thermosetting polymers (8). Since that introduction, dynamic mechanical analysis has become increasingly popular in the study of phenolic resin cure (9,10), since it provides information directly on the ultimate question of strength development. While these approaches are of unquestionable value, they suffer from one common shortcoming: without exception, the acquisition of data requires the carefully controlled environment of the sample chamber.

The recently introduced version of dielectric analysis known as microdielectric spectroscopy provides an additional means of thermoset resin cure analysis. Although dielectric analysis is an age-old technique, recent advances in microsensor technology have elevated it to the status of "routine" analytical instrumentation while simultaneously providing the luxury of remote data acquisition. The dielectric parameters, permittivity and dielectric loss factor, measured by this technique can be related to more common measures, such as viscosity, making it ideally suited for both fundamental and applied investigations of the chemorheology of the curing reaction. This has recently been demonstrated for epoxies (11) and urethane/acrylate interpenetrating networks (12). This paper describes some introductory investigations on the application of μ-dielectric spectroscopy for characterizing the cure properties of phenolic resins as influenced by both pH of the cure environment and chemical structure of the prepolymer.

Experimental

Resin Synthesis. The phenol-formaldehyde prepolymer studied is an experimental laboratory preparation. The reaction kettle was charged with 115.12 g of an 88% phenol solution and 200.7 g of 37% formaldehyde

(F/P= 2.3), and the reaction mixture was slowly heated. As seen in Figure 1, two samples were isolated from the reaction prior to the addition of 0.44 moles of NaOH (PF-I, PF-II). Four resins were sampled after the addition of caustic (PF-III through PF-VI).

Chemical Structure Analysis. The phenol-formaldehyde prepolymers were freeze-dried, and a portion of each was acetylated using excess pyridine-acetic anhydride for 48 hours at ambient temperature. The acetate derivative was prepared for NMR analysis by extraction into dichloromethane from water, followed by repeated evaporation of toluene to remove residual acetic acid and pyridine and was then dissolved in deuterated chloroform. Both ^{1}H and ^{13}C NMR analyses were performed on a Varian FT-80 spectrometer at 80 MHz and 20 MHz, respectively. Spectra were analyzed according to published procedures (6).

Curing Reaction Analysis. A Perkin-Elmer DSC-II differential scanning calorimeter was used to monitor the heat evolution of the curing reaction. Approximately 5–7 mg of the freeze-dried, phenol-formaldehyde prepolymer was sealed in a large volume, stainless steel capsule and heated at a programmed rate of 10 deg/min under a dry nitrogen purge. Reaction peak areas were determined from a digitized image obtained with an Omnicon Image Analyzer, and the heat of reaction (ΔH_{rxn}) calculated using standard procedure. Indium was used as the calibration standard. Additionally, reaction peak temperatures were determined for samples analyzed at heating rates of 2, 5, 10, 20 and 40 deg/min and activation energies (E_a) calculated from the relationship (13):

$$E_a = -\left(\frac{R}{0.457}\right)\left(\frac{\Delta \log \phi}{\Delta 1/T_p}\right) \qquad (1)$$

where R is the gas constant, ϕ is the programmed heating rate, and T_p is the temperature of the reaction peak maximum.

Dielectric analysis was performed with a Eumetric System III Microdielectrometer used in conjunction with a Eumetric Programmable Oven manufactured by Micromet Instruments, Inc. An IBM-PC personal computer was interfaced to the system for instrument control and data acquisition. Measurements were made at a constant frequency of 1000 Hz using a high conductivity microsensor (active face dimensions = 0.4 × 0.3 × 0.006 in.; L × W × T). This particular dielectric sensor consists of two shielded electrodes which measure the current passing between them for a given excitation voltage. From the measured magnitude and phase of the output voltage, the Microdielectrometer calculates the conductance and capacitance of the material, corrected for electrode polarization effects, between the electrodes. Since the sampling area is fixed by the sensor, this information is readily converted to the microscopic parameter, conductivity.

The sampling procedure simply involved uniformly distributing the sample on the sensor face and applying a slight pressure to ensure adequate contact throughout the analysis. The dielectric response of the curing prepolymer was studied both as a function of temperature from 30°C

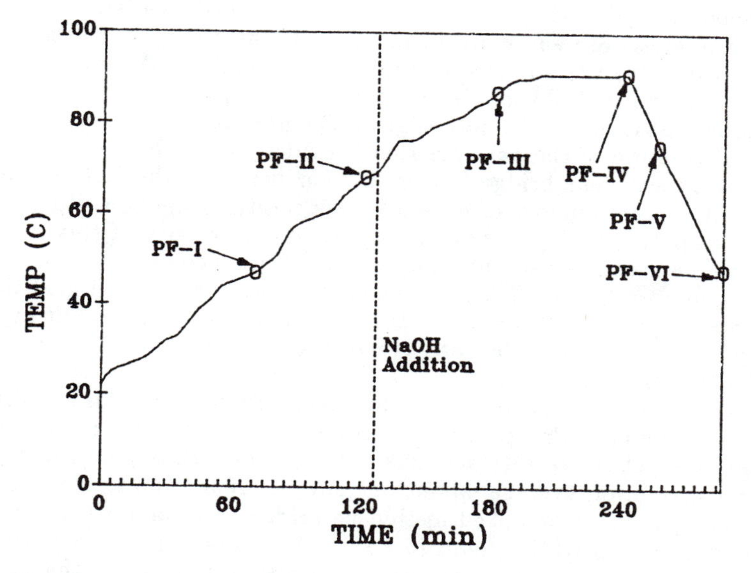

Figure 1. Temperature schedule of phenolic resin synthesis showing the origin of the six prepolymers studied.

to 200°C (heated at 5°C/min) and as a function of time at 150°C. A dry nitrogen atmosphere was used in all experiments. It should be noted that no direct influence of reaction water on the electrode-resin interface was observed; and, although the plasticizing effect of water undoubtedly influences the ionic mobility of the resin, no attempt to resolve this contribution was made.

Results and Discussion

Information on the total number of methylol and methylene functions is available from ^{1}H-NMR data, while substituent position (*ortho, para*) to the phenolic hydroxyl can be derived from ^{13}C-NMR spectra. The results of this analysis are summarized in Table I. Interestingly, the two resins isolated prior to the addition of caustic, PF-I and PF-II, have a substantial hydroxymethyl phenol (HMP) content, occupying about 1.7 sites per phenolic repeat unit with only slight differences between them. The preferential reaction of the *para* position is evidenced by the almost exclusive *para-para* methylene bridge formation. Furthermore, the fact that almost 80% of the total number of available *para* positions are substituted in the PF-II prepolymer emphasizes the preferential reactivity of this ring position, which is somewhat surprising given the absence of any catalyst. It is important to note that NMR revealed the presence of small quantities of hemiformal in both PF-I and PF-II that were not quantified, but highlight the acidic conditions of the early stage reaction (which will be carried over to the cure environment).

After the addition of NaOH to the reaction, a similar substitution pattern is found. The presence of residual *p*-HMP functionality all but disappears, yet the *o*-HMP remains fairly constant throughout the synthesis, occupying, on average, about one site per repeat unit. No significant amount of *ortho-ortho* linked methylene bridge appears until the complete consumption of *p*-HMP groups in PF-IV, but then remains relatively constant as the reaction progresses. Also, it is not until this point that any substantial increase in the number average degree of polymerization, $<X_n>$, is observed.

The influence of these structure variations and the pH of the system on the curing properties of the prepolymers is readily apparent from the results of differential scanning calorimetry analysis (Table II). It is interesting to note that very little variation in the activation energy is observed for the prepolymers, even though the pH conditions of the curing reactions are very different between resins PF-I and PF-II, and PF-III to PF-VI. There is, however, an inexplicably large difference in reaction peak temperature found for the two acidic prepolymers (PF-I and PF-II). Overall, the resins cured in an alkaline environment consistently exhibit a reaction peak temperature around 130°C, and an E_a of 20 kcal/mol. In spite of the different curing environments (*e.g.*, acid *vs.* alkaline), the onset of the curing reaction increases consistently from PF-I to PF-VI as the degree of polymerization of the prepolymer increases with cook time. Also, as would be expected, there is a gradual decrease in the heat of reaction as the cook

Table I. Summary of number average repeat unit structure and molecular size of phenol-formaldehyde prepolymers determined by nuclear magnetic resonance spectroscopy (^{1}H- and ^{13}C-NMR)

	Average Repeat Unit Structure									Molecular Size	
			$Ar\text{-}CH_2\text{-}O\text{-}H$			$Ar\text{-}CH_2\text{-}Ar$					
Resin ID	Ar-O-H	Ar-H	*para*	*ortho*	Total	*p-p*	*o-p*	*o-o*	Total	$\langle X_n \rangle$	$\langle M_n \rangle$
PF-I	1	3.19	0.45	1.22	1.67	0.10	0.04	0.00	0.14	1.07	156.1
PF-II	1	2.91	0.43	1.27	1.70	0.27	0.07	0.05	0.39	1.24	182.8
PF-III	1	3.03	0.23	1.18	1.41	0.32	0.21	0.03	0.56	1.39	199.6
PF-IV	1	3.07	0.04	0.90	0.94	0.36	0.44	0.19	1.00	2.00	270.4
PF-V	1	2.94	0.09	0.91	0.97	0.41	0.48	0.20	1.09	2.20	301.9
PF-VI	1	2.80	0.04	0.91	0.95	0.44	0.50	0.21	1.25	2.67	370.5

Table II. Analysis of phenol formaldehyde prepolymer curing reaction by differential scanning calorimetry (scan rate = 10 °C·min^{-1})

RESIN ID	TEMPERATURE (° C) ONSET	PEAK	ΔH_{rxn} (cal/g)	E_a (kcal/mol)
PF-I	60	140	5.65	23.7*
PF-II	65	127	5.50	19.2*
PF-III	78	134	4.54	21.2
PF-IV	80	129	2.25	23.2
PF-V	98	131	1.91	20.3
PF-VI	98	131	1.85	21.4

*Does not include programmed scan of 40°·min^{-1}.

time is increased. Again, although cured under dramatically different pH conditions, the residual cure of the resin closely parallels the hydroxymethyl phenol content determined by NMR. The conclusion to be drawn from the differential scanning calorimetry results is that while the degree of cure of the resins is closely related to residual HMP content, the characteristics of the curing reaction depend primarily on pH conditions. The extent to which this distinction impacts the chemorheology of phenolic resin cure is of considerable interest.

The application of dielectric spectroscopy to thermoset cure typically focuses on changes in the dielectric loss factor, ϵ'', rather than permittivity. The dielectric loss can be further resolved into its component dipole and ionic terms through the relation (11,14):

$$\epsilon'' = \left[\frac{\sigma}{\epsilon_o \omega}\right] + \left[\frac{(\epsilon_r - \epsilon_u)(\omega\tau)}{1+(\omega\tau)^2}\right] \tag{2}$$

where ϵ_r and ϵ_u = relaxed and unrelaxed permittivity, respectively; ϵ_o = permittivity of free space (8.85×10^{-14} F/cm); ω = applied field frequency; τ = dipole relaxation time; and σ = ionic conductivity $(\Omega/\text{cm})^{-1}$. The ionic conductivity is of particular utility in evaluating thermoset cure since it is a measure of the mobility of ions in a medium. Consequently, this term is inversely proportional to viscosity prior to gelation and matrix rigidity (*i.e.*, distance from the glass transition temperature) after gelation.

The dielectric response (log conductivity) as a function of temperature is compared with the reaction exotherm obtained by DSC in Figure 2 for the PF-II prepolymer. As the resin is heated, the conductivity increases rapidly until a slight slope change is observed at 65°C, corresponding to the onset of reaction as observed calorimetrically. There continues a net increase in conductivity until a maximum value is reached at about 110°C. Analagous to a viscosity minimum, this point appears to be related to a low temperature shoulder in the thermogram. It is particularly interesting that the conductivity remains relatively stable from this point up to about 140°C. It is not until the DSC exotherm nears completion that a rapid decrease in conductivity occurs, indicative of increased crosslinking and network development. Beyond this point, a final rapid drop in conductivity occurs as the temperature exceeds the network polymer's glass transition temperature (T_g), observed at slightly lower temperatures in the thermogram, and a final cure advance is achieved.

This same dielectric curing profile, relative to the other phenolic prepolymer systems, is presented in Figure 3. One of the more surprising aspects is the dramatic difference in the response of PF-I and PF-II which were both sampled prior to the addition of caustic to the cook. The initial difference in conductivity can be attributed to the slight molecular weight difference between the two prepolymers and its influence on the physical state of the material (*i.e.*, PF-I is liquid and PF-II is a gel). With the onset of cure at about 60°C, there is an immediate decline in conductivity, again suggesting a viscosity increase resulting from the increase in molecular

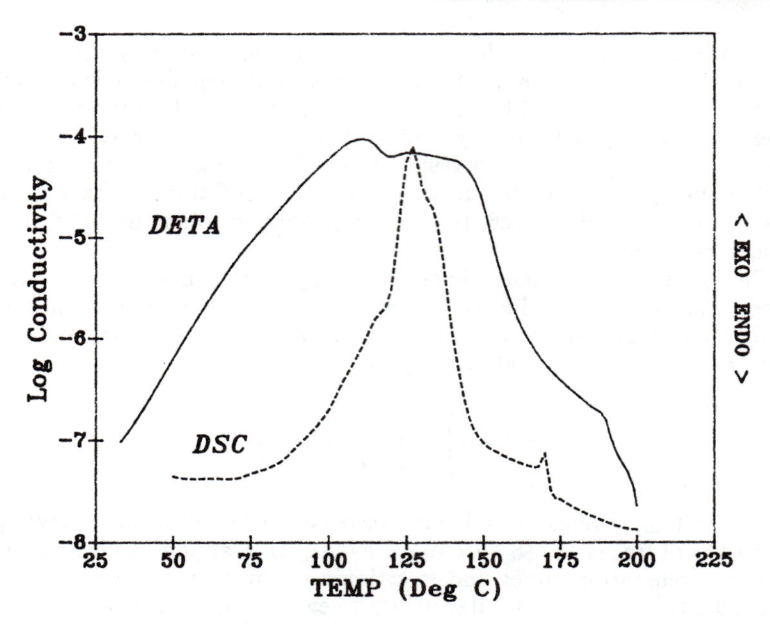

Figure 2. Comparison of the dielectric response (log conductivity) with the curing exotherm from DSC for prepolymer PF-II.

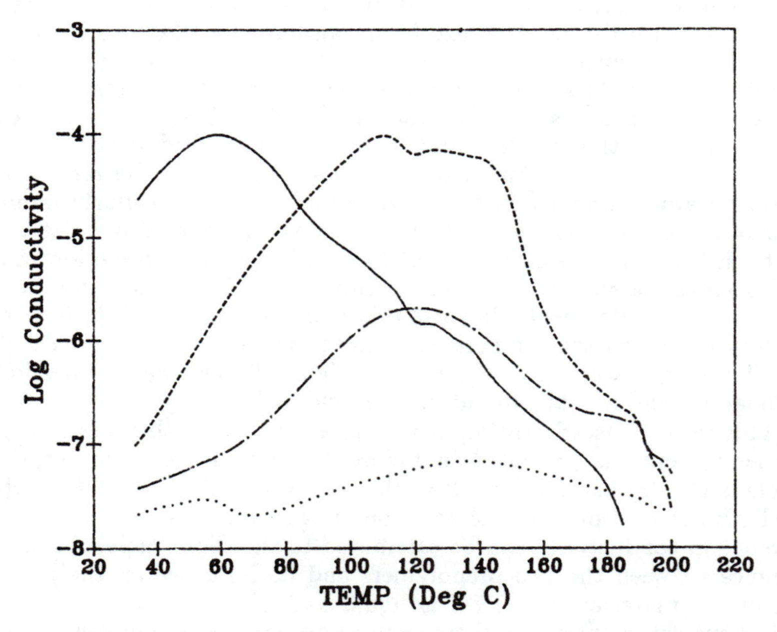

Figure 3. The variation in conductivity as a function of temperature for the phenolic prepolymers: PF-I (——), PF-II (— — —), PF-III (— · —), and PF-VI (···).

weight. A similar profile is found in both resins over the temperature region 110°C to 140°C, possibly attributable to the conversion of dimethylene ether linkage intermediates. It is not possible to account for the difference in chemorheological behavior in terms of residual functionality alone, as the hydroxymethyl phenol content of these two prepolymers is essentially identical. The primary distinction between these resins is the number of aromatic protons at the *para* position available for reaction which could account for the observed discrepancies in the curing profile.

Further comparison between PF-II and the alkaline cure systems reveals surprisingly similar profiles despite the difference in pH. The conductivity maximum in PF-III is observed at 120°C, essentially identical in position as PF-II; however, this characteristic point is shifted some 20° higher in temperature for the final resin, PF-VI. Generally, the maximum conductivity value that is achieved decreases from −4.0 to −7.2 and is primarily a reflection of the level of advancement, or cure stage, of the phenol-formaldehyde prepolymer. It is of interest, as well, that an additional peak is found at low temperatures for PF-VI which arises from the initial softening of this solid powder resin, emphasizing the importance of molecular mobility on the crosslinking reaction.

Often, the question of greater practical consequence is that of curing properties under isothermal conditions. In addressing this consideration, the prepolymers were rapidly heated to 150°C and the conductivity monitored over time. The results of this temperature regime are shown in Figure 4 for all six phenolic resins studied. Resins PF-II and PF-VI exhibit remarkably similar cure profiles under these conditions. In all of these materials, the peak maximum (viscosity minimum) corresponds to the point where the isothermal temperature (150°C) was reached. The time required for completion of the curing reaction decreases from about 30 minutes for PF-II to about 15 minutes for resins PF-V and PF-VI, and appears to be related to the degree of crosslinking that occurred prior to analysis. Similarly, with the decline in HMP content of the resin the intensity of the conductivity peak falls off considerably. The relationship between residual cure as determined by the difference between the conductivity maximum and minimum is illustrated in Figure 5. Although failing to account for contributions from reaction water or slight temperature variations, this simplified analysis provides a relatively convenient measure of residual cure. A similar relationship can be generated from the DSC data (ΔH_{rxn}), as shown. The increasing discrepancy at lower functionality between the two methods is to be expected since the dielectric data are generated at 150°C, and the DSC data are obtained at temperatures up to 210°C. Consequently, a higher residual cure is observed at the higher temperature, the effects of which are magnified for the more advanced resins.

Consistent with observations made under the dynamic temperature schedule, the PF-I prepolymer again distinguishes itself. Relative to PF-II a similar conductivity maximum is obtained; however, this peak is reached at an earlier time than the remaining resins. Also, the initial stage of the curing reaction proceeds much more quickly as almost 75% of the reaction

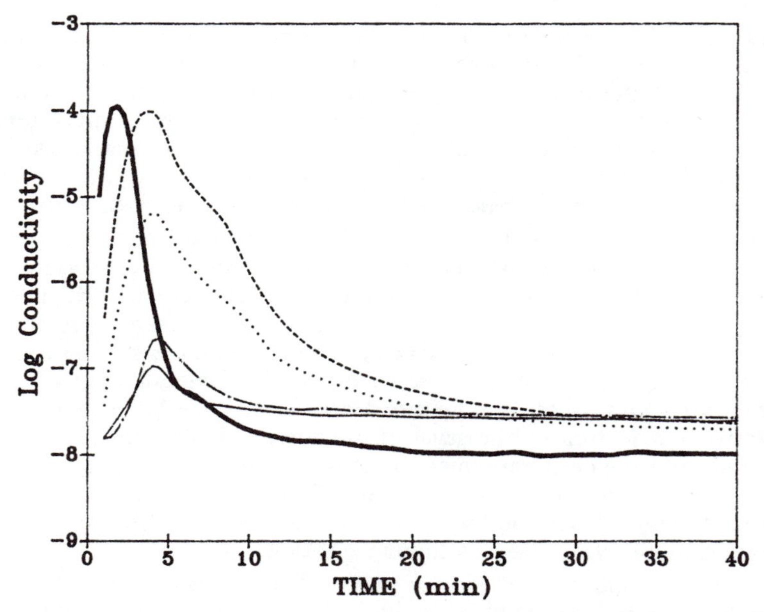

Figure 4. The variation in conductivity as a function of time at 150°C for the phenolic prepolymers: PF-I (▬▬), PF-II (— — —), PF-III (···), PF-IV (— · —), and PF-VI (——).

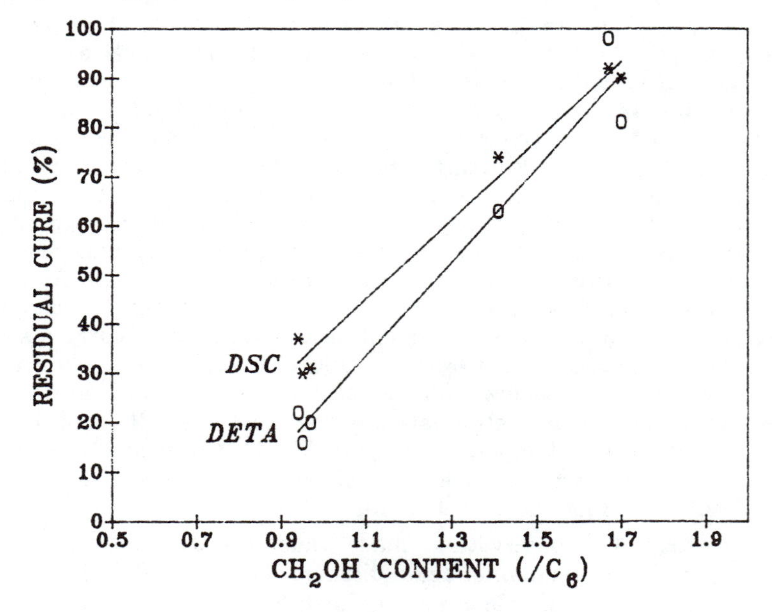

Figure 5. The relationship between residual cure of the phenolic prepolymer determined by DSC and dielectric analysis, and hydroxymethyl phenol content.

occurred at 5 minutes and completion was reached in only 20 minutes. The PF-I resin is also distinguished by a substantially lower conductivity for the cured material, suggesting a much higher level of crosslinking in the final network structure. Although the influence of pH conditions cannot be ruled out, its contribution to the overall chemorheological response appears to be of secondary importance, at least in distinguishing the early stage resin PF-I. Rather, the difference in the curing behavior of this system, as revealed under both dynamic (Figure 3) and isothermal conditions, would appear more readily attributable to chemical structure. Both PF-I and PF-II contain similar levels of residual HMP functionality, but the number of available *para* positions on the aromatic ring is about 20% greater for PF-I. Since a methylene is produced more readily by reacting directly with an unsubstituted *ortho* or *para* position on the phenolic ring rather than through formation and decomposition of a dimethylene ether, this would lead to a more substantial molecular weight increase at the early stages of cure and a more rapid viscosity increase (15,16). Consideration of this point, coupled with the initially higher molecular mobility of this liquid resin, could account for the distinctive curing profile, particularly at the early stage of the curing reaction.

Conclusions

Phenol-formaldehyde resins isolated from a synthesis employing a slow temperature ramp and delayed addition of catalyst provided materials varying considerably in their functionality, as well as the pH of the cure environment. Analysis by μ-dielectric spectroscopy substantiated thermal analysis results as clear differences in the rheological profile of the resins cured under different pH conditions were observed. Surprisingly, the greatest disparity was found to occur between the two acid-cured resins. PF-I, with a higher proportion of unsubstituted *para* positions, gave a faster and more extensive reaction. Although similar in terms of hydroxymethyl phenol functionality, the number of *para* positions available for reaction was substantially lower (about 20%) for the PF-II prepolymer. From these results, it appears that the type of residual functionality of the prepolymer has a greater influence on the net chemorheological response than does pH.

The limited data presented here leave considerable room for speculation regarding the origin of certain events observed by μ-dielectric spectroscopy. It does, however, illustrate the utility of this method to the study of phenol-formaldehyde resin cure rheology. Research in this laboratory is continuing to focus on the comprehensive definition of phenolic resin cure using microdielectric spectroscopy.

Acknowledgments

The author would like to thank Dr. R. W. Hemingway and Dr. G. W. McGraw for their assistance in NMR data analysis, as well as some very helpful discussion.

The use of trade, firm, or corporation names in this publication is for the information and convenience of the reader. Such use does not constitute an official endorsement or approval by the U.S. Department of Agriculture of any product or service to the exclusion of others that may be suitable.

Literature Cited

1. Knop, A.; Scheib, W. Springer-Verlag: Berlin, Heidelberg, 1979.
2. Katovic, Z. *J. Appl. Polym. Sci.* 1967, **11**, 85-93.
3. Chow, S. *Holzforschung* 1972, **26**, 229-232.
4. Kim, M. G.; Tiedeman, G. T.; Amos, L. W. In *Phenolic Resins: Chemistry and Applications*; Weyerhaeuser Science Symp. 2, Weyerhaeuser Co.: Tacoma, WA, 1981; pp 263-289.
5. Maciel, G. E.; Chuang, I.-S.; Gollob, L. *Macromolecules* 1984, **17**, 1081-1087.
6. McGraw, G. W.; Landucci, L. L.; Ohara, S.; Hemingway, R. W. *J. Wood Chem. and Technol.* 1989, **9**, 201-217.
7. Gillham, J. K.; Lewis, A. F. *J. Appl. Polym. Sci.* 1963, **7**, 2293-2306.
8. Gillham, J. K. *Polym. Eng. Sci.* 1979, **19**, 676-682.
9. Kelley, S. S.; Gollob, L.; Wellons, J. D. *Holzforschung* 1986, **40**, 303-308.
10. Steiner, P. R.; Warren, S. R. *Holzforschung* 1981, **35**, 273-278.
11. Day, D. R. *Dielectric Properties of Polymeric Materials*; Micromet Instruments, Inc.: Cambridge, MA, 1987.
12. Holmes, B. S.; Tiask, C. A. *J. Appl. Polym. Sci.* 1988, **35**, 1399-1408.
13. Muller, P. C.; Kelley, S. S.; Glasser, W. G. *J. Adhesion* 1984, **17**, 185-206.
14. Danial, V. *Dielectric Relaxation*; Academic Press: New York, 1977.
15. Ishida, S. In *Phenolic Resins: Chemistry and Applications*; Weyerhaeuser Science Symp. 2, Weyerhaeuser Co.: Tacoma, WA, 1981; pp 7-16.
16. Megson, N. J. L. *Phenolic Resin Chemistry*; Butterworth's Scientific Publications: London, 1958.

RECEIVED December 16, 1991

Relaxation Phenomena

Chapter 19

Conformation and Dynamics of (1→3)-β-D-Glucans in the Solid and Gel State

High-Resolution Solid-State ^{13}C NMR Spectroscopic Study

Hazime Saitô[1]

Biophysics Division, National Cancer Center Research Institute, Tsukiji 5-chome, Chuo-ku, Tokyo 104, Japan

Gel network of $(1 \rightarrow 3)$-β-d-glucans turned out to be highly heterogeneous from its motional state, from liquid-like, through intermediate, to solid-like domains. They were analyzed by a variety of NMR experiments. We found that conformations of these domains in a gel of a linear glucan (curdlan) exhibits an identical single helix conformation with a low proportion of a triple helix with reference to the data obtained in the solid state. Cross links of curdlan gel are thus ascribed to the triple helical junction as well as pseudo-cross-links formed by association of the single helical chains. In contrast, it was found that the network structure of branched glucans arose mainly from the triple helical chains. Gelation of the branched glucans thus proceeds from partial association of the triple helical chains.

The network structure of gels is generally highly heterogeneous from structural and dynamic points of view. The existence of solid-like domains from cross-linked region is characteristic of a gel formation but tends to prevent one from performing a structural characterization of gels in detail. Gelation of polysaccharides occurs as a result of physical association of chains adopting an ordered conformation. Thus, conformational elucidation of solid-like domains as well as liquid-like domains is essential for better understanding of network structures in polysaccharide gels.

We have previously demonstrated that ^{13}C-NMR spectra of liquid-like domains of gels consisting of synthetic polymers and $(1 \rightarrow 3)$-β-d-glucans are readily visible by a conventional high-resolution ^{13}C NMR spectrometer (1-4). Conformational features of the solid-like domains, however, were inevitably inferred from the data of the liquid-like domains, because ^{13}C NMR signals were completely suppressed. Naturally, conformation of these

[1]Current address: Department of Life Science, Himeji Institute of Technology, Harima Science Garden City, Kamigori-cho, Hyogo 678–12, Japan

0097–6156/92/0489–0296$06.00/0

domains can be elucidated by x-ray diffraction studies as in the cases of (1 → 3)-β-d-glucan (5-8), agarose (9), and carrageenans (10,11). It is not always guaranteed, however, that conformation of the solid-like domains is identical to that of the liquid-like domains. In fact, many of the gelation mechanisms of polysaccharide gels based on x-ray diffraction studies have been questioned (4, 12-14): the role of multiple-stranded helices has been overemphasized in the cases of curdlan, agarose, and carrageenan.

It is pointed out that direct NMR observation of the solid-like domains has recently been made possible (14,15) by the development of a highly stable magic-angle spinning system using a rotor of the double air bearing type. In the present article, we summarize our previous results on conformation and dynamics of (1 → 3)-β-d-glucans in the solid and gel states (15-22) by conventional and high-resolution solid-state ^{13}C NMR, with emphasis on revealing gelation mechanisms.

Liquid-Like Domains of Gels from Linear and Branched Glucans

Curdlan (I, Figure 1), a bacterial linear (1 → 3)-β-d-glucan of high molecular weight from *Alcaligenes faecalis*, is unique in its ability to form an elastic gel when its aqueous suspension is heated to a temperature above 54°C (23). This polysaccharide is now under production of > 100 ton/year by Takeda Chemical Industries, Osaka, to be used for food additives. Lentinan, from *Lentinus edodes* (Ajinomoto Co., Inc., Tokyo) (24), and schizophyllan, from *Schizophyllum commune* (Taito Co., Tokyo) (25), are fungal branched (1 → 3)-β-d-glucans (II, Figure 1), with branches of single glucose for every one to three glucose residues, and are clinically used in Japan as an anti-cancer drug in combination with other therapeutic agents. Lentinan, schizophyllan, HA-β-glucan (26), and other branched glucans are also equally able to form gels of brittle types under a higher concentration in aqueous media (~10%).

It is pointed out that rather sharp ^{13}C signals are visible from the liquid-like domains of curdlan gel by broad band decoupling on a conventional high-resolution NMR spectrometer (Figure 2B) (1,4), whereas high-resolution NMR signals are almost completely suppressed for brittle gels of the branched glucans (Figure 3, bottom trace) (27,28). The ^{13}C NMR signals of the C-1 and C-3 carbons at the glycosidic linkages from the liquid-like domains are appreciably displaced from those of oligomers (numerically averaged degree of polymerization, DPn, 14) taking random coil conformation (1,2,4). Such a displacement of peaks was ascribed to the presence of single helix conformation on the basis of the conformation-dependent displacement of peaks (29,30). The existence of the single helix conformation is also based on the following results: (1) the absorption maximum of Congo Red in the visible region was largely shifted to a longer wavelength by the addition of (1 → 3)-β-d-glucan of high molecular weight (31); (2) the annealed film of curdlan was considerably less stained with Congo Red than the original curdlan (32); and (3) about 70% of the heated curdlan gel (120°C/4 h) became resistant to treatment by endo(1 → 3)-β-d-glucanase or 32% sulfuric acid, although only 0-16% was resistant by

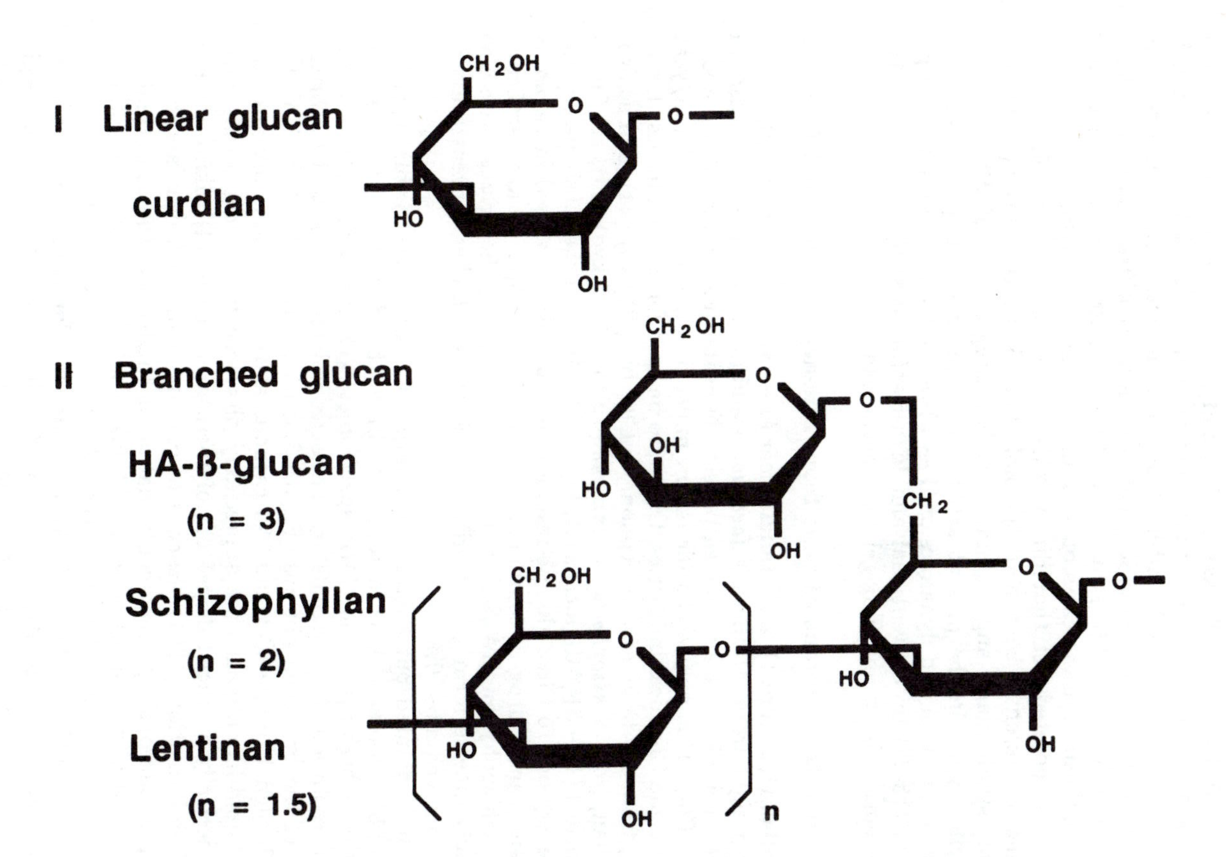

Figure 1. Chemical structures of linear and branched (1 → 3)-β-d-glucans.

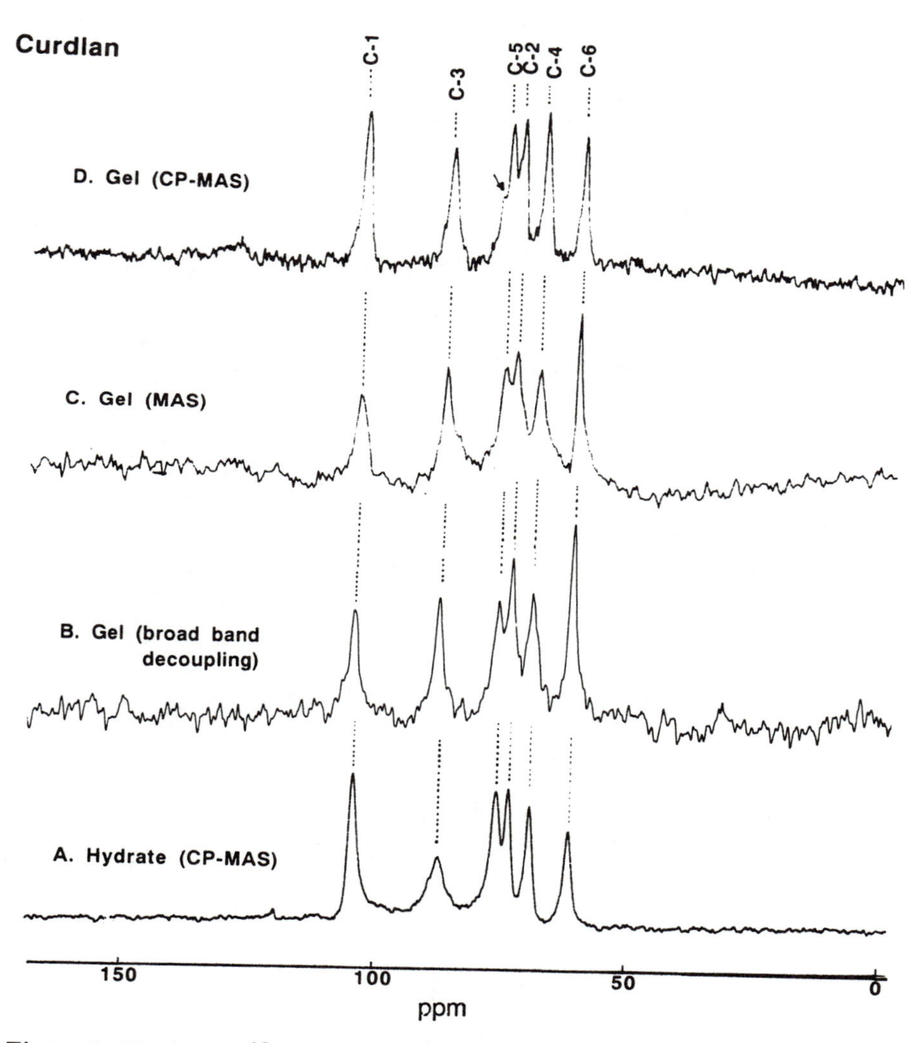

Figure 2. 75.46 MHz ^{13}C NMR spectra of curdlan gels recorded by a variety of experimental conditions (15). A and D, CP-MAS NMR technique; B, conventional NMR using broad-band decoupling; C, MAS NMR together with high power decoupling.

(Reproduced with permission from reference 15. Copyright 1990 Wiley.)

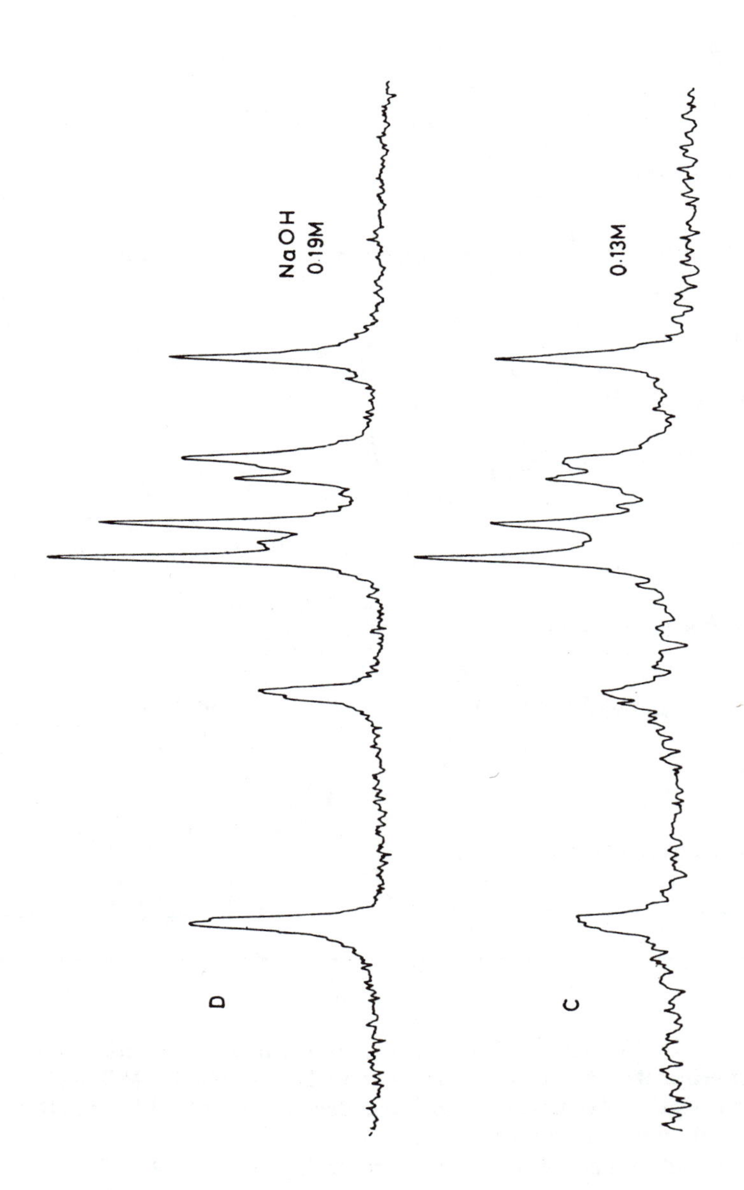
D
NaOH
0.19M
C
0.13M

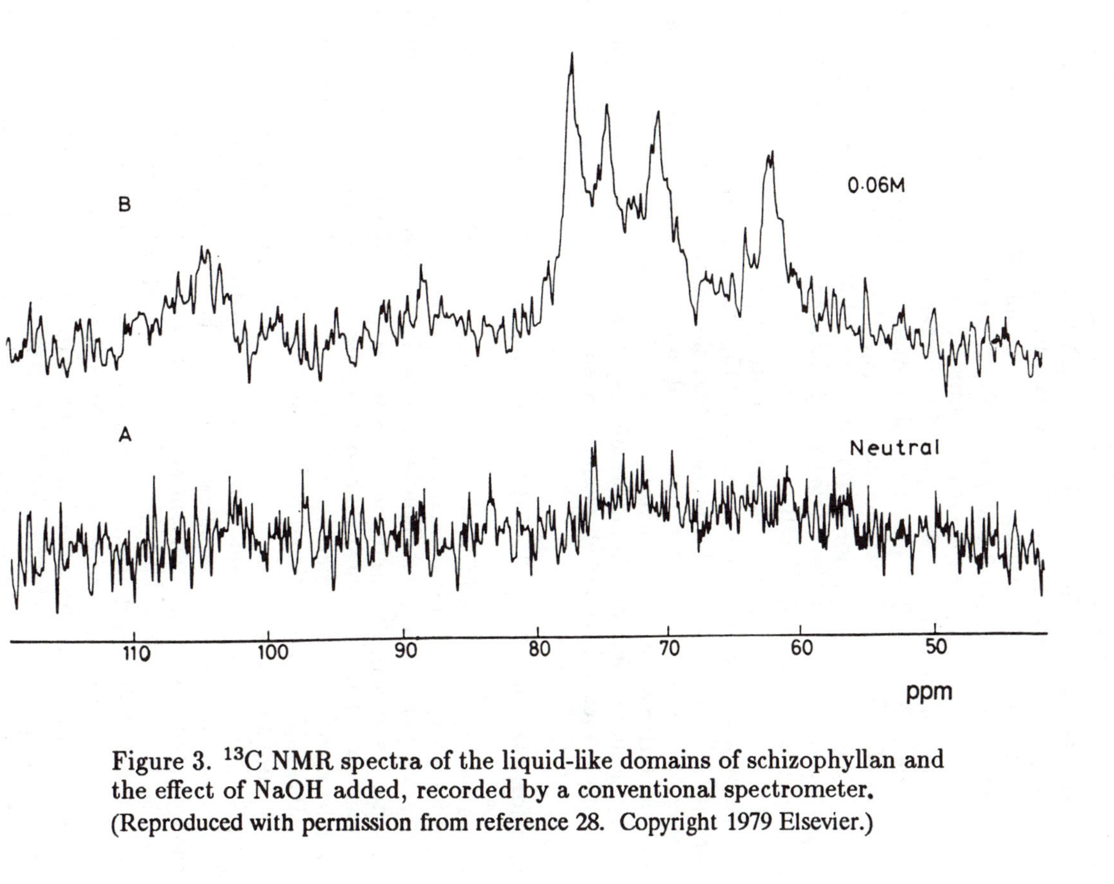

Figure 3. ^{13}C NMR spectra of the liquid-like domains of schizophyllan and the effect of NaOH added, recorded by a conventional spectrometer. (Reproduced with permission from reference 28. Copyright 1979 Elsevier.)

heating at a lower temperature (60°C/30 min) (33). Nevertheless, the single helix conformation was reported to be very unstable by a calculation of conformational energies corresponding to the anhydrous state (34, 35). This is not true, however, for the hydrated state, as will be discussed later.

Peak intensities of high-resolution ^{13}C NMR signals from the branched glucans, however, were gradually recovered by the addition of NaOH, in proportion to its concentration (between 0.06 and 0.2 M), and peaks were substantially narrowed at 0.2 M NaOH (4, 27-28) (Figure 3). This observation can now be well explained in terms of the following two steps of conformational changes, in the light of our recent data of high-resolution solid-state ^{13}C NMR: The first gradual recovery of the peaks is explained in terms of partial conformational change from the triple to the single helix form initiated by the addition of NaOH up to 0.2 M NaOH. The final conformational change from the single helix to the random coil form is achieved by the addition of 0.2 M NaOH.

These distinct spectral changes indicate that the gelation mechanism differs between the two types of glucans (15): the network of curdlan gel consists mainly of the flexible single helical chains, whereas branched glucans consist of the rather rigid triple helical chains. Crosslinks of the former and the latter are a small proportion (ca. 10%) of the triple helical junctions together with hydrophobic association of the single helical chains, and partially associated triple helical chains, respectively.

Conformations of (1 → 3)-β-d-Glucans in the Solid State

It has been demonstrated (29, 30) that ^{13}C chemical shifts of backbone carbons for a variety of biopolymers such as polypeptide and polysaccharides in the solid state are substantially displaced (up to 8 ppm but 12 ppm in some instances), depending on their respective conformations, and can be used as a diagnostic means for conformational elucidation for the solid and gel states. Such distinctions could be obscured or lost in solution by a plausible rapid conformational isomerism, however.

Figure 4 illustrates how the ^{13}C NMR spectral feature of curdlan varies with the condition of sample preparation: anhydrous, hydrated, and annealed samples (21). It is noteworthy that the ^{13}C NMR linewidths of the hydrated sample are substantially narrowed from those of the annealed sample. In general, broadened peaks in the solid state are ascribed to the superposition of slightly different conformations in view of the conformation-dependent displacement of peaks (29, 30). On the contrary, the peak narrowings in hydrated and annealed samples are ascribed to the presence of better crystalline packing or molecular ordering. At first glance, it appears that the ^{13}C chemical shifts of the hydrated sample are similar to those of the annealed curdlan. However, distinction of the spectra between the hydrated and annealed curdlan is straightforward on closer examination of the C-5 chemical shifts (75.8 for the former and 77.5 ppm for the latter, respectively), and the peak-separation between the C-5 and C-2 carbons (2.0 and 3.2 ppm, respectively) (21).

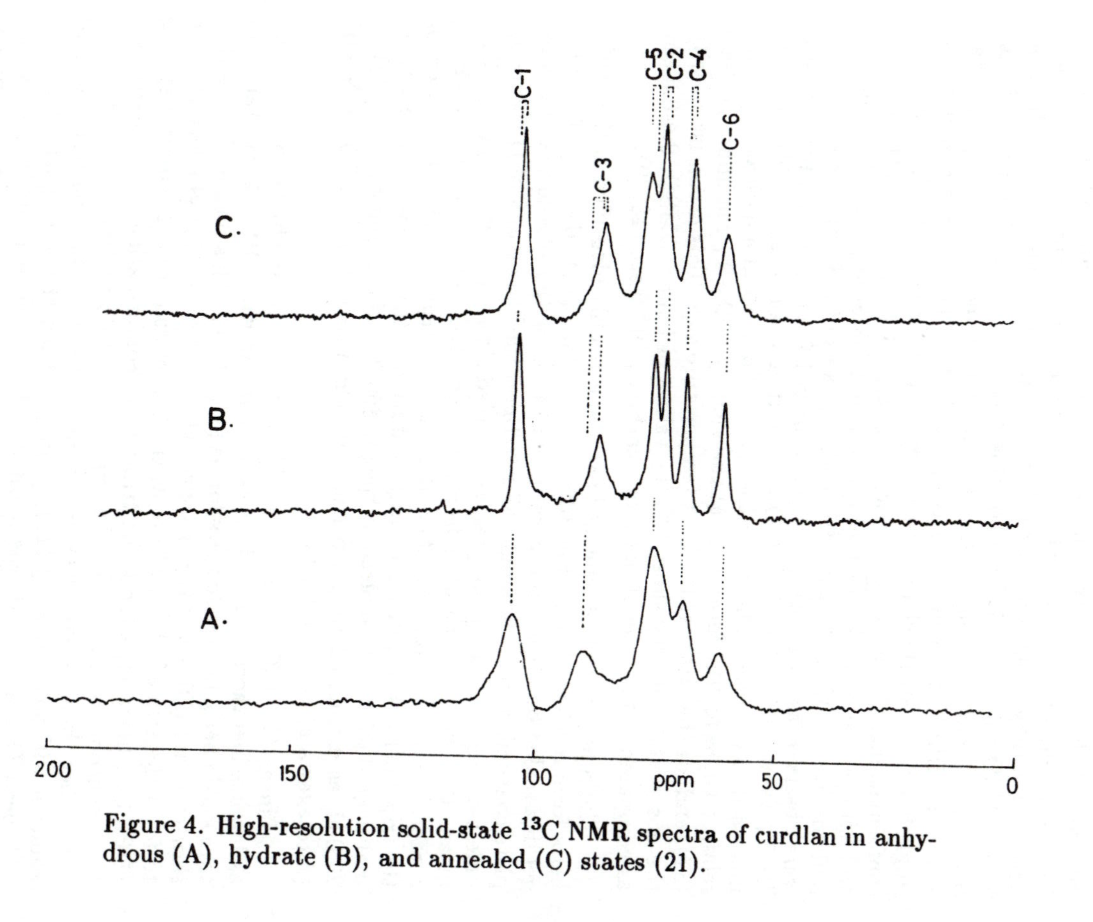

Figure 4. High-resolution solid-state ^{13}C NMR spectra of curdlan in anhydrous (A), hydrate (B), and annealed (C) states (21).

Surprisingly, the ^{13}C chemical shifts of the hydrated sample are in good agreement with those of the liquid-like domain of curdlan gel (with ± 0.1 ppm) (Figures 2A and 2B) (15,21). Thus, the conformation of the hydrated sample can be unequivocally identified as the single helix. It is emphasized that the spectral change between the anhydrous and hydrated forms is reversible upon hydration or dehydration. A similar spectral pattern to the anhydrous form of curdlan was also obtained by lyophilization from DMSO solution (17,20). For this reason, the anhydrous form is ascribed to a "single chain" form (20). In this connection, it appears that water and DMSO molecules are loosely bound to the polymer chains of the single helix and single chain forms, respectively, to stabilize the respective conformations. To prove this view, we analyzed 46 MHz ^2H NMR spectra and spin-lattice relaxation times of D_2O and DMSO-d_6 bound to curdlan and amylose (36). In all cases, ^2H NMR signals of the former and the latter were observed as single peaks of line widths 1.9 and 2.4 kHz, respectively. ^2H T_1 values of D_2O and DMSO-d_6 are 10 and 44 msec at 22°C and are increased to 68 msec at 50°C for the latter. These results indicate that correlation times for the former are at the vicinity of the T_1 minimum (10 sec) (37), and for the latter at the high temperature site. This is consistent with the observation of very short ^{13}C T_1 values of bound DMSO, 0.4-0.6 sec, as compared with those of backbone carbons, 10-20 sec (18,20).

The conformation of annealed curdlan is readily identified as the triple-helix form in view of the x-ray diffraction pattern (18). Heating of curdlan gel in a sealed bomb at a temperature of 150°C or 180°C resulted in incomplete conversion from the single helix to the triple helix (ca. 50%) (Figure 5), unless otherwise heated samples were cooled slowly (Figure 4C) (18). This means that slow cooling after heating is essential to the completion of the conformational change. Further, it is rather surprising that formation of the triple helix is facilitated by fragmentation of the polymer chain by thermal degradation from molecular weight 860,000 of original curdlan (38) to 20,000, as estimated by gel permeation chromatography in 0.3 M NaOH (Aketagawa *et al.*, unpublished).

Linear (1 → 3)-β-d-glucans of low molecular weight oligomers, such as laminariheptaose and acid-degraded curdlan with DPn 14, can take neither single helix nor triple helix forms (17,18,21). This means that neither shorter chain (DPn $<$ 14) nor longer chain (DPn $>$ 250) glucans are able to form the triple helix conformation. Laminaran from *Laminaria* species (DPn 39), however, adopts the triple helix conformation by lyophilization from aqueous solution (18). The triple helix conformation of branched glucans other than lentinan was also achieved by lyophilization from aqueous solution without annealing (19), as judged from the peak positions of the C-3 peak (Figure 6). It was pointed out that lentinan undergoes a conformational change from the single chain to the triple helix by complete dissolution in 8 M urea followed by dialysis against water and lyophilization (19). Such a triple helix conformation, however, is readily converted to the single chain by lyophilization from DMSO solution (17,20,21). Thus, sufficient hydration and dehydration of (1 → 3)-β-d-glucans are necessary

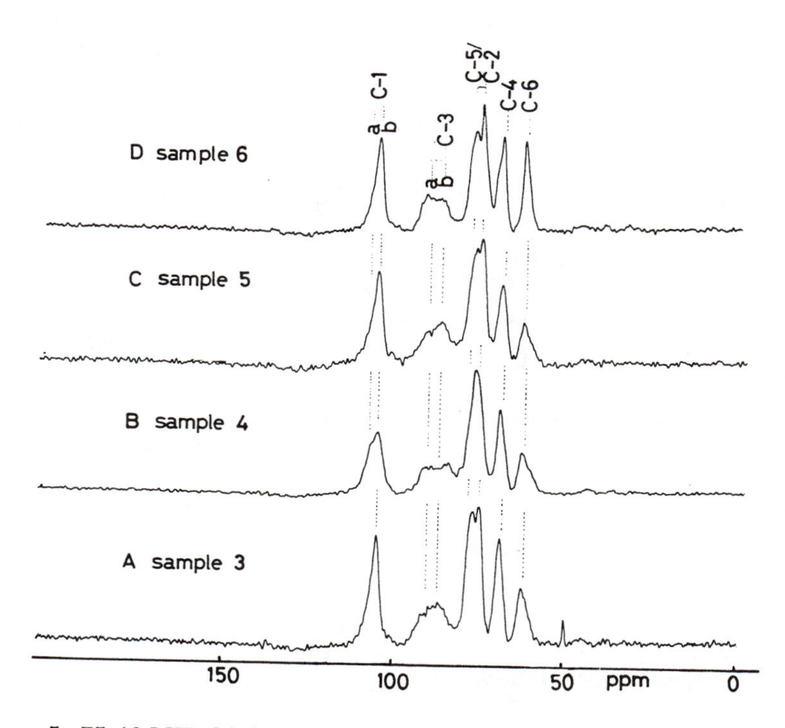

Figure 5. 75.46 MHz high-resolution solid-state ^{13}C NMR spectra of curdlan annealed at various temperatures followed by rapid cooling, except for sample 6 (18). Sample 3, annealed at 180°C; Sample 4, annealed at 150°C; Sample 5, annealed at 120°C; Sample 6, annealed at 80°C followed by slow cooling.

(Reproduced with permission from reference 18. Copyright 1987 Chemical Society of Japan.)

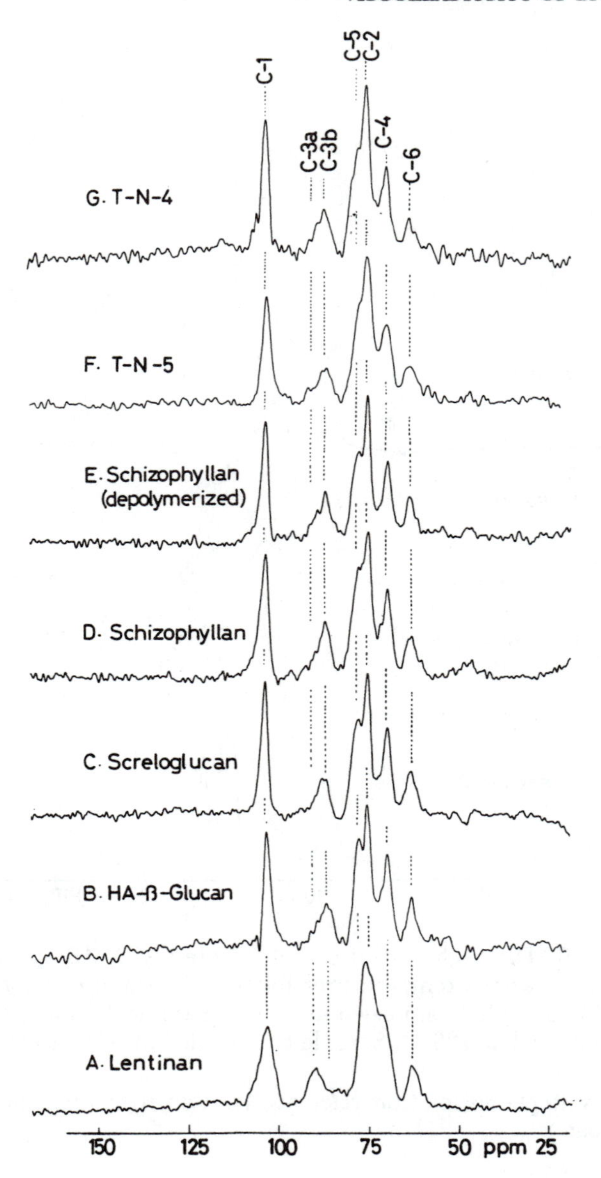

Figure 6. 75.46 MHz high-resolution solid-state ^{13}C NMR spectra of a variety of branched (1 → 3)-β-d-glucans (19). C-3a and C-3b peaks are characteristic of the presence of the single chain and triple helix forms, respectively.

(Reproduced with permission from reference 18. Copyright 1987 Chemical Society of Japan.)

conditions to achieve the triple helix and single chain conformations, respectively.

Solid-Like Domains of Gel Networks

The ^{13}C NMR spectra of the intermediate and solid-like domains in gel samples can be selectively recorded by a series of experiments with magic angle spinning (MAS) and cross polarization-magic angle spinning (CP-MAS), respectively. Interestingly, the resultant ^{13}C NMR spectra of curdlan gel are identical among the spectra by CP-MAS (Figure 2D), MAS only (Figure 2C), and broad-band decoupling (Figure 2B) in spite of a significant difference in the experimental conditions. This result indicates that the single helix form is dominant in elastic curdlan gel, although a low-intensity peak of the triple helical form is visible as shown by the arrowed peak (C-5 signal from the triple helix) in the spectra by CP-MAS experiment (Figure 2D). In this connection, it is worthwhile to relate the present view of gel network with the gel strength as a function of heating temperature presented by Harada *et al.* (23, 39): the gel strength is almost the same at a temperature between 60-80°C (low set gel), and thereafter increases with heating temperature until 120°C (high-set gel). Obviously, the first step corresponds to the formation of pseudo-crosslinks arising from hydrophobic association of the single helical chains as viewed from decrease of transmittance and shrinkage of gel (23). The increased gel strength in the high set gel is well explained in terms of increased proportion of the triple helical conformation together with heating temperature.

On the contrary, the ^{13}C NMR spectrum of lentinan gel (Figure 7C) exhibits characteristic peak positions of the triple helix conformation (15), although the single chain is dominant in the anhydrous and hydrated states (Figures 7A and 7B). Similar results were observed for polyol and aldehyde derivatives of HA-β-glucan (15). These results are consistent with the data of the complete suppression of peaks by conventional NMR spectrometer (27, 28), x-ray diffraction studies (34, 40), and hydrodynamic studies (41). Thus, it is concluded that gelation of branched (1 → 3)-β-d-glucans proceeds from partial association of the triple helical chains. It appears that gels consisting of higher proportion of the triple helical chains as in these branched glucans are of brittle type, in contrast to the case of curdlan gel consisting mainly of the single helical chain. In this connection, it is worthwhile to point out that the schizophyllan gel exhibits rather elastic property. In fact, we found that schizophyllan gel consists of both the single helical (60%) and triple helical (40%) chains (22).

Concluding Remarks

We demonstrated that combined use of conventional and solid-state high-resolution ^{13}C NMR is essential to delineate the complex network structures of polysaccharide gels. In particular, we showed that the network structures of gels substantially differ between linear and branched (1 → 3)-β-d-glucans. It is also pointed out that many biological responses such as

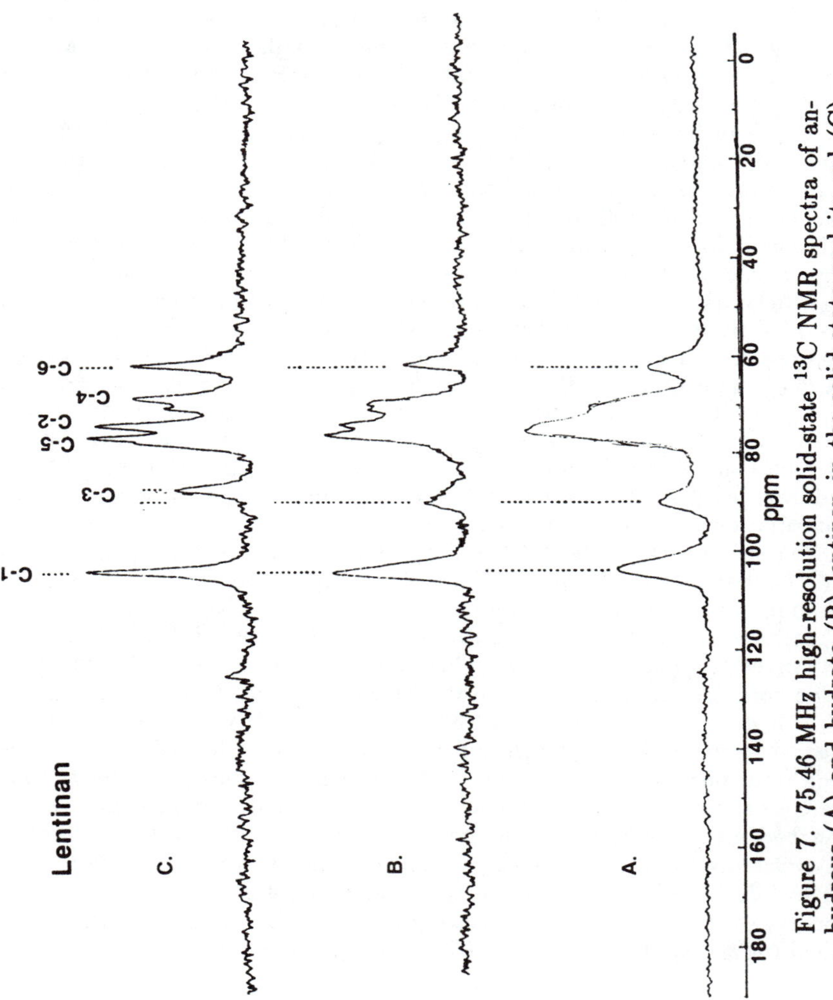

Figure 7. 75.46 MHz high-resolution solid-state ^{13}C NMR spectra of anhydrous (A) and hydrate (B) lentinan in the solid state and its gel (C). (Reproduced with permission from reference 15. Copyright 1990 Wiley.)

antitumor activity, activation of the coagulation system of horseshoe crab amebocyte lysate (LAL) (22), and endo(1 → 3)-β-d-glucanase are mainly initiated by recognition of the single helix conformation.

Acknowledgments

The authors are indebted to many collaborators listed in the references, especially to Yuko Yoshioka.

Literature Cited

1. Saitô, H.; Ohki, T.; Sasaki, T. *Biochemistry* 1977, **16**, 908.
2. Saitô, H.; Miyata, E.; Sasaki, T. *Macromolecules* 1978, **11**, 1244.
3. Yokota, K.; Abe, A.; Hosaka, S.; Sakai, I.; Saitô, H. *Macromolecules* 1978, **11**, 95.
4. Saitô, H. In *ACS Symp. Ser. No. 150*; American Chemical Society: Washington, DC, 1981; pp 125-147.
5. Marchessault, R. H.; Deslandes, Y.; Ogawa, K.; Sundararajan, P. R. *Can. J. Chem.* 1977, **55**, 300.
6. Deslandes, Y.; Marchessault, R. H.; Sarko, A. *Macromolecules* 1980, **13**, 1466.
7. Chuah, C. T.; Sarko, A.; Deslandes, Y.; Marchessault, R. H. *Macromolecules* 1983, **16**, 1375.
8. Fulton, S.; Atkins, E. D. T. In *ACS Symp. Ser. No. 141*; American Chemical Society: Washington, DC, 1981; p 385.
9. Arnott, S.; Fulmer, A.; Scott, W. E.; Dea, I. C. M.; Moorhouse, R.; Rees, D. A. *J. Mol. Biol.* 1974, **90**, 269.
10. Arnott, S.; Scott, W. E.; Rees, D. A.; McNab, C. G. A. *J. Mol. Biol.* 1974, **90**, 253.
11. Millane, R. P.; Handrasekaran, R. C.; Arnott, S.; Dea, I. C. M. *Carbohydr. Res.* 1988, **182**, 1.
12. Foord, S. A.; Atkins, E. D. T. *Biopolymers* 1989, **28**, 1345.
13. Smidsrod, O. *Proc. 27th Intl. Congress Pure Appl. Chem.*, 1980, p 315.
14. Saitô, H.; Yokoi, M.; Yamada, J. *Carbohydr. Res.*, 1990, **199**, 1.
15. Saitô, H.; Yoshioka, Y.; Yokoi, M.; Yamada, J. *Biopolymers*, 1990, **29**, 1689.
16. Saitô, H.; Tabeta, R.; Harada, T. *Chem. Lett.* 1981, 571.
17. Saitô, H.; Tabeta, R.; Sasaki, T.; Yoshioka, Y. *Bull. Chem. Soc. Jpn.* 1986, **59**, 2093.
18. Saitô, H.; Tabeta, R.; Yokoi, M.; Erata, T. *Bull. Chem. Soc. Jpn.* 1987, **60**, 4259.
19. Saitô, H.; Tabeta, R.; Yoshioka, Y.; Hara, C.; Kiho, T.; Ukai, S. *Bull. Chem. Soc. Jpn.* 1987, **60**, 4267.
20. Saitô, H.; Yokoi, M. *Bull. Chem. Soc. Jpn.* 1989, **62**, 392.
21. Saitô, H.; Yokoi, M.; Yoshioka, Y. *Macromolecules* 1989, **22**, 3892.
22. Saitô, H.; Yoshioka, Y.; Aketagawa, J.; Tanaka, S.; Shibata, Y. *Carbohydr. Res.*, in press.

23. Harada, T. In *ACS Symp. Ser. No. 45*; American Chemical Society: Washington, DC, 1977, p 265.
24. Chihara, G.; Suga, T.; Hamuro, J.; Takasuka, N.; Maeda, Y. Y.; Sasaki, T.; Shiio, T. In *Cancer Detection and Prevention, Suppl. 1*; Nieburgs, H. E.; Bekesi, J. G., Eds.; Alan R. Liss, Inc.: New York, 1987; pp 423-443.
25. Kikumoto, S.; Miyajima, T.; Kimura, K.; Ohkubo, S.; Komatsu, N. *Nippon Nogei Kagaku Kaishi* 1971, **45**, 162.
26. Yoshioka, Y.; Tabeta, R.; Saitô, H.; Uehara, N.; Fukuoka, F. *Carbohydr. Res.* 1985, **140**, 93.
27. Saitô, H.; Ohki, T.; Takasuka, N.; Sasaki, T. *Carbohydr. Res.* 1977, **58**, 293.
28. Saitô, H.; Ohki, T.; Sasaki, T. *Carbohydr. Res.* 1979, **74**, 227.
29. Saitô, H. *Magn. Reson. Chem.* 1986, **24**, 835.
30. Saitô, H.; Ando, I. *Ann. Rep. NMR Spectrosc.* 1989, **21**, 209.
31. Ogawa, K.; Tsurugi, J.; Watanabe, T. *Chem. Lett.* 1972, 689.
32. Ogawa, K.; Hatano, M. *Carbohydr. Res.* 1978, **67**, 527.
33. Kanzawa, Y.; Harada, T.; Koreeda, A.; Harada, A.; Okuyama, K. *Carbohydr. Polym.* 1989, **10**, 299.
34. Bluhm, T. L.; Sarko, A. *Can. J. Chem.* 1977, **55**, 293.
35. Bluhm, T. L.; Sarko, A. *Carbohydr. Res.* 1977, **54**, 125.
36. Saitô, H.; Yamada, J.; Yukumoto, T.; Yajima, H.; Endo, R. *Bull. Chem. Soc. Jpn.*, in press.
37. Davis, J. H. *Biochim. Biophys. Acta* 1983, **737**, 117.
38. Lawford, H. G. *Polymer Preprints, ACS* 1988, **29**, 633.
39. Maeda, I.; Saitô, H.; Masada, M.; Misaki, A.; Harada, T. *Agr. Biol. Chem.* 1967, **31**, 1184.
40. Bluhm, T. L.; Deslandes, Y.; Marchessault, R. H.; Perez, S.; Rinaudo, M. *Carbohydr. Res.* 1983, **100**, 117.
41. Yanaki, T.; Norisue, T.; Fujita, H. *Macromolecules* 1980, **13**, 1462.

RECEIVED December 16, 1991

Chapter 20

Solid-State ^{13}C NMR Spectroscopy of Multiphase Biomaterials

Roger H. Newman

Chemistry Division, Department of Scientific and Industrial Research, Private Bag, Petone, New Zealand

Proton spin relaxation parameters can be determined for particular phases in multiphase materials through indirect measurements on signals in carbon-13 NMR spectra. The results can be used to explore spatial relationships between phases or to edit spectra so that carbon-13 spin relaxation parameters can be determined for overlapping signals from different phases. These methods have been used to show that adding moisture to wood softens the hemicelluloses, with weaker effects on the lignin and cellulose.

Nuclear spin relaxation time constants can provide information about structure and molecular dynamics in polymers. Proton spin relaxation parameters can not be measured for individual functional groups because spin-spin "flip-flop" transitions result in proton spins diffusing over path lengths of a few nanometers during measurement of the rotating-frame relaxation time constant $T_{1\rho}^{H}$, or even a few tens of nanometers during the longer timescale required for measurement of the spin-lattice relaxation time constant T_1^H (1). The observation of a single time constant for a mixture of polymers has become established as an indicator of compatibility (1). Only 1.1% of carbon atoms contain ^{13}C nuclei, so the distances between ^{13}C nuclei inhibit spin diffusion and distinct time constants can be observed for different NMR signals provided the experiments are completed within a few seconds (2).

Proton spin relaxation time constants T_1^H, $T_{1\rho}^{H}$, and T_2^H can all be measured indirectly by ^{13}C NMR experiments with cross polarization (CP) and magic-angle spinning (MAS). This approach has the advantage of using the much larger dispersion of chemical shifts for ^{13}C compared with protons, increasing the probability of resolving NMR signals that can be assigned to each phase of a multi-phase sample. Experiments described in this paper

0097–6156/92/0489–0311$06.00/0

go one step further by using proton spin relaxation processes to edit ^{13}C NMR spectra, separating sub-spectra for different phases so that values of the ^{13}C spin-lattice relaxation time constant T_1^C can be measured for similar functional groups in different phases.

The new procedure was tested on samples of wood. Possible extensions to other multiphase biomaterials will be discussed at the end of the paper.

Methods

The NMR experiments described here were all run on a Varian XL-200 NMR spectrometer with a Doty Scientific MAS probe. Samples were packed in 7 mm diameter sapphire rotors and sealed with Kel-F caps. Poly(trifluorochloroethylene) grease was used to ensure watertight seals. Rotors were spun at 4 kHz.

Values of $T_{1\rho}^H$ were measured by the "delayed contact" pulse sequence in which the protons are spin-locked for a variable delay t_s before the contact time begins (3). The spin-locking field strength was maintained at 0.95 mT by adjusting the transmitter power after each sample change, compensating for variations in radiofrequency absorption due to variations in moisture contents.

The procedure for editing spectra has been described in detail elsewhere (4). The delayed-contact pulse sequence was used to generate a pair of spectra with $t_s = 0$ and T. If the sample contains just two phases labelled A and B, then it can be shown that the subspectrum of phase A can be isolated by multiplying the first spectrum by K and the second by K', then adding the spectra together with:

$$K = 1/\left\{1 - \exp\left[(T/T_{1\rho}^B) - T/T_{1\rho}^A)\right]\right\}$$

$$K' = -K \exp\left[T/T_{1\rho}^B\right]$$

Values of T= 6 or 8 ms were used, and transients from at least 5000 contacts were averaged. A similar method was used by VanderHart and Perez (5) to separate subspectra of crystalline and non-crystalline phases in polyethylene. If $T_{1\rho}^B$ is much less than a typical contact time of 1 or 2 ms then signals from phase B would be suppressed by inefficient cross polarization, but that was not the case in the present work where typical values of $T_{1\rho}^H$ were in the range 4 to 10 ms and values for different phases differed by no more than a factor of 2. Integrals over the two subspectra do not give an accurate estimate of the relative amounts of carbon associated with each phase unless allowance is made for differences in rotating-frame relaxation during the contact time.

Values of T_1^C were measured by the method of Torchia (6) in which cross-polarization is followed by two $\pi/2$ pulses separated by a variable relaxation interval t_r. In some experiments the delayed-contact pulse sequence was combined with Torchia's pulse sequence as in Figure 1. Pairs of spectra were generated by averaging 3000 transients for $t_s = 0$ and 9000 transients for $t_s = 6$ ms, and subspectra were separated for each value of t_r.

Proton Spin Relaxation

Figure 2 shows subspectra separated from the ^{13}C CP/MAS NMR spectrum of *Pinus radiata* wood moistened to 35% by weight water. A subspectrum associated with $T_{1\rho}^{H} = 7$ ms (Figure 2a) displayed signals that can be assigned to cellulose in both crystalline and non-crystalline domains (7), while a subspectrum associated with $T_{1\rho}^{H} = 4$ ms (Figure 2b) displayed signals that can be assigned to lignin and hemicelluloses (8).

The clean separation of two such subspectra forms the basis of a new method for estimating the degree of crystallinity of cellulose in wood (4). Such a clean separation is made possible by close similarities in the values of $T_{1\rho}^{H}$ for lignin and for hemicellulose (4). This point was explored further by measuring values of $T_{1\rho}^{H}$ for a collection of eight samples of eucalypt woods, using the signals at $\delta = 56$ and 22 ppm as representative of lignin and hemicellulose, respectively. The results are shown in Figure 3, along with published results for six assorted hardwood species (4). Values of $T_{1\rho}^{H}$ for wood are sensitive to the strength of the spin-locking field (9). The NMR instrument was modified to improve control of transmitter power in between the two sets of experiments, so the range of values of $T_{1\rho}^{H}$ for the eucalypts at least can be attributed primarily to differences in molecular motion in different samples rather than to variations in instrument performance.

The points plotted in Figure 3 are scattered around a line of unit slope, with no deviations greater than the experimental uncertainty. Such close similarities in $T_{1\rho}^{H}$ indicate mixing of the molecules on a scale of a few nanometers or less (1). This is consistent with the model of Fengel (10), in which lignin molecules are linked to hemicellulose fibrils in wound and coiled structures.

$T_{1\rho}^{H}$ values were not measured for hemicelluloses in softwoods because the signal at $\delta = 22$ ppm was too weak. Mean values of $T_{1\rho}^{H}$ for lignin were 5.1 ms for five softwoods and 5.4 ms for six hardwoods, with standard deviations of 0.9 and 1.2 ms, respectively (4). New data for eight eucalypts provided a mean of 6.8 ms and a standard deviation of 1.3 ms. The scatter is too large for any clear distinction between molecular motion in softwoods and hardwoods.

Carbon-13 Spin Relaxation

The pulse sequence of Figure 1 was used to estimate values of T_1^C for chemical functional groups in wood from *P. radiata* as an example of a softwood and *Eucalyptus delegatensis* as an example of a hardwood. Both samples were moistened to 31% moisture content. The values of T_1^C for both crystalline and non-crystalline cellulose were > 15 s and therefore too long for reliable estimation without extending the experiments beyond three days of data averaging. Results for the non-cellulosic material are given in Table I. Shortest values of T_1^C were observed for $-CH_2OH$ and $-OCH_3$ groups ($\delta = 62$ and 56, respectively), as might be expected for groups that are not part of the polymeric backbones. Signals at $\delta = 106$ and 114 ppm were assigned to CH groups in lignin syringyl and guaiacyl

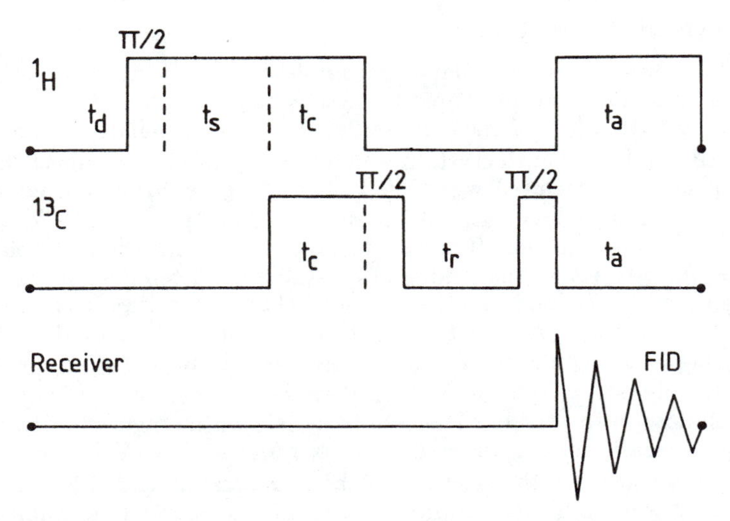

Figure 1. Pulse sequence for measuring T_1^C . Phases of the $\pi/2$ pulses have been omitted for clarity.

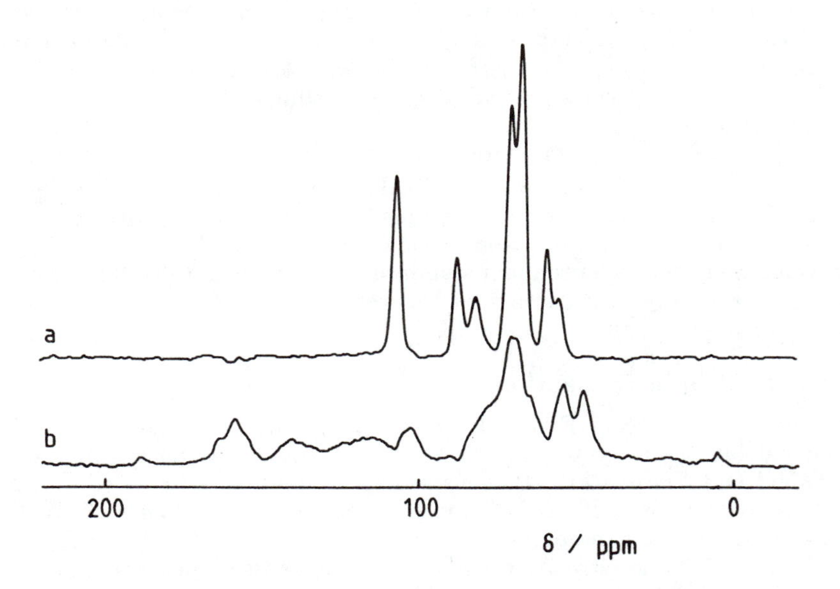

Figure 2. Subspectra separated from a ^{13}C CP/MAS NMR spectrum of *P. radiata* wood, selected for domains with (a) $T_{1\rho}^H = 7$ ms; (b) $T_{1\rho}^H = 4$ ms.

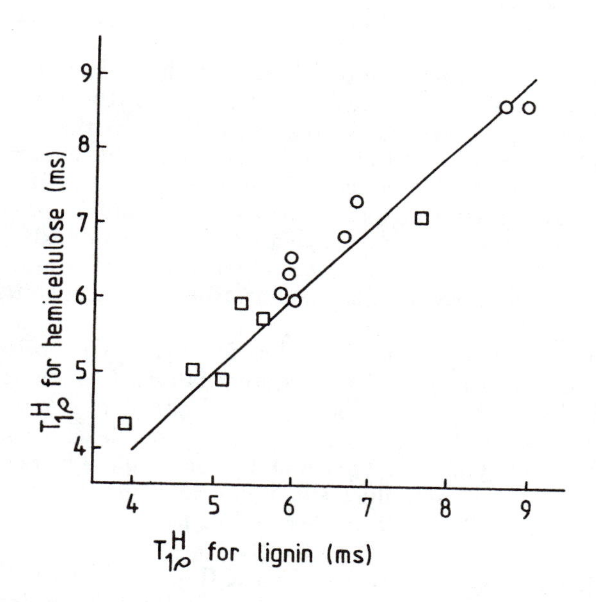

Figure 3. Values of $T^H_{1\rho}$ for lignin and hemicellulose in wood from eucalypts (○) and assorted hardwoods (□).

units, respectively. The signal at $\delta = 84$ was assigned primarily to CH groups in hemicellulose (8), with a minor contribution from CH groups in lignin.

Table I. Values of T_1^C (s) for specific signals in subspectra of non-cellulosic material

δ/ppm	114	106	84	62	56
P. radiata	5	–	3	0.3	1.5
E. delegatensis	–	3	3	0.6	1.5

When Torchia's pulse sequence was used without spectral editing, the relaxation curve for the signal at $\delta = 84$ was distinctly non-exponential. The spectral editing experiment showed that this was the result of overlapping signals from non-crystalline cellulose and hemicellulose, that the shorter time constant should be assigned to the latter, and that the ratios of contributing strengths were 59 : 41 for *P. radiata* and 54 : 46 for *E. delegatensis*. This information was used to analyze non-exponential relaxation curves from experiments without spectral editing. The results are shown in Figure 4, along with results obtained at other chemical shifts for which spectral editing was unnecessary.

The curves of Figure 4 show evidence for increasing motion involving polymeric backbones as the moisture content is raised to about 25%, with little further change beyond. The effects of added moisture were most pronounced for hemicellulose (square symbols) and least pronounced for cellulose (triangular symbols). Values of T_1^C for cellulose were so long that spin diffusion between crystalline and non-crystalline domains may have provided an important relaxation mechanism for the crystalline material (2).

Figure 4b shows no information about CH groups in lignin because the signal at $\delta = 106$ ppm was swamped by much stronger overlapping signals assigned to C-1 of cellulose in both crystalline and non-crystalline domains.

Glass transition temperatures have been published for isolated non-crystalline cellulose and hemicelluloses (11, 12). Water acts as a plasticizer, lowering T_g to room temperature at about 20% moisture content for both substances. The NMR results for wood can be explained by water interacting with hemicelluloses, softening them beyond the glass transition. The relatively small effect on T_g for non-crystalline cellulose in wood provides evidence for this component being less accessible to water. Winandy and Rowell (13) have suggested that: "Lignin is the most hydrophobic component of the wood cell wall. Its ability to act as an encrusting agent on and around the carbohydrate fraction, and thereby limit water's influence on that carbohydrate fraction, is the cornerstone of wood's ability to retain its strength and stiffness as moisture is introduced to the system."

Recent experiments on spruce wood have provided evidence for two distinct glass transitions (14). The lower-temperature transition occurs at room temperature for a moisture content of about 10%, and can therefore be identified with the softening of hemicellulose as observed in the NMR experiments. The higher-temperature transition does not drop below 50°C even at high moisture contents. The latter transition has been attributed to softening of lignin. The NMR results show some evidence for lignin being softened by water at room temperature, but this observation can be explained in terms of Fengel's model for interactions between lignin and hemicellulose, i.e., the NMR results for lignin may reflect softening of hemicellulose fibrils allowing greater freedom of movement of the lignin molecules in the interconnected structure.

Spin Diffusion and Spatial Relationships

Similarities and differences in proton spin relaxation time constants have been used in studies of spatial relationships between the chemical components of wood and pulp (15-18). Similarities could be the result of rapid spin diffusion between domains with very different relaxation time constants, or they could be the result of coincidental similarities in molecular motion in well-separated domains. This ambiguity can be resolved by measuring the time constants for spin diffusion between domains, e.g., using pulse sequences described by Zumbulyadis (19) or the pulse sequence "MOPS" (Mixing of Proton Spins) (20). MOPS has been used to estimate a time constant of 14 ms for mixing between the protons of lignin and crystalline cellulose in *Eucalyptus regnans* wood (20).

The averaging effect of such mixing on relaxation time constants was explored with a highly simplified model for a cell-wall microfibril. A cylinder of cellulose was surrounded by a mixture of lignin and hemicelluloses. A spin diffusion coefficient of 9.7×10^{-16} m^2s^{-1} was used (20). Computer simulations matched the experimental observations when the trial diameter of the cellulose microfibril was 24 nm. This is within the range of 9 to 29 nm previously estimated by electron microscopic examination of cellulose microfibrils from wood (21). Two versions of the model were tested; one with crystalline cellulose confined to a central core, and one with crystalline cellulose distributed in a non-crystalline matrix. The former model gave slightly less convincing results, because simulations predicted a difference of about 11% between apparent values of $T_{1\rho}^{H}$ for crystalline and non-crystalline cellulose. This difference would be large enough to upset the clean separation of cellulosic and non-cellulosic material that was demonstrated earlier. Simulations for both models predicted that apparent values of $T_{1\rho}^{H}$ for lignin are distorted by no more than 10% by proton spin diffusion between domains.

Other Biomaterials

Figure 5 shows a pair of subspectra separated from the CP/MAS NMR spectrum of ryegrass. The subspectrum associated with longer values of $T_{1\rho}^{H}$ (Figure 5a) shows signals assigned to crystalline and non-crystalline cellulose, while the subspectrum associated with shorter values of $T_{1\rho}^{H}$ is

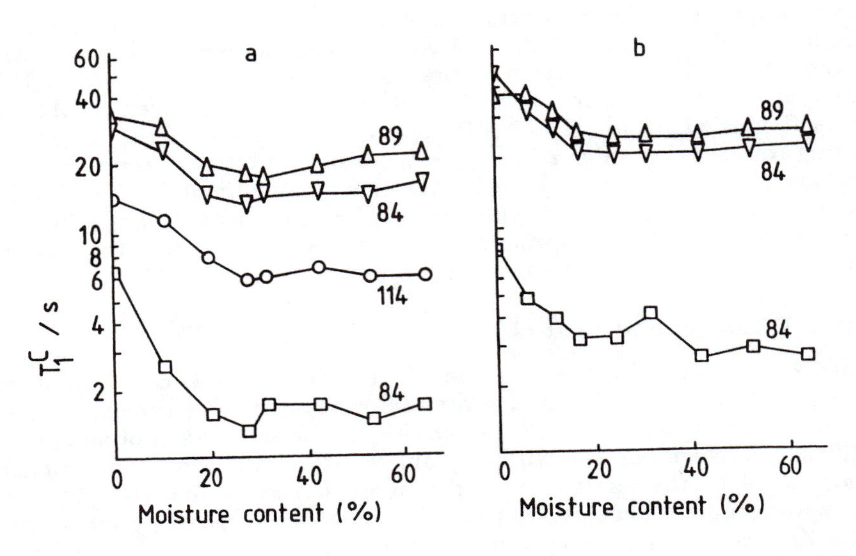

Figure 4. Values of T_1^C for (a) *P. radiata* and (b) *E. delegatensis*. The chemical shift is shown beside each curve.

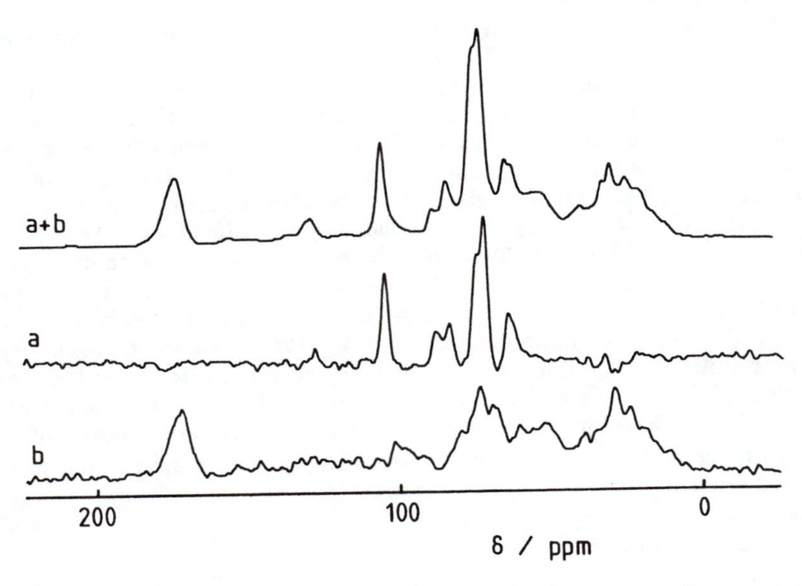

Figure 5. Subspectra separated from a ^{13}C CP/MAS NMR spectrum of ryegrass, selected for domains with (a) $T_{1\rho}^H = 4.6$ and (b) $T_{1\rho}^H = 3.6$ ms.

dominated by signals assigned to proteins and hemicelluloses; signals at δ = 173 and 10–40 are assigned to protein amide groups and sidechains, respectively.

Wool is thought to contain microfibrils of crystalline and non-crystalline protein, but attempts at separation of NMR subspectra failed.

Literature Cited

1. McBrierty, V. J.; Douglass, D. C. *Phys. Rep.* 1980, **63**, 61-147.
2. VanderHart, D. L. *J. Magn. Reson.* 1987, **72**, 13-47.
3. Alla, M.; Lippmaa, E. *Chem. Phys. Lett.* 1976, **37**, 260-264.
4. Newman, R. H.; Hemmingson, J. A. *Holzforschung* 1990, **44**, 351-355.
5. VanderHart, D. L.; Perez, E. *Macromol.* 1986, **19**, 1902-1909.
6. Torchia, D. A. *J. Magn. Reson.* 1978, **30**, 613-616.
7. Horii, F.; Hirai, A.; Kitamaru, R. *J. Carbohydr. Chem.* 1984, **3**, 641-662.
8. Kolodzieski, W.; Frye, J. S.; Maciel, G. E. *Anal. Chem.* 1982, **54**, 1419-1424.
9. Newman, R. H. *Proc. 4th Int. Symp. Wood and Pulping Chem., Paris* 1987, **1**, 195-199.
10. Fengel, D. *Svensk Papperstid.* 1976, **79**, 24-28.
11. Salmén, N. L.; Back, E. L. *Tappi* 1977, **60**(12), 137-140.
12. Irvine, G. M. *Tappi J.* 1984, **67**, 118-121.
13. Winandy, J. E.; Rowell, R. M. *Adv. Chem. Ser.* 1984, **207**, 211-255.
14. Kelley, S. S.; Rials, T. G.; Glasser, W. G. *J. Mater. Sci.* 1987, **22**, 617-624.
15. Gerasimowicz, W. V.; Hicks, K. B.; Pfeffer, P. E. *Macromolecules* 1984, **17**, 2597-2603.
16. Haw, J. F.; Maciel, G. E.; Schroeder, H. A. *Anal. Chem.* 1984, **56**, 1323-1329.
17. Teeäär, R.; Gravitis, J.; Andersons, B.; Lippmaa, E. *Teubner-Texte Phys.* 1986, **9**, 240-249.
18. Tekely, P.; Vignon, M. R. *J. Polym. Sci. C: Polym. Lett.* 1987, **25**, 257-261.
19. Zumbulyadis, N. *J. Magn. Reson.* 1983, **53**, 486-494.
20. Newman, R. H.; Leary, G. J.; Morgan, K. R. *5th Int. Symp. Wood and Pulping Chem., Raleigh, NC* 1989, **1**, 221-223.
21. Jayme, G.; Koburg, E. *Holzforschung* 1959, **13**, 37-43.

RECEIVED December 16, 1991

Chapter 21

Chemically Modified Wood

Solid-State Cross-Polarization–Magic-Angle Spinning NMR Spectroscopy

J. J. Lindberg[1], A. Björkman[2], L. Salmén[3], and K. Soljamo[1]

[1]University of Helsinki, Helsinki SF–00170, Finland
[2]Technical University of Denmark, Lyngby DK–2800, Denmark
[3]Swedish Pulp and Paper Research Institute, Stockholm S–11486, Sweden

Our investigations on birch, spruce, aspen, beech and eucalyptus wood show that solid state cross polarization/magic angle spinning nuclear magnetic resonance (CP/MAS NMR) spectroscopy can be used to check small changes in composition by treating *in situ* solid wood with various chemical agents. Mechanical strength data correlate with changes in intensities of NMR bands showing the crystallinity of fibers and especially with changes in T1 relaxation data. The advantages of NMR and Fourier transform infrared spectroscopy (FTIR) methods as tools for analysis of chemical and mechanical properties of solid wood are discussed in terms of current viscoelastic theories.

Nature has solved in the wood structure, with its many variants of connections between the cells, the problem of a lightweight and tough composite, consisting of strong fibers that are embedded in a cohesive matrix. Among drawbacks may be mentioned deformation by moisture and aging of the chemically untreated wood. However, as a renewable resource, wood is also one of the best and most competitive answers to the environmental problems involved in the production of construction materials and treating of their wastes.

To study the viscoelastic and mechanical properties of chemically modified wood, Anders Björkman, Technical University of Denmark (DTH), formed a group in collaboration with colleagues in Sweden, at the Swedish Pulp and Paper Research Institute (STFI), and in Finland, at the University of Helsinki (HU). The present paper deals with results of the investigations at HU using NMR chemical shifts and relaxation times to demonstrate the influence of chemical modification on

0097–6156/92/0489–0320$06.00/0

the wood structure. We know from earlier studies that the results relate most closely to phenomena on the submicroscopic scale. However, they are also connected with macroscopic mechanical and viscoelastic behavior. The latter parameters were measured on the same wood samples at DTH and STFI. In addition, we used FTIR spectroscopy to check the validity of the NMR data.

Materials

Air-dried samples from Danish spruce, birch, aspen and beech woods and eucalyptus wood were made resin-free by solvent extraction, and they were cut to small bars. The bars were treated chemically as described in (1) by reaction with hydrochloric acid, acetic acid anhydride, succinic acid anhydride, ammonia, or by carbanilation.

Methods

The solid-state carbon-13 spectra were recorded on a JEOL FX 200 FT NMR spectrometer at 50.1 MHz with CP-MAS technique. The contact time was 1 ms, and the delay between the pulse sequence repetitions was 1 s. In the experiments 1500–3000 scans were collected.

The samples could not be measured as whole cylindrical blocks owing to irregular spinning. They were therefore disintegrated to small chips in a mill. The chips were packed in MACOR-rotor cells and spun at speeds of 3 kHz. Chemical shifts of the spectra were calibrated with the benzene carbon resonance of solid hexamethylbenzene taken before each measurement. The T1-measurements were carried out by means of a modified inversion-recovery-T1 experiment in which the π-pulse was replaced for a normal cross-polarization (CP)-sequence with a following $\pi/2$-pulse. The relaxation curves were numerically resolved into their components. Curve fitting was done using an iteration technique. At most two components could be obtained.

The FTIR spectra were recorded in the 400–4000 1/cm region with a Nicolet Instruments Corp. spectrometer on finely milled wood powders using the KBr-pellet technique. The spectral resolution was 2 1/cm over the entire range. The computer program SX FTIR and a three-dimensional stick spectrum plot program developed by Mr. Matti Laanterä, Chem. Eng., were used to examine the difference spectra and present the results in a suitable form (Figure 1).

Results

The results of the NMR measurements of chemical shifts and relaxation times T1 are given in Tables I, II, and III, respectively.

We know from earlier investigations that the intensities of the NMR bands of the lignin structure are much less intense than the bands of the carbohydrate part and especially the cellulose part. Table

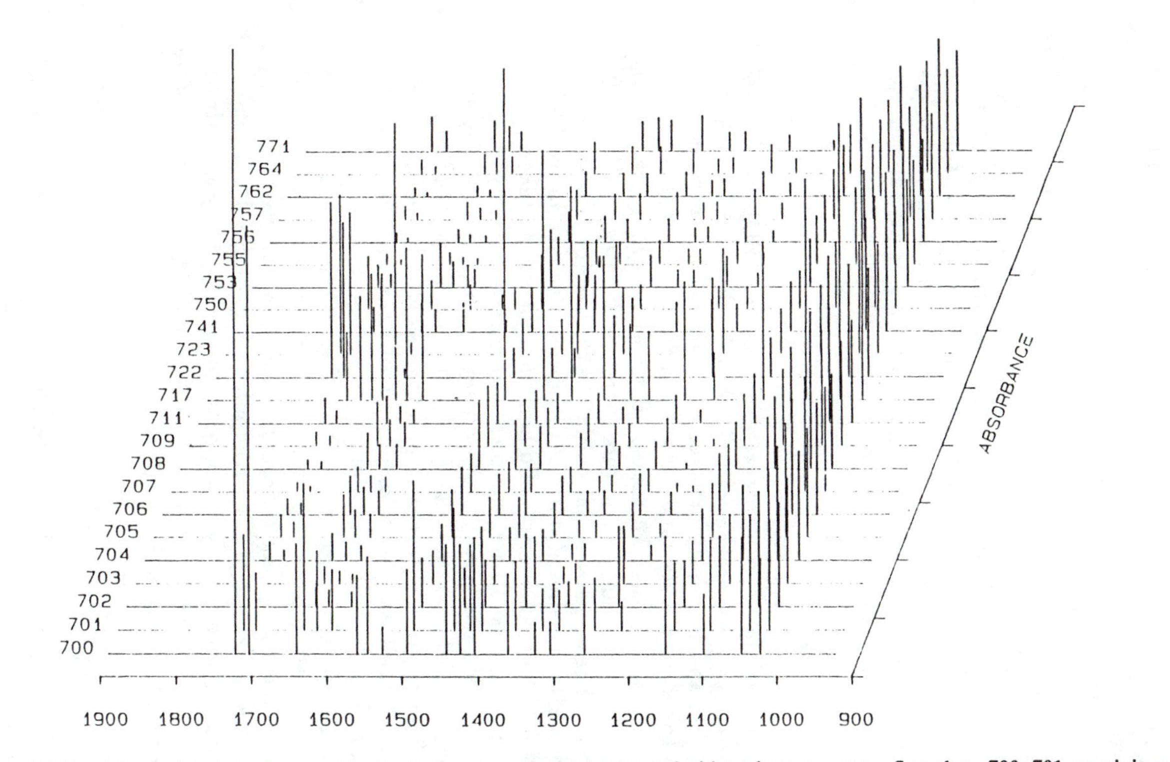

Figure 1. FTIR stick spectra (absorbance—1/cm) of spruce wood samples treated with various reagents. Samples: 700, 701, succinic anhydride (700:scale × 10); 702, 703, acetic anhydride; 704, 705, 706, liquid ammonia, 1h, 1d, 3d, respectively; 707, 708, 709, acid hydrolysis (HCl); 711, reference; 717, sodium in liquid ammonia; 722, 723, acetic anhydride, 3d, 6d, respectively; 741, 750, carbanilation, 2d, 4d, respectively; 753, liquid ammonia, 3d; 755, 756, 762, acid hydrolysis (HCl), 4d/40C, 4d/20C, 4d/60C, respectively; 757, 764, acid hydrolysis, HCl/LiCl/dimethylacetamide/60C/7d and 5d, respectively; and 771, hydrogen peroxide 40 mL/acetic acid 20 mL 50 mL/11d 60C.

Table I. ^{13}C CP/MAS NMR spectral shifts (ppm) of chemically treated solid spruce (first column) and birch (second column) woods at RT. G = glucose in carbohydrates; L = lignin; Cn = carbon atom n

	Chemical Reaction, ppm							
Atom or Group	Reference[1]		Acetylation		Carbanilation		Ammonia Reaction	
Methyl group	22.1	21.3	21.4	21.2	22.2	21.7	22.2	21.7
Alkyl group	—	—	—	—	40.0	—	—	—
L: Methoxyl	56.4	56.5	56.2	56.5	56.5	56.6	56.8	56.5
G: C6	63.4	—	62.8	—	—	—	—	63.3
G: C6	65.8	65.4	65.8	65.6	63.1	65.2	65.8	—
G: C2,C3,C5	72.9	73.2	72.8	73.0	75.4	72.9	73.0	—
	75.5	74.7	75.4	75.2	75.1	75.0	—	75.0
G: C4	84.7	84.0	84.5	84.1	84.0	83.5	83.4	83.1
	—	89.1	—	89.4	89.6	89.1	—	88.3
G: C1	105.6	106.0	105.6	105.6	105.8	105.3	105.4	105.3
L: C2,C5	114.0	—	114.4	—	118.6	—	—	—
	124.1	125.2	122.7	123.3	123.1	123.7	—	—
L: C-aromat.	129.7	—	—	—	130.2	129.7	129.0	135.1
L: C6	133.6	136.4	133.3	135.2	—	130.7	132.2	135.1
L: C-aromat.	149.5	—	—	—	—	—	—	—
L: C-aromat.	149.5	153.8	149.3	151.1	153.2	152.8	148.0	152.5
Carbonyl Groups	173.9	172.2	172.3	171.5	171.6	171.7	172.8	171.6

[1] Similar values are obtained for mildly acid (HCl) hydrolyzed wood.

I demonstrates that also the shift parameters show only small changes from sample to sample.

Major changes are found for ammonia-reacted wood where the change in the cellulose polymorph structure from I (72.9 and 75.5 ppm doublet for spruce and 73.2 and 74.7 ppm for birch, respectively) to III (73.0 ppm singlet for spruce and 75.0 ppm for birch, respectively) is clearly demonstrated. This has also been reported earlier by several authors (2,3).

Other major differences are only observed when a new group, *e.g.*, urethane (*e.g.*, the alkyl carbon at 40.0 ppm in carbanilation), or succinic acid or succinic ester is introduced. In the above respect the FTIR spectra are often much more informative, *e.g.*, by carbanilation or addition of succinic anhydride into the cell structure of spruce wood (Figure 1). Table III indicates that also for aspen and beech woods the same trends seem valid. The NMR spectral structure of eucalyptus wood is very similar to that of the other hardwood samples,

Table II. ^{13}C CP/MAS NMR T1 relaxation data at RT of chemically treated birch wood. G = glucose; L = lignin

	T1 Values, [s]			
Atom or Group	Hydrolysis with HCl	Carbani-lation	Ammonia	Succunic Anhydride[1,2]
Methyl[4]	2.77	3.25	5.06	8.81
Alkyl	—	—	—	1.17/30.18
L: Methoxyl	0.76	0.85	0.91	2.41
G: C6[3]	12.09	8.18	0.64	21.15
G: C2,C3,C5				
T11[4]	3.43	2.26	18.42	5.23
T12[4]	53.01	43.13	—	39.58
G: C4	5.36	1.56	12.19	2.78/34.23*
G: C4	1.21	1.63	—	7.06/27.75*
G: C1[3]	19.10	30.85	24.22	6.76*/26.11
Carbonyl Group	—	1.01	1.01	13.8–20.7

[1] Dry wood was impregnated for 4 days with a 1:2 by volume solution of succinic anhydride in DMSO using DMBA as a catalyst. Subsequently, the samples were washed for a long time with acetone and water and dried over phosphorus pentoxide.
[2] Two phases were observed in the spectra. Only data with an asterisk (*) were used in the calculation of the statistical correlation.
[3] Relaxation times correlate roughly (correlation factor > 0.7) with the relaxation enthalpy data, Ha, reported by Salmen (see text).
[4] Relaxation times correlate roughly (correlation factor > 0.7) with the softening point, T_g, reported by Salmen (see text).

although large differences are known for their mechanical strength behavior. Thus, a simple NMR spectroscopical shift analysis in the solid state provides information only on large chemical changes, especially in the cellulose fibers.

Owing to the very long NMR relaxation times T1 of the lignin component in comparison to the carbohydrates, the present technique gives satisfactory information with a reasonable amount of scans and instrumental time only for the latter substances and the various methoxyl and methyl groups in the materials. In the present study we have therefore given our primary attention to the changes of carbohydrate relaxation times by chemical treatments of wood.

In the same manner as for chemical shifts, the change in crystallinity of cellulose from the initial I-state to the III-state on treatment with ammonia is clearly demonstrated in the 75 ppm region: T1 = 3.43/53.01 to 18.42 s, respectively.

Table III. ^{13}C CP/MAS NMR spectra of chemically modified woods. G = glucose in carbohydrates; L = lignin; Cn = Carbon Atom n

Atom or Group	Sample and Chemical Reaction, ppm							
	Aspen[2]	Aspen Acet. Anh.	Aspen Succ. Anh.[1]	Beech[2]	Beech Succ. Anh.[1]	Birch[2]	Birch Dioxan	Eucalyptus[2]
Groups:								
Methyl	21.7	21.6	22.2	22.0	22.2	22.2	22.3	22.5
			31.2		30.3			
			40.8					
Methoxyl Groups	57.2	56.9	57.2	57.1	57.1	56.8	56.1	57.1
G: C6	64.1	61.5	64.4	64.0	63.5	64.0	63.9	64.2
G: C6	66.0	64.6	66.3	65.7	66.0	65.5	66.0	66.0
G: C2, C3,C5	73.7	73.3	73.6	73.6	73.4	73.6	73.6	73.9
	75.6	75.3	75.9	75.6	75.3	75.3	75.8	75.9
G: C4	83.9	83.2	84.3	84.8	84.4	83.7	84.7	85.0
	85.1							
	89.5	89.8	90.3	89.6	89.9	89.2	89.9	89.6
G: C1	106.2	106.0	106.0	106.0	106.2	105.9	106.2	106.0
L: C-arom.	130.0	129.0		129.5				130.0
L: C6	136.7	133.5	136.4	136.5	134.7	134.5	137.4	135.2
		141.36			138.3			138.2
L: C-arom.	149.7				141.0	148.5		
	154.1	154.5	154.7	154.1	153.9	151.0	154.2	154.4
C: Carbonyl Groups	173.1	172.0	173.9	172.0	174.2	173.8	172.6	173.2
			220.0		220.0			
			240.0		240.0			

[1] Wood treated 4 days with 37.5 succinic acid anhydride in 75 ml DMSO and 1.5 ml DMBA at RT. After reaction extraction with water and acetone.
[2] Untreated resin-free wood.

The reaction of succinic anhydride with the wood structure also gives marked changes in the relaxation behavior. As a general observation, the T1 spectrum shows two series of values. They can be interpreted as the formation of two different phases: one, with lower T1-values, is only weakly influenced by the reagent and nearly similar to untreated wood. The other, with higher T1-values, is looking more like a succinic ester structure. NMR shifts and FTIR spectra show that the formation of the two patterns increases step by step with reaction time.

Carbanilation gives smaller changes than the above. The region of C2, C3, and C5 glucose ring atoms (75 ppm region) shows decreased T1 values (T1 = 3.43/53.01 to 2.26/43.13 s). T1 (T1 = 19.1 to 30.85 and 12.09 to 8.18 s, respectively) is markedly changed for the C1 and C6 atoms of the glucose ring (105.5 and 65 ppm, respectively). A detailed interpretation of the effects is still difficult. The changes are probably caused by the bulky urethane groups introduced which induce changes in mobility of the surrounding structures. Also here, both NMR shifts and especially FTIR spectra show a gradual change with reaction time. Sundquist and Rantanen (4) observed similar changes for technical pulps.

Discussion

The viscoelastic behavior of wood differs significantly from the properties of synthetic polymer composites. Various theories have been proposed in the last 50 years to explain the complex aspects of stress–strain–moisture–history relationships in wood, including the much discussed anomaly known as mechano-sorptive creep. In the present discussion we rely especially on the thermodynamic theory of Barkas (6) and the dynamic lenticular trellis theory of Boyd (7). They explain most of the phenomena related to viscoelasticity of wood and its fibrillar structure. The mentioned theories have recently been unified in the relaxation theory of Hunt (8), which explains still more exactly the complicated dynamics of wood and, among others, its anomalies in moisture diffusion and dimensional hysteresis.

From our point of view, it is especially important that already Boyd (7) concluded that the statements proposed for the moisture dynamics of wood also can be applied to other solvents, especially Boyd's working concept that dynamics behavior of wood is determined by two parameters related to its fibrillar structure: a fibrous and comparatively rigid envelope composed of the crystalline part of wood, and a softer and gel-like matrix. The former is a rough model for the fibrillar network made up of intertwisting cellulose chains. The latter consists of the amorphous components, hemicellulose, the non-crystalline parts of cellulose, lignin and resins. On swelling with water or other solvents, the gel-like part is deformed within the framework of the stiffer envelope and makes the system more rigid.

Boyd concluded that for moistened wood, the resins are among the gel-like components, that contribute, through their partly hydrophilic character, to the viscoelastic behavior. In our experiments, we have removed the resins and thus obtained a non-collapsing void volume. The dried system seems to be in thermodynamic equilibrium in the sense of the Barkas theory. The void volume can be filled by the action of organic reagents.

If Boyd's views are correct, it is expected that by removal of resins, after a chemical substitution generally only small changes in NMR relaxation behavior should be observed. Exceptions are made by very bulky substituents, as in the case of succinic ester and phenylurethanes. Especially big changes should be noted in accordance with the model of Boyd and Hunt when the crystalline polymorph of the cellulosic fibrillar envelope is changed, e.g., as by ammonia treatment from cellulose I to III-type. Here also changes in chirality and semicrystalline areas may be observed.

In accordance with the above discussion, Salmen (5) concluded from macroscopic mechanical relaxation measurements on samples of spruce and birch wood that the reaction with ammonia decreases the softening point of lignin. Also, the introduction of phenylurethane and succinic ester or acetate groups into the wood gives a similar effect. Acid hydrolysis seems both to remove carbohydrates and to crosslink lignin, and make the system more stiff. He concluded also that (in the resin-free samples) lignin is the determining component in the softening behavior.

The remarkably small changes in macroscopic mechanical strength of chemically treated as compared to resin-free wood reported by Björkman and Lassota (1) are also in good agreement with the above theories.

To get some more insight into the quantitative connection between the parameters, a rough statistical correlation analysis was carried out. The NMR relaxation times, T1-values, were compared with the softening point values, T_g, and with the activation energies, Ha, of mechanical relaxations measured by Salmen (5). A significance higher than 0.7 was obtained for Ha only for the carbon atoms C1 and C6 of the glucose ring. For the softening point parameter, T_g, a significant correlation was noted with the NMR relaxation times of the glucose ring carbon atoms and with the methyl group (21.5 ppm).

It is generally difficult to locate exactly the positions of chemical coupling in the cell wall, and the hydrogen bonds, with the methods used. Indirect evidence is obtained from the FTIR spectra from changes in intensities in the whole "fingerprint" region and especially in the vibrations in the hydroxyl and ether bridge regions. Thus, the infrared spectra indicate that the chemical reactions partly break up and change the hydrogen bonds and association networks, and modify the composition of the fibrillar envelope of the viscoelastic units. These effects are also clearly indicated by the macroscopic and microscopic relaxation data discussed above.

The work of Boyd, and studies on the technical utilization of wood, indicate that the dimensional stability of wood is markedly increased by the removal of resins by extraction with organic solvents. The "static" data, NMR chemical shifts, and storage moduli of viscoelasticity, are therefore less sensitive and give weaker responses to viscoelastic changes for resin-free samples than for original wood. Thus, the weight changes on chemical modification of the samples found by Björkman and Lassota, and the non-polarized infrared spectra of the present work, seem generally to reflect the degree of chemical modification of the cell wall, and provide an indication of the extent of the reaction.

Acknowledgments

The authors are indebted to Nordisk Industrifond, Puututkimus OY, Pulp and Paper Research Institute, Otaniemi, and Chemical Instrument Center of the Faculty of Science, University of Helsinki, for financial and instrumental support.

Literature Cited

1. Björkman, A.; Lassota, H. *Abstr., Symp. on Viscoelasticity of Biomaterials, Boston*; American Chemical Society, April 22-27, 1990.
2. Lindberg, J. J.; Hortling, B. *Adv. Polymer Sci.* 1985, **66**, 1-22.
3. Isogai, A.; Usuda, M.; Kato, T.; Toshiyuki, U.; Atalla, R. H. *Macromolecules* 1989, **22**, 3168-72.
4. Sundquist, J.; Rantanen, T. *Paperi ja Puu* 1983, **65**, 733-737.
5. Salmen, L. *STFI-Kontakt* 1989, 14-15. Olsson, A.-M.; Salmen, L. *Abstr., Symp. on Viscoelasticity of Biomaterials, Boston*; American Chemical Society, April 22-27, 1990.
6. Barkas, W. W. *The Swelling of Wood Under Stress*; HMSO: London, 1949.
7. Boyd, J. D. In *New Perspectives in Wood Anatomy*; Haas, P., Ed.; The Hague, 1982; p 17. Cf. also *Nature* 1982, **299**, 775.
8. Hunt, D. G. *J. Mater. Sci. Lett.* 1989, **8**, 1474-1476.

RECEIVED December 16, 1991

Chapter 22

Molecular Relaxation of Cellulosic Polyelectrolytes with Water

T. Hatakeyama[1] and Hyoe Hatakeyama[2]

[1]Research Institute for Polymers and Textiles and [2]Industrial Products Research Institute, Tsukuba, Ibaraki 305, Japan

The molecular relaxation of the liquid crystalline state of the water-cellulosic polyelectrolyte system was investigated using nuclear magnetic resonance spectroscopy (NMR) and differential scanning calorimetry (DSC). DSC and polarizing light microscopic observation showed that the system forms the liquid crystalline state, when the water content (W_c) of the system ranges from 0.5 to 2.5 (W_c = grams of water/gram of dry sample) depending on the type of polyelectrolyte. The water-polyelectrolyte system forms a homogeneous mixture and has a regular molecular alignment. The temperature and the enthalpy of transition from the liquid crystalline state to the isotropic liquid state decreased with increasing W_c. The longitudinal relaxation time (T_1) and transverse relaxation time (T_2) of ^1H and ^{23}Na of a water-Na cellulose sulfate system were measured as functions of temperature and W_c in the range where the liquid crystalline state was formed. The ^1H T_2 value gradually decreased with decreasing temperature and showed a sudden decrease in the temperature range where the mobile fraction of water molecules in the system froze. The ^{23}Na T_{2slow} value showed a sudden decrease at the temperature where the liquid crystalline state was formed. NMR and DSC results suggested that the Na ion is associated with the main chain in the liquid crystalline state, and the increase of free and freezing bound water surrounding hydrophilic groups hinders the regular molecular alignment of polyelectrolyte molecules through the dissociation between the Na ion and the main chain.

0097–6156/92/0489–0329$06.00/0

The molecular motion of biopolymers is markedly affected by the presence of water. In the dry state, the main chain motion of biopolymers is generally not observed. The higher-order structure of proteins and polysaccharides is stable and the inter-molecular bonds are maintained at the thermal decomposition temperature of the main chain. However, if biopolymers sorb a small amount of water, the main chain motion is enhanced and can be detected by various measurement techniques (1-5).

We have reported that various types of water-polyelectrolyte systems form the liquid crystalline state when the water content of the systems is in a range from 0.5 to 2.5 W_c (6-9). This suggests that water molecules break the inter-molecular bonds of polyelectrolytes, and that the water-polyelectrolyte system forms a homogeneous mixture and is arranged in a regular molecular alignment in the temperature range from 280 to 310K, depending on W_c.

NMR and thermal studies suggest that the structure of water molecules is altered by the presence of ionic and hydrophilic groups of polymers (7-12). Water molecules tightly bound by ionic groups show no first-order phase transition. The number of the bound water molecules increased depending on valency and radius of each ion (13). It is also reported that the heat capacity values of proteins and saccharides which sorb water are smaller than the heat capacity values of absolute dry samples at the same temperature (14,15). On this account, it is necessary to take into consideration the role of water when we investigate the molecular relaxation of natural polymers.

In this study, we focus on the molecular relaxation of the liquid crystalline state of water-cellulosic polyelectrolytes using NMR and DSC.

Experimental

Samples. Sodium cellulose sulfate (NaCS) was obtained from Scientific Polymer Products, Inc. The degree of substitution (DS) was 2.26. Daiichi Seiyaku Co. supplied the carboxymethylcellulose (CMC) that was substituted by various ions: lithium (DS = 0.61), sodium (DS = 0.95), potassium (DS = 0.59), cesium (DS = 0.63), calcium (DS = 0.57). Na salt of xanthan was obtained from Kelco Co. The amount of the pyruvate group per tetrasaccharide unit of xanthan was ca. 0.4 (mol/mol).

The water content (W_c) of each sample was defined as follows: W_c = grams of water/gram of dry sample, (g/g).

Measurements. A Nicolet FT-NMR Model NT-200 WB and a Bruker FT-NMR Model MSL-300 were used for the NMR measurements. The longitudinal relaxation time (T_1) was measured by the 180-τ-90 degree pulse method. The transverse relaxation time (T_2) was measured by either the Meiboom-Gill variant of the Carr-Purcell method or the solid echo method.

A Leitz polarizing light micrograph, Orthoplan Pol, equipped with camera and temperature controller, was used for the observation of the higher-order structure of the sample.

A Perkin Elmer differential scanning calorimeter DSC-II was used for the measurement of phase transition. The sample weight was ca. 3 mg. An aluminum sample vessel for volatile samples was used. The scanning rate was 10K/min. The samples were weighed to the precision of $\pm$ 0.001 mg using a Sartorius microbalance. The sample preparation technique was as previously reported (3).

Results and Discussion

Elastic and transparent gel-like materials are formed when cellulosic polyelectrolytes sorb a small amount of water. At around room temperature (20°C) and in a W_c ranging from 0.5 to 2.5, an ordered structure was observed in the above systems under the polarizing light microscope. Figure 1 shows polarizing light micrographs of water-Na cellulose sulfate (Figure 1a), –Na xanthan (Figure 1b), –Na carboxymethylcellulose (Figure 1c), –K carboxymethylcellulose (Figure 1d), systems (W_c = ca. 1.0). A similar structure was also observed in the water-Cs carboxymethylcellulose and –Ca carboxymethylcellulose systems. When a color-sensitive plate was inserted in the microscope, bright portions of the photographs were observed as yellow or blue and both colors changed following rotation of the sample by 90°. This fact suggests that the molecular chains of the bright portions are aligned parallel to the glass plate. Figure 1 shows that the systems form liquid crystals. Liquid crystals were found to form not only in natural polyelectrolytes (4,6), but also in synthetic polyelectrolytes, such as in Na poly(styrene sulfonate) and in Na poly(hydroxystyrene sulfonate) derivatives, under the same conditions.

A transition from the liquid crystalline state to the isotropic liquid state was observed by DSC. The transition was thermally reversible, although the temperature difference between heating and cooling was large compared with that of the liquid crystalline transition observed in low molecular weight compounds. This large temperature difference can be attributed to the viscoelastic nature of the sample. Figure 2 shows the DSC heating curves of water-Na cellulose sulfate systems with various water contents. All of the samples were quenched from 330K to 150K and heated at the rate of 10K/min. The system shows a glass transition only until W_c reaches 0.38. Water molecules in the system are tightly bound to the ionic groups and show no first-order phase transition. In the W_c ranging from 0.38 to ca. 5.0, the system shows T_g and an endothermic peak (T^*). The peak temperature of T^* was higher than that of pure water. In the W_c exceeding 0.50, another endothermic peak appeared between T_{cc} and T^*. This was the melting peak of freezing water in the system (T_m). The temperature of the melting peak is lower than that of pure water. In W_c higher than 1.5, T_g, T_{cc} and T^* disappeared, and T_m was observed at the same temperature as pure water. The endotherm peak T^* was attributed to the transition from the liquid crystalline state to the isotropic liquid state. The T^* was also observed in the water-Li, Na, K, Cs, and Ca carboxymethylcellulose systems.

The above DSC data indicate that in the W_c range of 0.4 to 1.5, the water-NaCS forms a liquid crystalline state. If we assume that the hydroxyl group interacts with one water molecule (3,16) and the remaining water

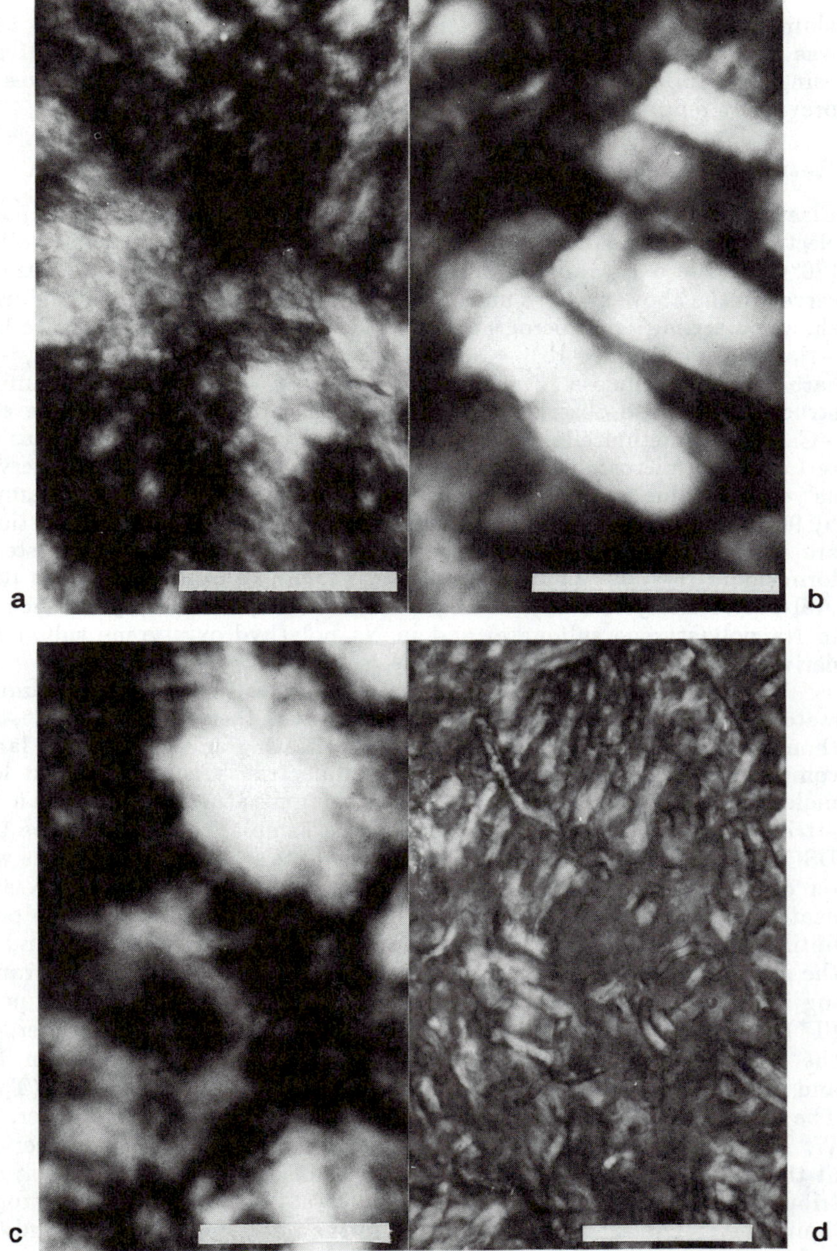

Figure 1. Polarizing light micrographs of water-polyelectrolyte systems. water-Na cellulose sulfate (a); –Na xanthan (b); –Na carboxymethylcellulose (c); K carboxymethylcellulose (d); (W_c = ca. 1.5), Scale bar = 100μm.

molecules interact with the ionic group, the number of water molecules attached to each repeating unit can be calculated. From 4 to 20 water molecules are strongly restricted by the hydrophilic group.

Figure 3 shows the phase diagram of water-NaCS. The transition temperature in the phase diagram was determined by DSC heating curves of the samples quenched from the isotropic liquid state to 150K and heated at 10K/min. The phase diagrams of the other water-polyelectrolyte systems showed a similar tendency (8,9), i.e., the temperature range of the liquid crystalline state decreased with increasing water content.

The enthalpy of the liquid crystalline transition (ΔH^*) was calculated from the peak area. It was found that the enthalpy decreased with increasing W_c when we used the summation of the dry sample and non-freezing water for the normalization of enthalpy value. Figure 4 shows the relationship between ΔH^* and W_c of the water-NaCS system at T^*.

To investigate the molecular relaxation of the liquid crystalline state of water-polyelectrolytes, the transition from the liquid crystalline state to the isotropic liquid state was investigated by ^1H and ^{23}Na nuclear magnetic relaxation.

Figure 5 shows ^1H T_1, ^1H T_2 and τ_c of the water-NaCS system (W_c = 0.86) as a function of inverse temperature. The τ_c was calculated on the basis of several assumptions as reported in our previous studies (4,7). From Figure 3, the liquid crystalline state can be observed from 242K to 293K at $W_c = 0.86$. The amount of freezing water can be calculated from the melting enthalpy as previously reported. When W_c is 0.86, 56% of the total amount of water in the system forms ice having an irregular crystalline structure in the water-NaCS system. The ice in the above system started to melt at around 253K, and a melting peak was observed at 270K. The peak temperature of melting (T_m) observed by DSC is indicated as a broken line in Figure 5. In both T_2 and τ_c curves, the inflection point can be observed at around 270K, although temperatures obtained from NMR measurements were not as precise as those obtained by DSC. It is also necessary to keep in mind that the thermal history of the sample used in the NMR study was not the same as that used in DSC. In spite of the above differences, the inflection point agreed fairly well with T_m values obtained by DSC. The inflection point observed for the water-NaCS system having various W_c's agreed well with each T_m value. The T_2 value in Figure 5 is considered to express the average molecular mobility. Figure 5 shows that the motion of ^1H of water molecules in the water-NaCS system is enhanced at temperatures higher than T_m.

Figure 6 is a three-dimensional diagram showing the relationship between ^1H T_2, inverse absolute temperature and water content (W_c). The liquid crystalline state is shown as a triangular peak in the figure. The T^* and T_m lines show the transition temperatures observed by DSC. It can be seen that the liquid crystalline state does not easily form when the mobility of water molecules increases. DSC data shown in Figure 3 suggest that the melting peak temperature (T_m) of water-NaCS approaches the T_m of pure water at around $W_c = 1.2$. The T_2 value shown in Figure 6 starts to

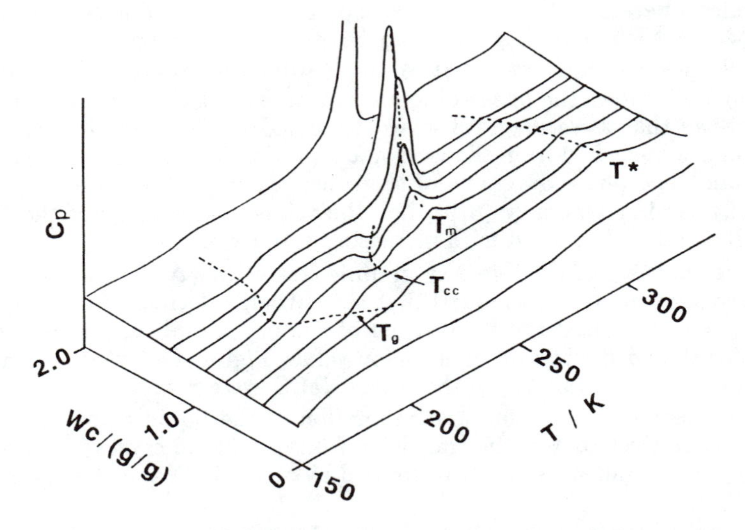

Figure 2. DSC curves of water-NaCS with various water contents.

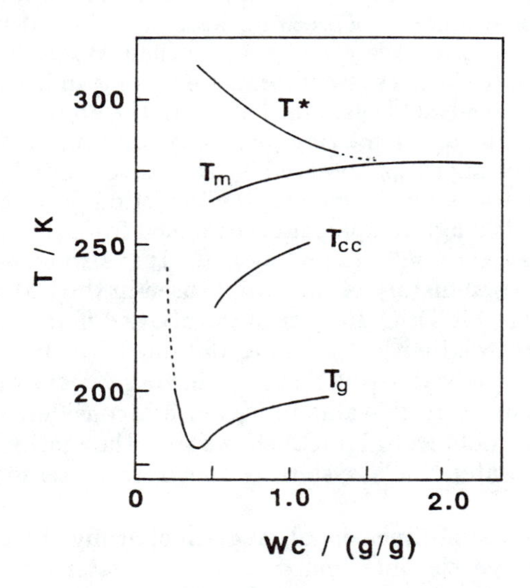

Figure 3. Phase diagram of water-NaCS system. T_g, glass transition; T_{cc}, cold-crystallization; T_m, melting; T*, liquid crystalline transition.

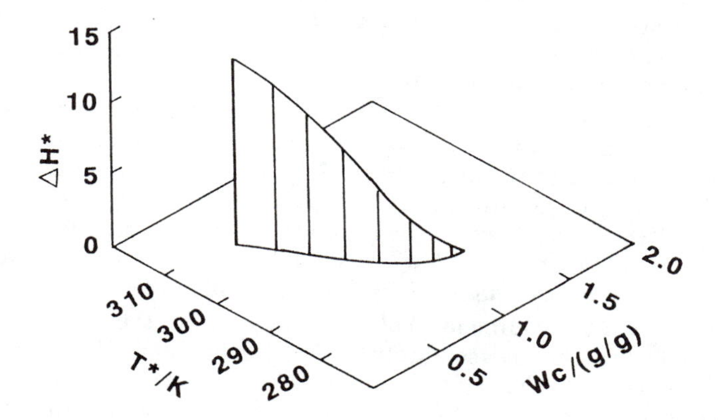

Figure 4. Three-dimensional diagram showing the relationship between enthalpy of liquid crystal transition (ΔH^*), temperature and water content (W_c) of water-NaCS system.

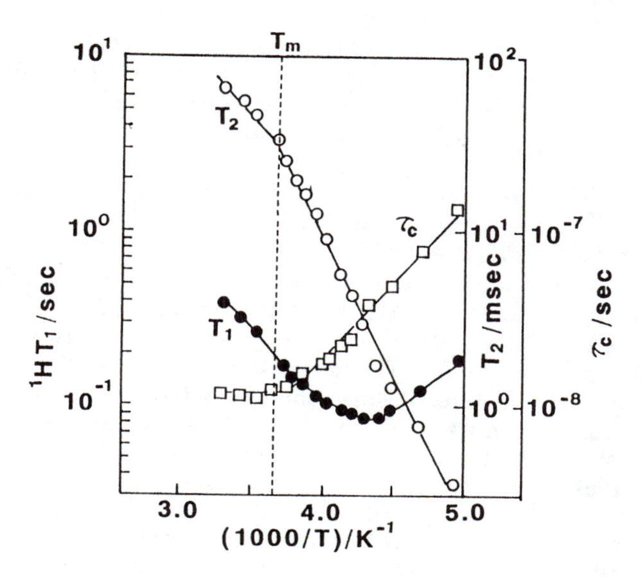

Figure 5. Longitudinal relaxation time (T_1), transverse relaxation time (T_2) and correlation time τ_c as a function of temperature of the water-NaCS system with water content 0.86.

increase markedly at around $W_c = 1.2$, and the area that forms the liquid crystalline state is reduced. In contrast, the T_2 value at $W_c = 0.86$ is almost the same in the broad temperature range, suggesting that the system has a stable liquid crystalline state.

Figure 7 is a three-dimensional diagram showing the relationship between ^{23}Na T_1, the inverse absolute temperature and water content (W_c). The hatched area shows the liquid crystalline state. The symbols C, LC and L show the crystalline state, the liquid crystalline state and the isotropic state, respectively. As shown in the diagram, T_1 values gradually decrease at temperatures higher than T^* when W_c is between 0.5 and 1.5, and also decrease at temperatures higher than T_m when W_c exceeds 1.5.

Figure 8 is a three-dimensional diagram showing the relationship between ^{23}Na T_{2slow}, the inverse absolute temperature, and W_c. The symbols are the same as those shown in Figure 7. The temperature and W_c where the T_2 value shows the inflection point agreed well with T^* and T_m measured by DSC. The fast and slow components exist in T_2, because the ^{23}Na nucleus shows quadrapole relaxation, although the decrease of the fast component with temperature is not shown in Figure 8. The fast component in the T_2 value decreases with temperature in the same manner as the slow component.

The results of Figures 7 and 8 suggest that ^{23}Na T_1 and T_2 are closely related to the mobility of polyelectrolyte molecules. When the system forms the liquid crystalline state, the molecular motion of NaCS in the system is restricted. When the Na ion is dissociated from the cellulose sulfate group in NaCS, the T_2 value is low. The above results agree well with the enthalpy of transition from the liquid crystalline state to the isotropic liquid state (ΔH^*).

Figure 9 is a three-dimensional diagram showing the relationship between ΔH^*, ^{23}Na T_{2slow} at T^*, and W_c. From this figure, it can be seen that the ΔH value is high when T_2 and W_c are low.

The above results obtained from ^{1}H and ^{23}Na NMR relaxation agreed well with those obtained by DSC. From the variation of ^{1}H T_2 with temperature, the inflection point where the T_2 values started to decrease was obtained for each system W_c. The inflection point observed in ^{23}Na T_{2slow} was also calculated. Figure 10 shows the phase diagram of water-NaCS in the temperature region where the liquid crystalline state is observed. Figure 10 also shows the temperatures at the inflection points of the ^{1}H and ^{23}Na T_2 curves vs. inverse absolute temperature. As shown in Figure 10, the inflection point obtained from ^{23}Na T_2 is located in the temperature range where the liquid crystalline state is observed, while the inflection point obtained from ^{1}H T_2 is located in the crystalline region. As already mentioned in the above section, the transition temperature observed by DSC does not correspond well with the temperature observed by NMR. In spite of the above discrepancy, Figure 10 shows that the temperature ranges obtained from DSC and NMR agree well. At a higher W_c range, both ^{1}H and ^{23}Na T_{2slow} show the same tendency, suggesting that the Na ion is dissociated from the main chain of polyelectrolytes and moves co-

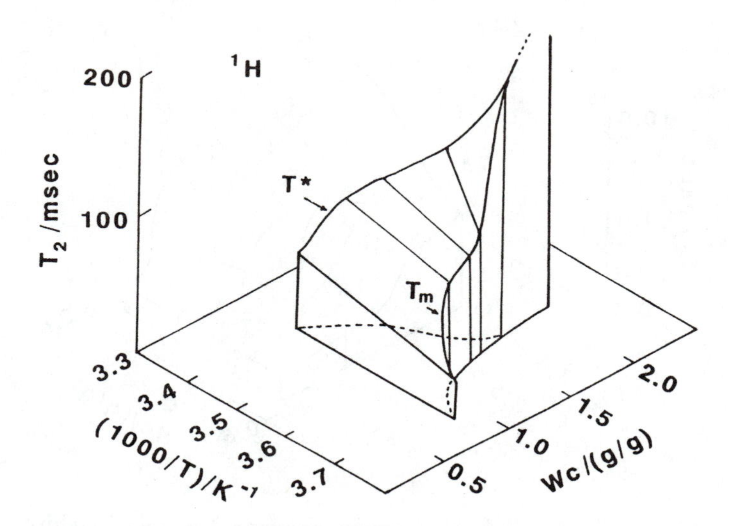

Figure 6. Three-dimensional diagram showing the relationship between ^{1}H T_2, temperature and W_c of water-NaCS system in the liquid crystalline state. T_m, melting of water; T*, liquid crystalline transition.

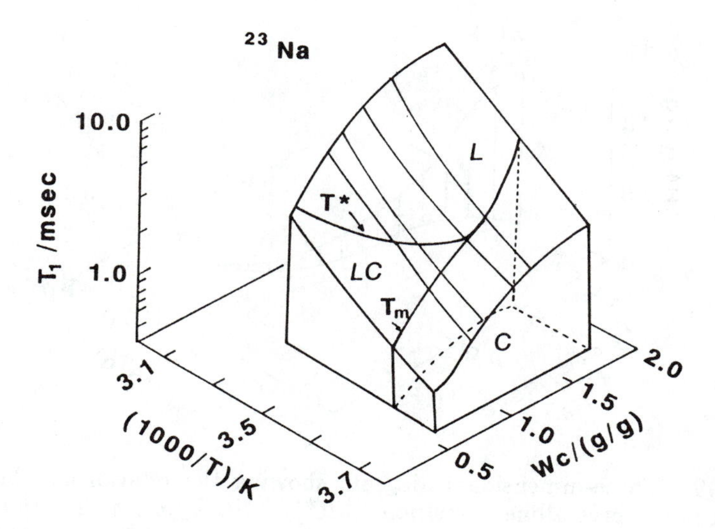

Figure 7. Three-dimensional diagram showing the relationship between ^{23}Na T_1, temperature and W_c of water NaCS system. L, isotropic liquid; LC, liquid crystal; C, crystal; T_m, melting of water; T*, liquid crystalline transition.

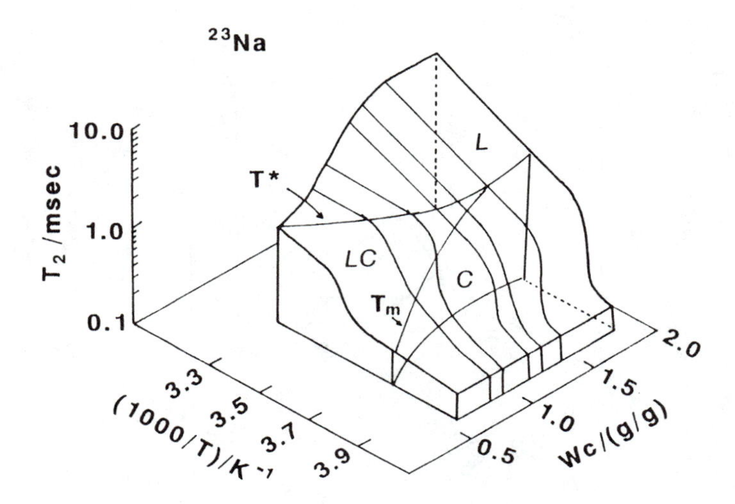

Figure 8. Three-dimensional diagram showing the relationship between ^{23}Na T_{2slow}, temperature and W_c of water-NaCS system. L, isotropic liquid; LC, liquid crystal; C, crystal; T_m, melting of water; T*, liquid crystalline transition.

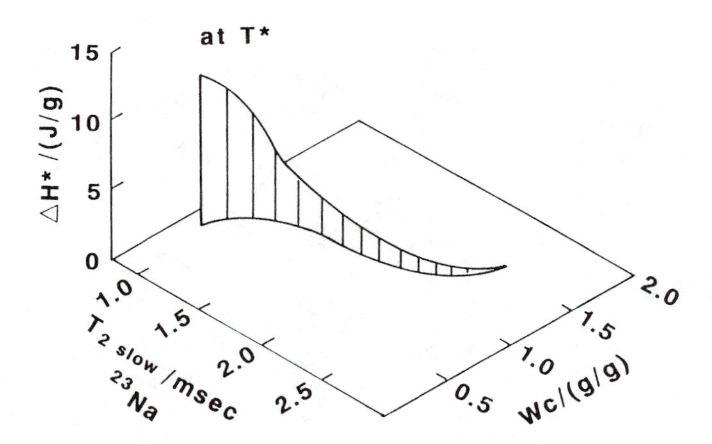

Figure 9. Three-dimensional diagram showing the relationship between heat of liquid crystalline transition (ΔH*), ^{23}Na T_{2slow} at T* and water content (W_c) of water-NaCS.

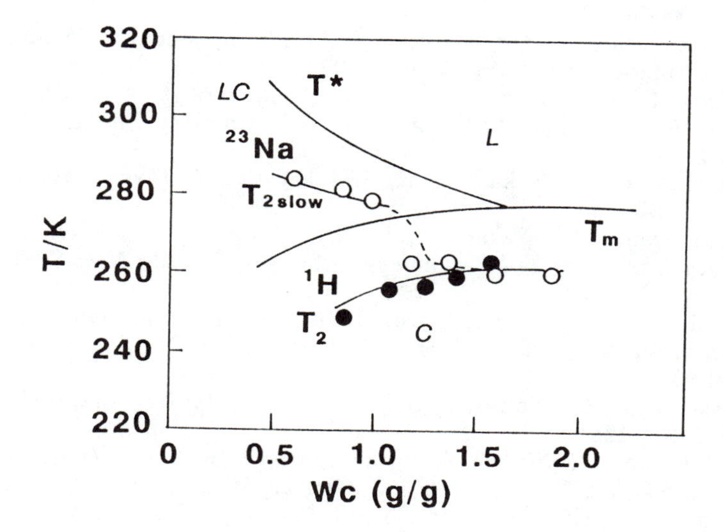

Figure 10. Relationship between transition temperature and water content of water-NaCS systems. ○, ^{23}Na T_{2slow}; ●, ^{1}H T_2; solid line, DSC endothermic peak.

operatively with water molecules. The results shown in Figure 10 suggest that the molecular motion of cellulose sulfate is reflected by the variation of ^{23}Na T_2 and the molecular mobility of water surrounding the main chain is reflected by ^{1}H T_2.

NMR and DSC results suggest that the Na ion is associated with the main chain in the liquid crystalline state, and that the increase of free and freezing bound water surrounding hydrophilic groups inhibits the regular molecular alignment of polyelectrolytes through the dissociation of the Na ion from the main chain by the influence of water molecules.

Literature Cited

1. Rowland, S. P. *Water in Polymers*; ACS Symp. Ser. 127; Washington, DC: American Chemical Society, 1980.
2. Kuntz, I. D. Jr. In *Proteins at Low Temperatures*; Adv. Chem. Ser. 180; Fennema, O., Ed.; Washington, DC: American Chemical Society, 1981.
3. Hatakeyama, T.; Nakamura, K.; Hatakeyama, H. *Thermochimica Acta* 1988, **123**, 153.
4. Hatakeyama, H.; Nakamura, K.; Hatakeyama, T. In *Cellulose and Wood Chemistry and Technology*; Schuerch, C., Ed.; New York: John Wiley, 1989; p 419.
5. Horii, F.; Hirai, A.; Kitamaru, R. In *Polysaccharide Solutions*; ACS Symp. Ser. 260; Arthur, J. Jr., Ed.; Washington, DC: American Chemical Society, 1984; p 27.
6. Hatakeyama, T.; Nakamura, K.; Yoshida, H.; Hatakeyama, H. *Polymer* 1987, **28**, 1281.
7. Hatakeyama, H.; Hirose, S.; Hatakeyama, T. In *Lignin: Properties and Materials*; ACS Symp. Ser. 397; Glasser, W. G.; Sarkanen, S., Eds.; Washington, DC: American Chemical Society, 1989; p 274.
8. Hatakeyama, T.; Nakamura, K.; Yoshida, H.; Hatakeyama, H. *Food Hydrocolloids* 1989, **3**, 301.
9. Yoshida, H.; Hatakeyama, T.; Hatakeyama, H. *Polymer*, in press.
10. Samoilov, O. Ya. (trans. D. J. G. Ivens). *Structure of Aqueous Electrolyte Solutions and the Hydration of Ions*; New York: Consultants Bureau, 1965.
11. Gekko, K. *Food Hydrocolloids* 1989, **3**, 289.
12. Komoroski, R. A. In *Ions in Polymers*; ACS Symp. Ser.; Eisenberg, A., Ed.; Washington, DC: American Chemical Society, 1980.
13. Nakamura, K.; Hatakeyama, T.; Hatakeyama, H. In *Wood and Cellulosics: Industrial Utilization, Biotechnology, Structure and Properties*; Kennedy, J. F.; Phillips, G. O.; Williams, P. A., Eds.; Chichester: Ellis Horwood Ltd., 1987; p 97.
14. Yang, R.-H.; Ruply, J. A. *Biochem.* 1979, **18**, 2654.
15. Hatakeyama, T. In *Cellulose, Structural and Functional Aspects*; Kennedy, J. F.; Phillips, G. O.; Williams, P. A., Eds.; Chichester: Ellis Horwood Ltd., 1989; p 43.
16. Nakamura, K.; Hatakeyama, T.; Hatakeyama, H. *Polymer* 1983, **21**, 871.

RECEIVED December 16, 1991

Chapter 23

Solution Properties of a Hydrophobically Associating Cellulosic Polymer

Electron Spin Resonance Spectroscopy

P. A. Williams, J. Meadows, Glyn O. Phillips, and R. Tanaka

North East Wales Institute of Higher Education, Connah's Quay, Clwyd CH5 4BR, Wales

Electron spin resonance spectroscopy has been used to investigate the solution properties of a hydrophobically associating cellulosic polymer. Nitroxide spin labels covalently attached to the cellulosic backbone have given information with regard to the segmental motion of the polymer chains, whereas nitroxide spin probes have demonstrated the formation of regions of hydrophobicity above a critical polymer concentration. The data is consistent with the formation of an extensive three-dimensional network in which the hydrophilic cellulosic backbones are effectively cross-linked by the intermolecular association of neighbouring hydrophobic side chains. The work has been extended to study the interaction of the polymer with sodium dodecyl sulphate surfactant and the electron spin resonance data has been used to elucidate the mechanism of interaction, and to explain the unusual rheological behaviour.

Hydrophobically associating polymers are finding increasing industrial use due to their ability to impart improved rheological behavior to particulate dispersions (1,2). Consequently, there have been a number of research publications concerning such polymers over recent years (3–13). Essentially, these polymers consist of a hydrophilic backbone and possess a small number of hydrophobic side chains, usually in the range of 8 to 40 carbon atoms in length. Whilst the nature of their backbone usually renders the polymer soluble in aqueous media, intermolecular association of the hydrophobic groups leads to the formation of a weak three-dimensional network structure giving rise to solutions of very high viscosity at low shear rates (8). Addition of surfactants to solutions of hydrophobically associating polymers has also been shown to have a dramatic effect on the rheological properties

0097–6156/92/0489–0341$06.00/0

(8–11). For example, Gelman (11) found that the addition of 0.04% sodium oleate to a 0.7% aqueous solution of hydrophobically modified hydroxyethyl cellulose produced an approximate hundred-fold increase in the Brookfield viscosity of the solution.

Nitroxide spin labels and spin probes are ideally suited to studying the dynamics of these systems since the nitroxide free radical is able to monitor molecular events in the 10^{-7} to 10^{-11} s timescale. Spin labels, which (by definition) are covalently attached to the backbone, can give information regarding the segmental motion of the polymer chains, whilst specially selected spin probes present in solution (but not covalently attached) can monitor the formation of intermolecular hydrophobic associations.

This paper reports on the use of electron spin resonance spectroscopy (ESR) to study the solution properties of hydrophobically modified hydroxyethyl cellulose (HMHEC) and its interactions with the anionic surfactant, sodium dodecyl sulphate (SDS).

Materials

HMHEC was kindly supplied by Aqualon (UK) Ltd., Warrington, UK, under the trade name Natrosol Plus Grade 330. The manufacturer reports the polymer to have a molecular mass of approximately 250,000, a molar substitution of 3.3, and to contain approximately 1–2% of chemically grafted C_{12}–C_{24} alkyl side chains.

A portion of the HMHEC was spin labelled with 4-amino Tempo (Sigma Chemicals Ltd., Poole, UK) as previously described (9). The spin label attaches to hydroxyl groups of the glucose residues and it was estimated that there was 1 spin label per 7,000 residues. This low degree of labelling ensures minimum perturbation of the polymer characteristics.

The spin probe used was 5-doxyl stearic acid (5-DSA; Sigma Chemicals Ltd.), and was used as supplied.

Methods

ESR Spectroscopy. The nitroxide free radical gives rise to a well characterized three-lined ESR spectrum and the relative shapes and intensities of the lines are a reflection of the mobility of the nitroxide moiety. If the motion of the nitroxide radical is unrestricted, then the three lines are relatively narrow and are of similar intensities. However, as the mobility of the free radical is reduced, line broadening occurs due to anisotropic effects. In spin label experiments where the nitroxide moiety is covalently attached to the polymer chain, it is argued that the shape of the ESR spectrum closely reflects the segmental motion of the polymer (14). This is illustrated in Figure 1, which shows the ESR spectra for spin labelled hydroxyethyl cellulose in 80% v/v aqueous glycerol as a function of temperature. At high temperatures (spectrum a) where the solution viscosity is lowest, the motion of the polymer segments is relatively unrestricted, resulting in a motionally narrowed isotropic spectrum. As the temperature is reduced and the solution viscosity increases, the segmental motion of the polymer is reduced

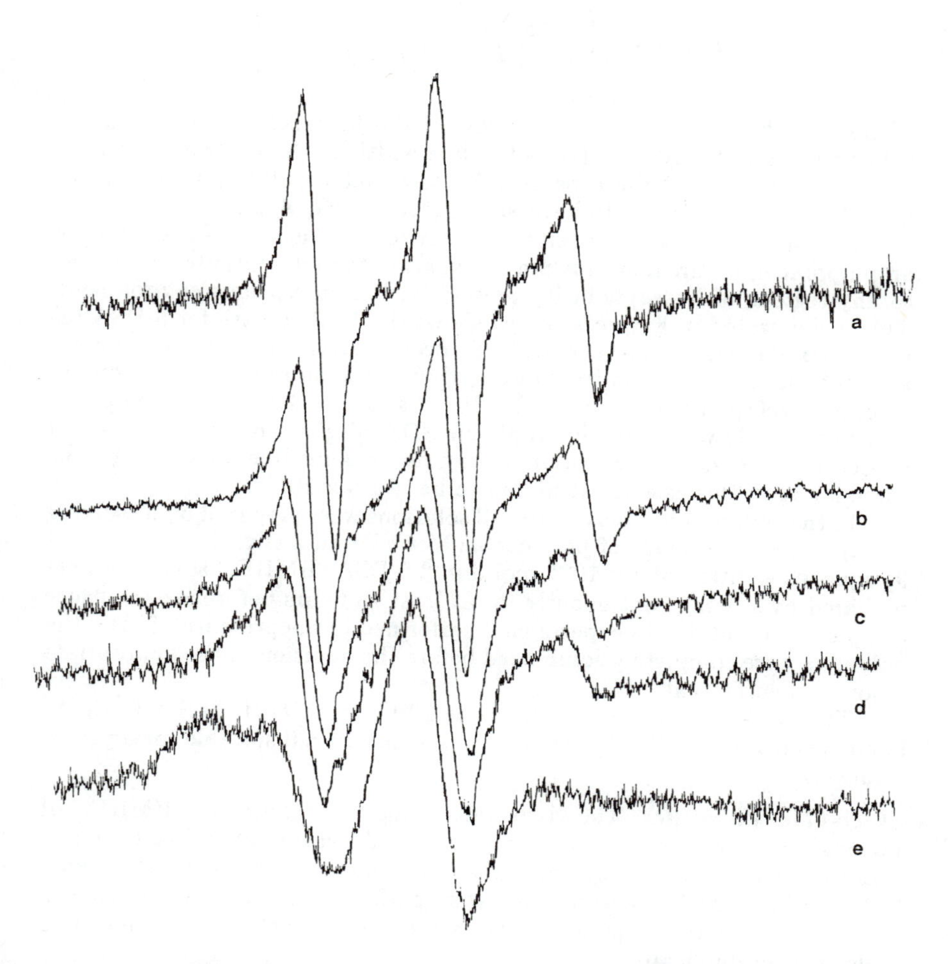

Figure 1. ESR spectra of a 2% solution of spin labelled HMHEC in 80% glycerol at (a) 80°C; (b) 60°C; (c) 40°C; (d) 30°C; and (e) 5°C.

resulting in line broadening (spectra b–e). For isotropic spectra the rotational correlation time τ_c of the label can be calculated using the equation of Stone *et al.* (15), which shows that:

$$\tau_c = KW_0 \left\{ \left(\frac{h_0}{h_{+1}}\right)^{0.5} + \left(\frac{h_0}{h_{-1}}\right)^{0.5} - 2 \right\}$$

where h_{-1}, h_0 and h_{+1} are the heights of the high field, central and low field lines respectively, and W_0 is the line width of the central line. Zhao *et al.* (16) calculated the constant, K, to be 6.08×10^{-10}, assuming the hyperfine coupling tensor to have values $A_{zz} = 32G$, A_{xx} and $A_{yy} = 5G$.

The spin probe experiments were carried out using 5-DSA which is amphiphilic in nature and thus can be expected to preferentially reside close to any regions of hydrophobicity present within an aqueous environment. This is illustrated in Figure 2, which shows the ESR spectra for 5-DSA (a) in water and (b) in the presence of SDS micelles. In the former environment, the probe undergoes rapid tumbling, giving rise to an isotropic spectrum of typical correlation time $\tau_c = \sim 2.3 \times 10^{-10}$ s. In the latter, the spin probe prefers to reside within the hydrophobic SDS micelles and this results in a reduction in its molecular motion, giving rise to some line broadening. The value for τ_c in this case is calculated to be $\sim 1.5 \times 10^{-9}$ s.

In the spin probe experiments, all solutions were prepared by dissolving the appropriate amount of polymer and/or SDS in a slightly alkaline (pH 9) aqueous solution of 5×10^{-6} mol dm^{-3} 5-DSA. HMHEC solutions were prepared by stirring continuously for at least 18 hours before use to ensure complete dissolution. In experiments performed in the presence of SDS, the polymers were completely solubilized before the addition of the appropriate amount of surfactant.

The ESR spectra were recorded at 20°C on a JEOL JES ME 1X X-band spectrometer (JEOL, Japan) using a quartz cell suitable for aqueous solutions.

Shear-Flow Viscosities. The viscosities of aqueous solutions of HMHEC of various concentrations were determined over the shear rate range 0–10 s^{-1} using a Carrimed CS100 controlled stress rheometer (Carrimed Instruments Ltd., Dorking, UK). Measurements were performed at 20°C using either a 4 cm 2° or a 2 cm 2° cone and plate attachment. Each measurement was performed in duplicate.

Oscillatory Measurements. The storage and loss moduli (G′ and G″ respectively) of 2% aqueous solutions of HMHEC at 20°C containing various amounts of added SDS were recorded at an amplitude of 6×10^{-3} radians over the frequency range $10^{-2} - 10$ Hz using a Carrimed CS100 controlled stress rheometer fitted with either a 4 cm 2° or a 2 cm 2° cone and plate attachment.

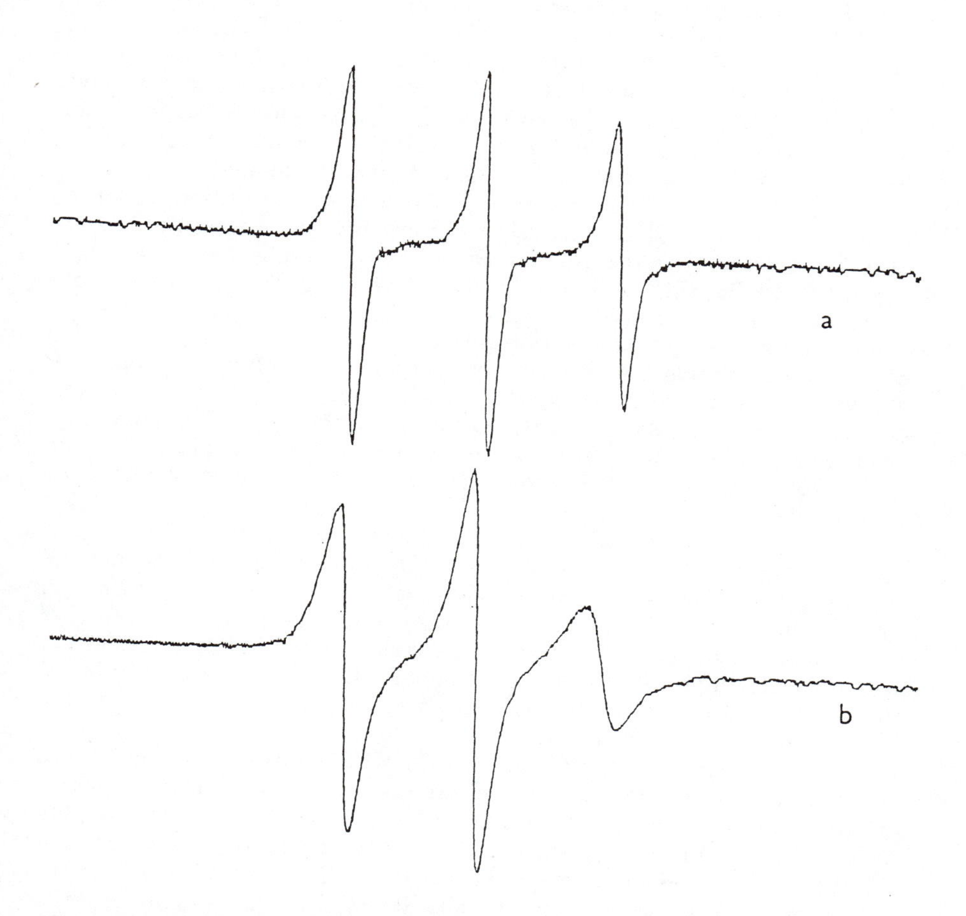

Figure 2. ESR spectra of 5×10^{-6} mol dm^{-3} 5-DSA in (a) water and (b) 10^{-2} mol dm^{-3} aqueous SDS solution.

Results

The viscosities of aqueous solutions of HMHEC at a shear rate of 1 s^{-1} are given as a function of polymer concentration in Figure 3. The viscosities of the solutions are seen to increase exponentially over the concentration range studied. The ESR spectra of spin labelled HMHEC in aqueous solution over a similar concentration range are given in Figure 4. The letters on the spectra correspond to the letters on the viscosity/concentration curve given in Figure 3. The spectra obtained were all motionally narrowed indicating a high degree of segmental motion. τ_c for all the spectra was calculated to be $1.2 \pm 0.2 \times 10^{-9}$ s, irrespective of the polymer concentration.

Figure 5 gives the ESR spectra of 5-DSA in aqueous solutions of various concentrations of HMHEC. At relatively low polymer concentrations the observed spectra (spectrum a) indicates the spin probe has a very high degree of mobility with τ_c having a value of $\sim 2.3 \times 10^{-10}$ s. However, above a polymer concentration of approximately 0.2% (9), the spectra is seen to contain both isotropic and anisotropic components, with the proportion of the latter increasing with increasing polymer concentration. Computer analysis indicates that at a polymer concentration of 1.5% the anisotropic component corresponds to approximately 70% of the signal. The composite spectra are believed to arise from the partitioning of the spin probe into hydrophobic regions created through the intermolecular association of the polymer molecules.

The effect of added SDS on G′ (at a frequency of 1 Hz) of 2% aqueous solutions of HMHEC is given in Figure 6. The value of G′ is seen to increase markedly with increasing surfactant addition up to an SDS concentration of approximately 8×10^{-3} mol dm^{-3}. Above this value, however, further additions of surfactant produce a progressive decrease in the values of G′. At sufficiently high concentrations of added SDS, the storage modulus of the polymer/surfactant system is actually lower than that of 2% HMHEC in the absence of any surfactant.

The variation of G′ and G″ of a 2% HMHEC solution alone, and in the presence of 8×10^{-3} mol dm^{-3} SDS are given as a function of frequency of oscillation in Figure 7. The values of both G′ and G″ are considerably increased in the presence of surfactant and furthermore G′ is greater than G″ over a wider frequency range. This closely reflects the difference in the appearance of the samples. In the absence of SDS, the sample is fluid, whereas in the presence of this concentration of SDS, the sample appears gel-like. However, the ESR spectrum of a 2% aqueous solution of spin labelled HMHEC in the presence of 8×10^{-3} mol dm^{-3} SDS is given in Figure 8 and is almost identical to that for the polymer in the absence of SDS indicating that the segmental motion of the polymer is virtually unchanged.

The ESR spectra of 5-DSA in the presence of 0.2% HMHEC and various amounts of SDS are given in Figure 9. This polymer concentration corresponds to the maximum concentration which will not itself affect the observed ESR spectrum of 5-DSA (9). It is seen that the mobility of the spin probe is progressively reduced, indicating the formation of regions of

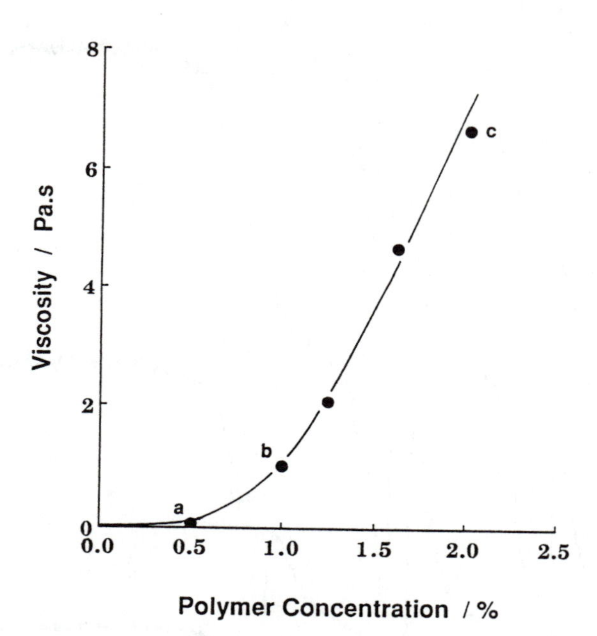

Figure 3. The effect of polymer concentration on the viscosities of aqueous solutions of HMHEC at a shear rate of 1 s^{-1}.

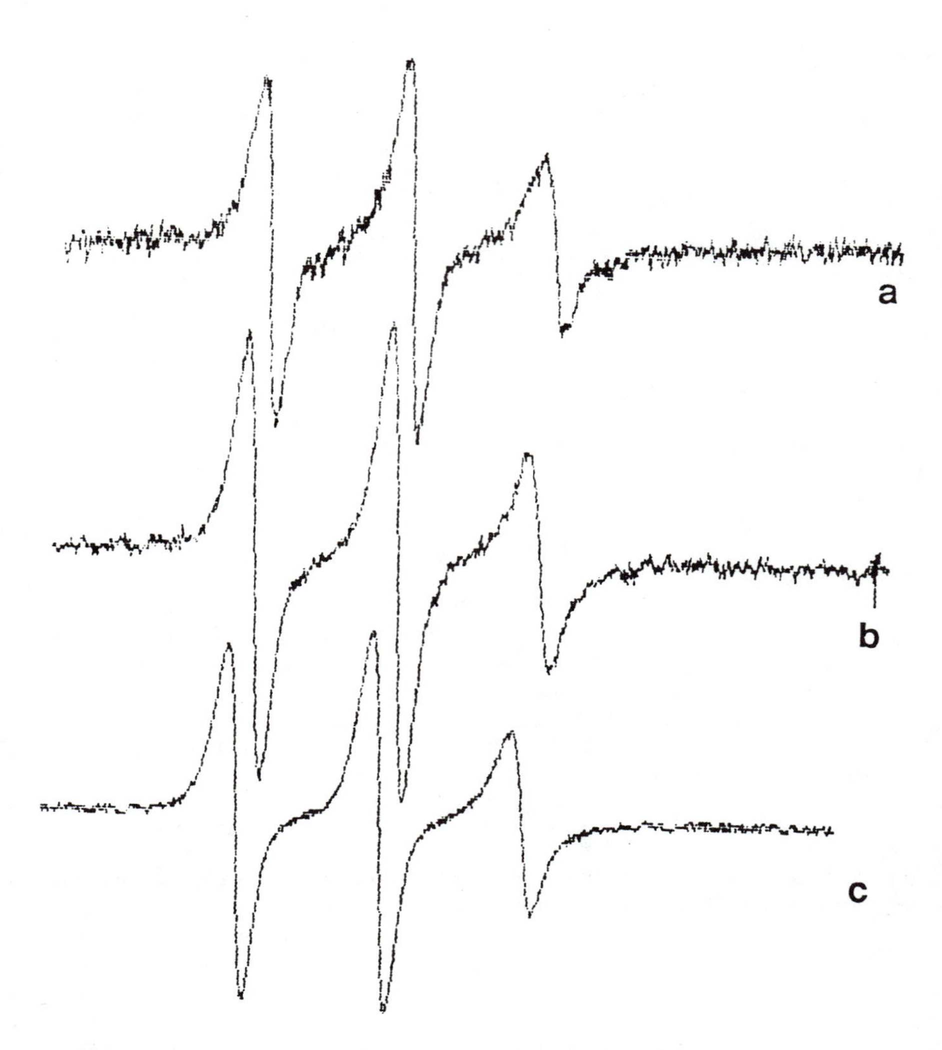

Figure 4. ESR spectra of spin labelled HMHEC in aqueous solution at concentrations of (a) 0.5%; (b) 1.0%; and (c) 2.0%.

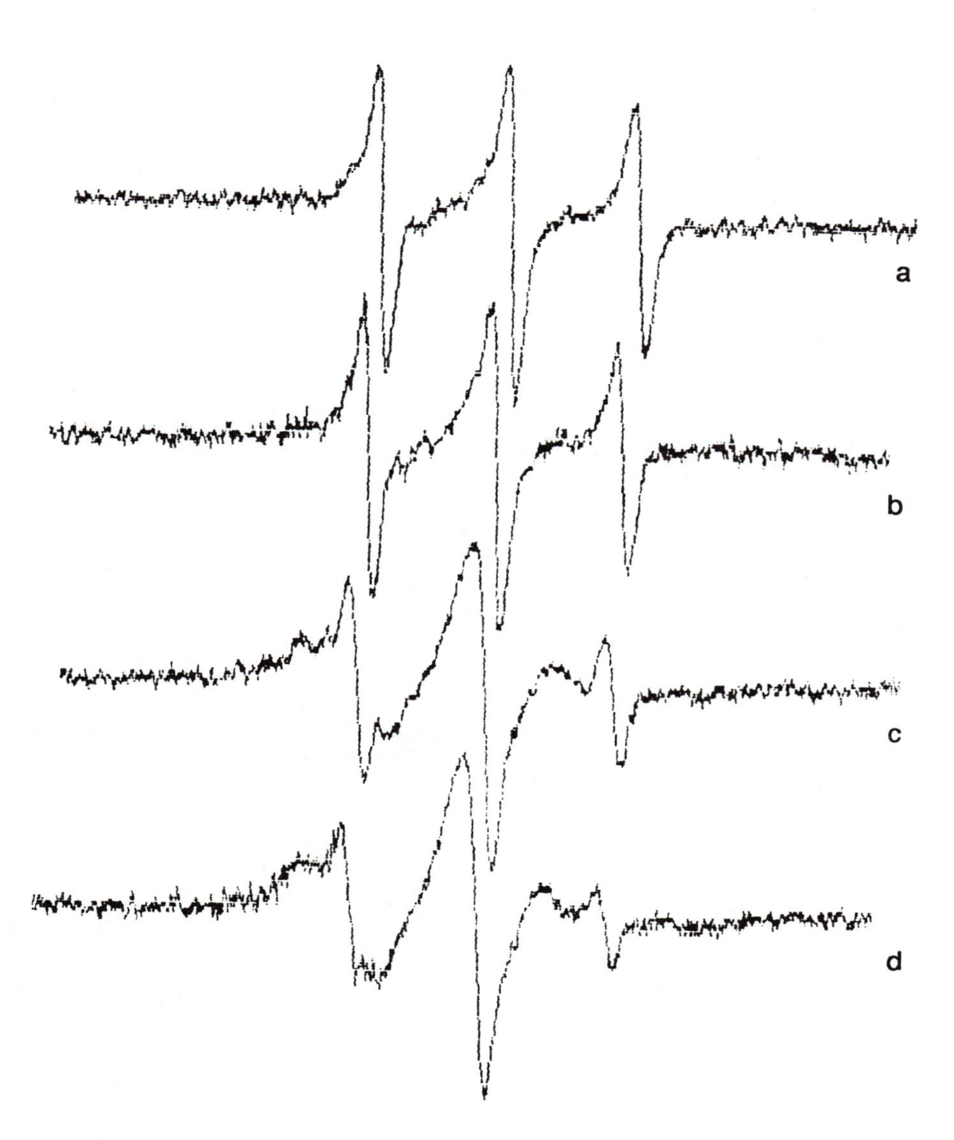

Figure 5. ESR spectra of 5×10^{-6} mol dm^{-3} 5-DSA in aqueous solutions of (a) 0.1% HMHEC; (b) 0.4% HMHEC; (c) 0.75% HMHEC; and (d) 1.5% HMHEC.

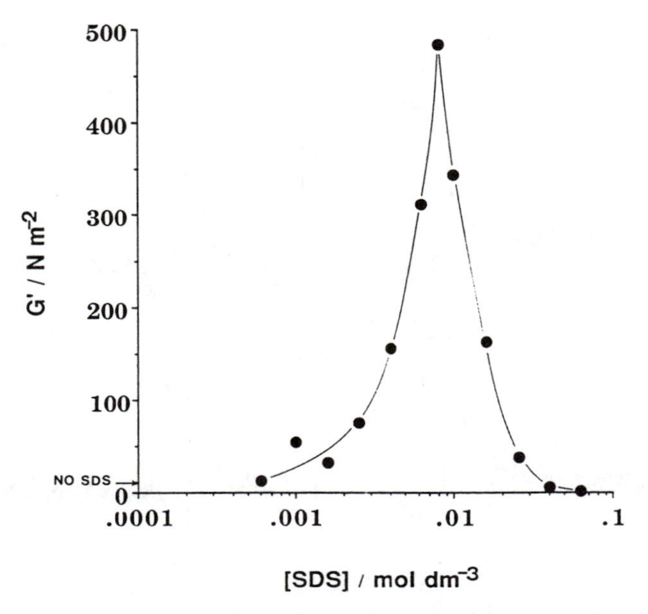

Figure 6. The effect of added SDS on the storage moduli, G′ (at a frequency of 1 Hz) of 2% aqueous solutions of HMHEC.

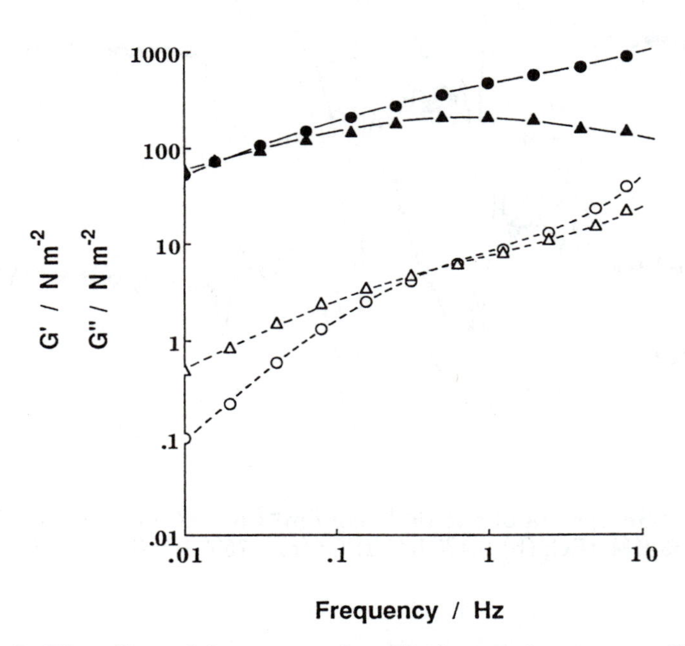

Figure 7. The effect of frequency of oscillation on the storage, G′ (○ and ●) and loss, G″ (△ and ▲) moduli of a 2% aqueous solution of HMHEC, alone (open symbols) and in the presence of 8×10^{-3} mol dm^{-3} SDS (closed symbols), respectively.

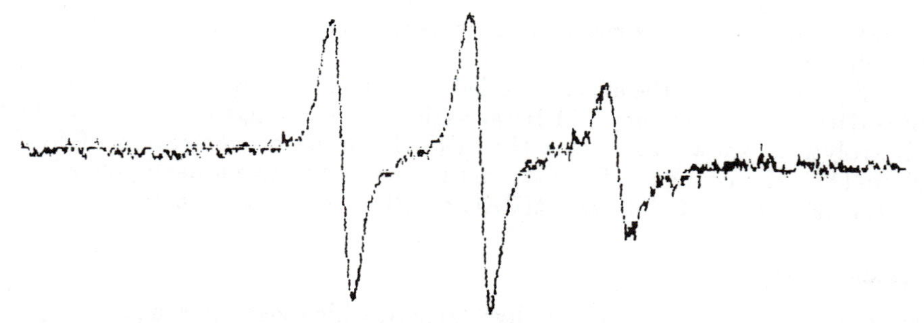

Figure 8. ESR spectra of a 2% aqueous solution of spin labelled HMHEC in the presence of 8×10^{-3} mol dm^{-3} SDS.

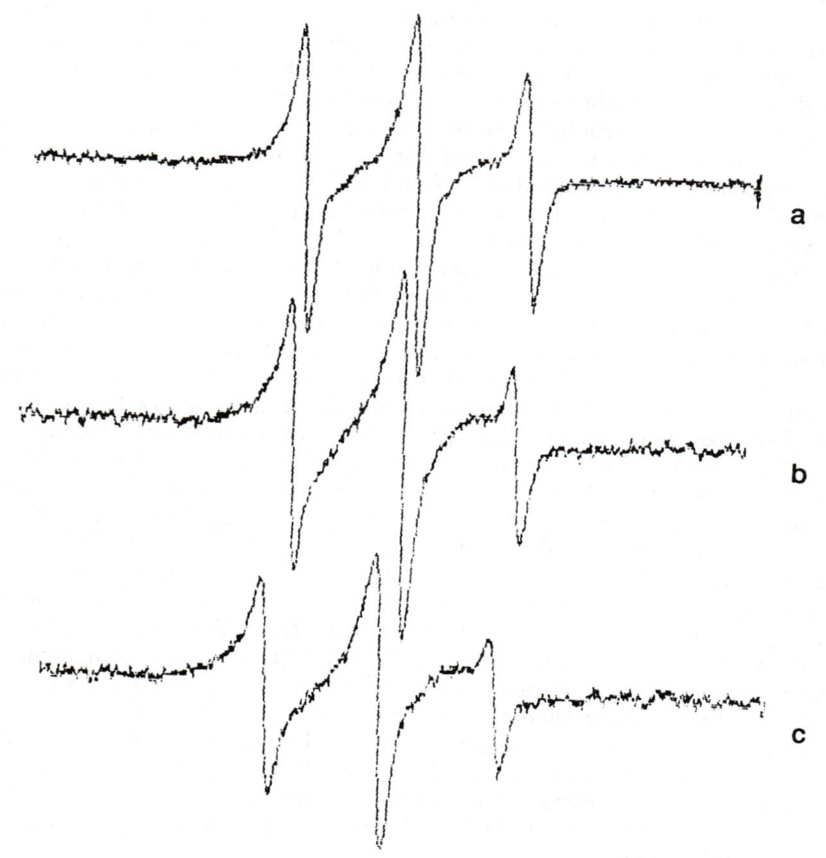

Figure 9. ESR spectra of 5×10^{-6} mol dm^{-3} 5-DSA in a 0.2% aqueous solution of HMHEC containing (a) $10^{-2.8}$ mol dm^{-3} SDS; (b) $10^{-2.5}$ mol dm^{-3} SDS; and (c) $10^{-2.2}$ mol dm^{-3} SDS.

hydrophobicity, above surfactant concentrations of approximately 2×10^{-3} mol dm^{-3}.

Figure 10 shows the effect of added SDS on G′ for 2% aqueous solutions of HMHEC at a frequency of 1 Hz at various ionic strengths. Although the electrolyte is known to reduce the critical micelle concentration of SDS alone (17), the change in ionic strength appears to have a negligible effect on the variation of G′ of the HMHEC solutions with surfactant addition.

Discussion

In contrast to hydroxyethyl cellulose (HEC), which viscosifies aqueous systems through chain entanglements, the comparatively high solution viscosities exhibited by HMHEC solutions are thought to reflect the formation of an extensive three-dimensional polymer network in which the hydrophilic cellulosic backbones of HMHEC are effectively cross-linked through the intermolecular association of neighboring hydrophobic side chains (8). The ESR data presented in Figure 5 show distinct changes in line shape of the 5-DSA spin probe above polymer concentrations of approximately 0.2% suggesting that hydrophobic associations occur at a critical polymer concentration analogous to the critical micelle concentration (cmc) of free surfactants as originally suggested by Landoll (18). This value is considerably less than the expected coil overlap concentration (C*) of approximately 1%, as calculated from the relationship proposed by Morris *et al.* (19), (*i.e.*, $C^* = 4/[\eta]$), but is of comparable magnitude to that observed viscometrically by ourselves (9) and by Goodwin *et al.* (20) for a similar HMHEC sample.

The spin label studies (Figure 4) indicate that despite the dramatic increase in the bulk viscosities of aqueous HMHEC solutions over the concentration range 0–2% (Figure 3), the segmental motion of the cellulosic chain remains relatively high and constant throughout, suggesting that the backbone segments are not directly involved in the interchain associations.

The marked increases in the rheological parameters (especially G′) upon the addition of 8×10^{-3} mol dm^{-3} SDS is indicative of a far greater degree of structure within the system. Despite the gel-like characteristics of the 2% solutions of HMHEC in the presence of 8×10^{-3} mol dm^{-3} SDS, as borne out by the oscillatory measurements given in Figure 7, spin label studies (Figure 8) indicate the mobility of the cellulosic backbone remains largely unaltered by the presence of the surfactant. This suggests that the dramatic increase in structure within the HMHEC/SDS is a consequence of the interaction of the surfactant molecules with the associated hydrophobic side chains rather than the cellulosic backbone. This inference is not unexpected in light of the relatively limited interaction of HEC and SDS observed by ourselves (9) and Goddard and Hannan (21), and the results of Steiner and co-workers (12,13) which demonstrate the ability of SDS to solubilize water-insoluble HMHEC's bearing a relatively high proportion of hydrophobic groups.

From the variation of G′ with added SDS illustrated in Figure 6, it can be seen that the interaction of SDS with HMHEC is relatively complex, with any suggested mechanism needing to explain first an increase,

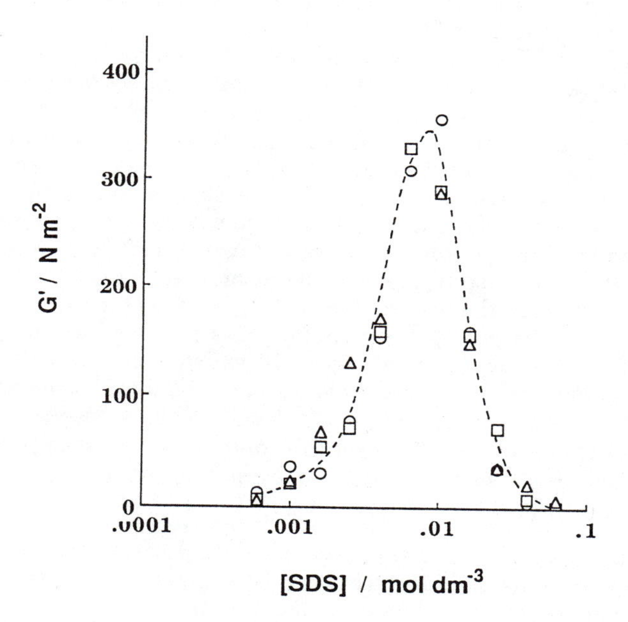

Figure 10. The effect of added SDS on the storage moduli, G′ (at a frequency of 1 Hz) of 2% aqueous solutions of HMHEC in (○) water, (□) 0.01 mol dm^{-3} NaCl; and (△) 0.1 mol dm^{-3} NaCl.

then a decrease, and finally a complete disruption of the network structure, as indicated by the significantly lower value of G′ at high surfactant additions compared to that of the polymer solution in the absence of any surfactant. On first inspection of Figure 6, it is evident that the maximum value of G′ (G'_{max}) is obtained at a concentration approximately equivalent to the expected cmc of the free surfactant (8.2×10^{-3} mol dm^{-3}). However, the cmc of SDS in the presence of polymer may be changed significantly (22,23). Numerical analysis (9) of the ESR spectra given in Figure 9 indicates that, in the presence of 0.2% HMHEC, surfactant-hydrophobe interactions occur at an SDS concentration of approximately 2×10^{-3} mol dm^{-3}, a factor of four lower than the expected cmc. Intuitively, in the presence of the higher 2% concentration of HMHEC used in the determination of the rheological data given in Figure 6, surfactant-hydrophobe interactions may occur at even lower SDS concentrations. The exact nature and composition of such polymer-surfactant "aggregates" are still matters for conjecture, but it is possible that such "aggregates" may differ markedly in parameters such as surfactant aggregation number in comparison with the micellisation behavior of the free surfactant in solution. It is, however, reasonable to assert that polymer-surfactant "aggregates" are present in the HMHEC/SDS/water system prior to the attainment of G'_{max} in Figure 6, and consequently that the occurrence of G'_{max} at the cmc expected for free SDS in water may be purely coincidental. This point is further illustrated by the rheological data given in Figure 10 which indicates that the SDS concentration at which G'_{max} occurs is largely indifferent to the addition of electrolyte at least up to 0.1 mol dm^{-3} NaCl. This is despite the known reduction in the cmc of free SDS over this ionic strength range (8.2, 5.6 and 1.5×10^{-3} mol dm^{-3} in water, 0.01 mol dm^{-3} and 0.1 mol dm^{-3} NaCl, respectively (17)).

From the rheological and spectroscopic data described above, the complex rheological behavior of HMHEC can be explained along the lines proposed by Gelman (11) as follows: in the absence of surfactant, HMHEC adopts an extensive three-dimensional structure in which the cellulosic polymer backbones are effectively and reversibly cross-linked by intermolecular hydrophobic associations between neighboring alkyl side chains. At levels of surfactant addition considerably lower than the expected cmc of free SDS, micellar-type aggregates of surfactant molecules form around the hydrophobic "bridges." In addition, it is possible that the presence of the surfactant aggregates may also enable alkyl side chains which, for geometrical reasons, were previously unassociated to become involved in the "bridging" network. The overall effect of this polymer-surfactant interaction would be to increase the degree and strength of the structure within the HMHEC network as indicated by the initial portion of the curves for G′ given in Figures 6 and 10. This micellar association with the hydrophobic "bridges," and hence the increase in G′ with increasing surfactant addition would continue until the optimum number of micellar bridges had been formed. This may correspond to a situation where each surfactant micelle incorporates a single grafted alkyl side chain from each of two neighboring polymer molecules. At concentrations of added surfactant above this optimum value, additional surfactant micelles would be available for the "solubilization" of the polymer-bound hydrophobes. Consequently, the average number of bound hydrophobes per surfactant micelle would decrease, thereby reducing the number of effective micellar "bridges" and thus producing a decrease in the observed G′ with increasing surfactant addition. At sufficiently high levels of surfactant addition, each polymer-bound hydrophobe would be encapsulated in its own individual surfactant micelle, thereby breaking all interpolymer "bridges" and causing the complete disruption of the network structure, as indicated by the very low values of G′ observed at high surfactant additions.

The validation of the above model necessitates the ability to accurately describe the micellisation behavior of the added surfactant in the presence of the polymer in terms of the cmc, the micellar aggregation number, and the number of bound hydrophobes per polymer-surfactant "aggregate." Our current investigations are aimed at resolving this problem.

Literature Cited

1. Shaw, K. G.; Leipold, D. P. *J. Coatings Technol.* 1985, **57**, 63.
2. Hall, J. E.; Hodgson, P.; Krivanek, L.; Malizia, P. *J. Coatings Technol.* 1986, **58**, 65.
3. Glass, J. E., Ed. *Water Soluble Polymers: Beauty with Performance*; Advances in Chemistry Series Vol. 213; American Chemical Society: Washington, DC, 1986.
4. Glass, J. E., Ed. *Polymers in Aqueous Media: Performance through Association*; Advances in Chemistry Series Vol. 223; American Chemical Society: Washington, DC, 1989.
5. Ananthapadmanabhan, K. P.; Leung, P. S.; Goddard, E. D. ACS Symposium Series 384; American Chemical Society: Washington, DC, 1989; p 297.
6. Schulz, D. N.; Kaladas, J. J.; Maurer, J. J.; Bock, J.; Pace, S. J.; Schulz, W. W. *Polymer* 1987, **28**, 2110.
7. Wang, K. T.; Iliopoulos, I.; Audebert, R. *Polym. Bull.* 1988, **20**, 577.
8. Sau, A. C.; Landoll, L. M. *J. Coatings Technol.* 1986, **58**, 344–364.
9. Tanaka, R.; Meadows, J.; Phillips, G. O.; Williams, P. A. *Carbohydr. Polym.* 1990, **12**, 443.
10. Tanaka, R.; Meadows, J.; Phillips, G. O.; Williams, P. A. In *Cellulose: Structural and Functional Aspects*; Kennedy, J. F.; Phillips, G. O.; Williams, P. A., Eds.; Ellis Horwood Ltd.: Chichester, UK, 1989; pp 323-327.
11. Gelman, R. In *TAPPI Proc., 1987 International Dissolving Pulps Conference*; Geneva, 1987; pp 159-165.
12. Steiner, C. A. In *Cellulosics Utilization*; Inagaki, H.; Phillips, G. O., Eds.; Elsevier Appl. Sci.: London, 1989; pp 132-140.
13. Dualeh, A. J.; Steiner, C. A. *Macromolecules* 1990, **23**, 251.
14. Fox, K. K.; Robb, I. D.; Smith, R. *J. Chem. Soc. Faraday Trans. I* 1974, **70**, 1186.

15. Stone, T. J.; Buckman, T.; Nordio, P. L.; McConnell, H. M. *Proc. Natl. Acad. Sci. USA* 1965, **54**, 1010.
16. Zhao, F.; Rosen, M. J.; Yang, N. L. *Colloids Surfaces* 1984, **11**, 97.
17. Stellner, K. L.; Scamehorn, J. F. *Langmuir* 1989, **5**, 70.
18. Landoll, L. M. *J. Polym. Sci. Polym. Chem. Ed.* 1982, **20**, 443.
19. Morris, E. R.; Cutler, A. N.; Ross-Murphy, S. B.; Rees, D. A.; Rice, J. *Carbohydr. Polym.* 1981, **1**, 5.
20. Goodwin, J. W.; Hughes, R. W.; Lam, C. K.; Miles, J. A.; Warren, R. C. H. In *Polymers in Aqueous Media: Performance through Association*; Advances in Chemistry Series Vol. 213; American Chemical Society: Washington, DC, 1986, pp 365–378.
21. Goddard, E. D.; Hannon, D. B. In *Proc. International Symp. on Micellisation, Solubilisation and Microemulsions—2*; Mittal, K. L., Ed.; Plenum Press: New York, 1977; pp 835-845.
22. Witte, F. M.; Buwalda, P. L.; Engberts, J. B. F. N. *Colloid Polym. Sci.* 1987, **265**, 42.
23. Witte, F. M.; Engberts, J. B. F. N. *J. Org. Chem.* 1988, **53**, 3085.

RECEIVED December 16, 1991

Chapter 24

Molecular Motions in Cellulose Derivatives

K. Nishinari[1], K. Kohyama[1], N. Shibuya[1], K. Y. Kim[1,2], N. H. Kim[1,2], M. Watase[3], and A. Tsutsumi[4]

[1]National Food Research Institute, Tsukuba 305, Japan
[2]Department of Chemistry, College of Natural Sciences, Sungshin Women's University, Sungbuk-ku, Seoul, Korea
[3]Chemical Research Laboratory, Shizuoka University, Ohya, Shizuoka 422, Japan
[4]Department of Applied Physics, Hokkaido University, Sapporo 060, Japan

Broad-line nuclear magnetic resonance (NMR) and viscoelastic measurements were carried out for hydroxyethyl cellulose (HEC) and hydroxypropyl cellulose (HPC) with various average numbers of substituents per anhydroglucose unit in the temperature range from liquid nitrogen temperature to 150°C at various moisture levels. The motional narrowing was observed at lower temperatures (around −100°C), and its origin was attributed to the rotational motion of CH_2OR where $R = H$ or $(CH_2CH_2O)_mH$ in HEC and CH_2OH in HPC attached to the C_5 atom in the glucose residue, and/or of $(CH_2CH_2O)_mH$ in HEC and CH_2OH in HPC which substituted at the second and/or the third hydroxyl groups. The mechanical loss peak at about −130°C was attributed to the same molecular motion observed at around −100°C in the broad-line NMR. A peak or a shoulder was observed at around −50°C in mechanical loss for a slightly humid specimen, and this was attributed to the commencement of motion of frozen bound water molecules. Mechanical loss peaks around room temperature were attributed to the melting of ice.

Cellulose is a linear polysaccharide in which glucosidic residues are linked by β-1,4 (glucosidic) linkages. Because a glucosidic residue has hydroxyl groups attached to C_2, C_3, and C_6 carbons, many derivatives of cellulose have been developed for industrial utilization (1,2). Figure 1 shows the structure of cellulose derivatives. Both hydroxyethyl cellulose (HEC) and hydroxypropyl cellulose (HPC) are non-toxic and edible, and they have been used in medical and related fields (3). Although cellulose, amylose, pullulan and dextran consist of the same structural units, i.e., glucosidic residues,

0097–6156/92/0489–0357$06.00/0

their physicochemical properties and molecular motion in the solid state are quite different from each other because of the different modes of glucosidic linkages (4–11). The study of molecular motion in HEC and HPC in the solid state helps to clarify the relation between structure and properties of these α- and β-glucans systematically, not only for scientific interest, but also for further industrial application. Films of HEC and HPC of various molar substitutions, MS (= the average number of substituents per anhydroglucose unit), were prepared, and broad-line NMR and viscoelastic studies were carried out in order to clarify the relation between structure and molecular motion in this work.

Experimental

Materials. HEC and HPC of various MS were supplied from Fuji Chemical Company and Nippon Soda Company, respectively. Films of these polymers were prepared by a conventional casting method. MS of HEC was determined by the Zeisel gas chromatographic method (12,13) as 1.7 and 2.5. MS of HPC was determined by the Morgan method (14) as 2.4, 3.0, and 4.1.

Evaluation of the Amount of Substituent Introduced in Each Glucosidic Residue. HEC or HPC preparations were methylated by the method of Hakomori (15) using CD_3Cl. Methylated polysaccharides were hydrolyzed, reduced, and acetylated as previously reported (16). The amount and position of substituent introduced in each glucosidic residue was evaluated by gas chromatography-mass spectrometry (GC/MS) of these partially methylated alditol acetates. Detailed conditions of this experiment and the analysis of the mass spectra will be reported elsewhere.

NMR Measurement. The film samples were put into glass tubes and evacuated (10^{-2}mm Hg) at 120°C for 2 hours. After evacuation, the glass tubes were sealed off. Broad-line, ^{1}H-NMR derivative spectra were measured by means of a continuous wave method with a JEOL-JNM-W-40 spectrometer operated at 40 MHz as a function of temperature from −160°C to 110°C, and the second moments were calculated from the spectra. Samples were kept at each temperature for one hour before the measurement. The field modulation frequency was 35 Hz, and in most cases the amplitude was kept low enough to neglect the modulation broadening. A radio frequency field below the limit of saturation was chosen by comparing the curves at different power levels.

Viscoelastic Measurements. A piezotron (Toyo Seiki Seisakusho Ltd.) was used for the measurement of the temperature dependence of the viscoelastic coefficients. The sinusoidal strain and stress at both ends of the film were detected by nonbonded strain gauges. The operational circuit calculated immediately the real and imaginary parts of the complex response function from strain and stress. The temperature was raised from −180°C at the rate of 2°C/min in the dried nitrogen gas flow. The storage and loss moduli, c' and c'', of the complex Young's modulus c^* at 10 Hz, were plotted on a two-pen X-Y recorder as a function of temperature.

Results and Discussion

The observed second moment values ΔH^2 for HPC samples of different moisture levels are shown in Figure 2 as a function of temperature. The decrease of ΔH^2 is proportional to the number of protons contained in the moving regions of molecules, and to the degree of motional intensity. ΔH^2 decreased remarkably above 0°C, and it was attributed to the motional narrowing due to a motion of polymeric main chains. The fact that the decrease of ΔH^2 in a humid sample is more remarkable than that in a dehydrated sample is due to the plasticizing effect of water molecules at this temperature range. Here, a humid (non-dried) sample is kept at 70% R.H. for two weeks. Because the magnitude of ΔH^2 is larger in a humid sample than in a dehydrated sample at lower temperatures (around −120°C), the molecular packing is considered to be better in a humid sample than in a dehydrated sample. The denser molecular packing in humid polymers than in dehydrated ones has been suggested from viscoelastic measurements for many biopolymers such as cellulose (6), amylose (7), poly-α-amino acids (17), and collagen (18).

The regularity of higher order structure of cellulose I was shown to increase in the presence of a small amount of water molecules by mechanical (19), thermal (20), and x-ray analysis (21). This increase in structural order has been observed with a water content of less than 0.2 g/g dry weight. Intermolecular hydrogen bonds which are randomly formed in a highly dehydrated state may be broken and rearranged to form a more ordered structure by the introduction of a small amount of water molecules.

The temperature dependence of ΔH^2 for dried HPC samples of three different MS is shown in Figure 3. For HPC of MS = 2.4, ΔH^2 decreased rapidly in the temperature range from −150°C to −130°C, and then it decreased gradually in the range (II) from −130°C to −80°C. Then, it decreased rapidly in the range (III) from −75°C to −40°C. In temperature range IV, from −40°C to −10°C, it decreased slightly, and then it decreased monotonously above −10°C (temperature range V).

For HPC of MS = 3.0, ΔH^2 decreased rapidly in the range (I) from −150°C to −120°C, and then it was almost constant in the range (II) from −120°C to −80°C. It decreased rapidly in the range (III) from −80°C to −50°C, and then was constant in the range (IV) from −50°C to −20°C. It decreased monotonously above −20°C (V).

For HPC of MS = 4.1, ΔH^2 decreased only slightly in the range (I) from −145°C to −130°C, and then it decreased in the range (II) from −130°C to −90°C. It decreased gradually in the range (III) from −90°C to −50°C, and then decreased monotonously above −20°C (V). The rapid decrease of ΔH^2 in the temperature range (II) from −130°C to −80°C in HPC of MS = 2.4, from −120°C to −80°C in HPC of MS = 3.0, and that from −130°C to −90°C in HPC of MS = 4.1, was attributed to the motional narrowing due to the rotational motion of CH_2OR, where R=H or $(CH_2CH(CH_3)O)_m$-H (m = 1,2,...), and/or of $(CH_2CH(CH_3)O)_m$-H which substituted the second and/or the third hydroxyl groups. However, the contribution of the rotation of hydroxymethyl groups seems to be larger

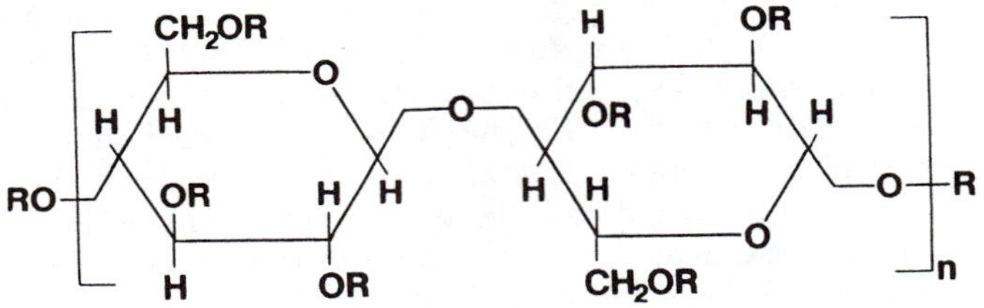

Figure 1. Structure of cellulose derivatives. R=H or $(CH_2CH_2O)_m$-H (m = 1, 2, ...) for HEC, R=H or $(CH_2CH(CH_3)O)_m$-H (m = 1, 2, ...) for HPC.

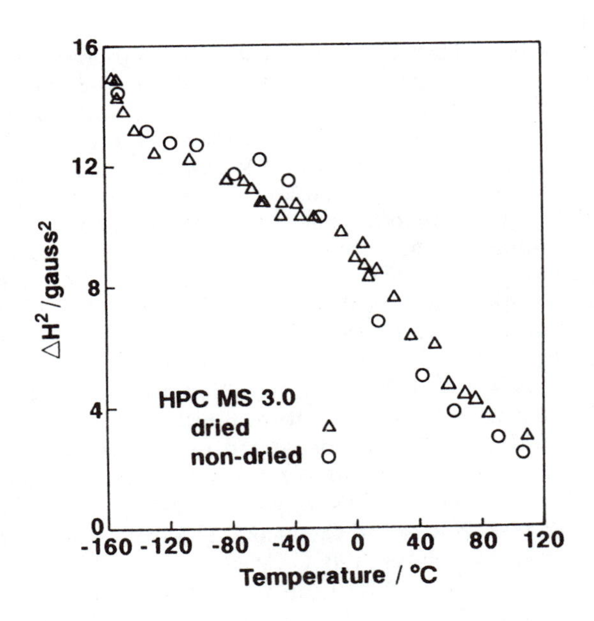

Figure 2. The temperature dependence of the second moment for dried and non-dried HPC (MS = 3.0).

than that of hydroxypropyl groups because the decrease of ΔH^2 at the temperature range (II) is smaller in HPC of higher MS than in HPC of lower MS. This corresponds well to the fact that this mechanical loss peak is more pronounced in HPC of higher MS than in HPC of lower MS. The finding that this temperature region shifted to lower temperatures with increasing value of MS corresponds well to the experimental results obtained by dielectric and viscoelastic measurements. This suggests that the hydroxypropyl groups in HPC with higher MS are more mobile than those in HPC with lower MS.

The rapid decrease of ΔH^2 at the temperature range (I) below −120°C or −130°C in HPC was attributed to the motional narrowing due to rotational motion of methyl groups. The steep decrease of ΔH^2 in temperature range I is attributed to the rotational motion of methyl groups which could begin to move at lower temperatures than −160°C. The rapid decrease of ΔH^2 above −10°C in HPC of MS = 2.4, and that above −20°C in HPC of MS = 3.0, and that above −50°C in HPC of MS = 4.1, was attributed to the motional narrowing caused by the oscillational motion of longer side chains derived from polymerized hydroxypropyl groups. The number of these longer side chains increases with increasing value of MS, which was shown clearly from the methylation/GC-MS analysis (Table 1). Therefore, the decrease of ΔH^2 at these temperatures was most remarkable in HPC of MS = 4.1.

Table 1. The ratio (%) of various numbers of hydroxyethyl or hydroxypropyl groups per glucose residue [a]

		Number of hydroxyethyl or hydroxypropyl groups/glucose residue						
	MS[b]	0	1	2	3	4	5	MS[c]
		Ratio (%)[a]						
HEC,	1.7	28.1	26.1	21.5	15.5	7.2	1.6	1.5
	2.5	18.7	20.6	22.6	22.5	12.2	3.4	2.0
HPC,	2.4	2.3	5.9	14.6	35.9	30.7	10.6	3.2
	3.0	0.4	2.4	8.8	37.5	35.1	15.9	3.5
	4.1	1.3	1.3	3.3	23.2	38.9	32.0	3.9

[a] The ratio is calculated by:

$$\frac{Sn}{\sum_{n=0}^{5} Sn} \times 100$$

where Sn = peak area for glucose residues with n hydroxyethyl or hydroxypropyl groups.

[b] MS of HEC by Zeisel method, MS of HPC by Morgan method.

[c] MS estimated by GC/MS.

Figure 4 shows the observed second moment for dried HEC samples of different MS as a function of temperature. The sample of higher MS (2.5) shows a more remarkable decrease than that of lower MS (1.7) with increasing temperature. This decrease was attributed to the rotational motion of CH_2OR, where R=H or $(CH_2CH_2O)_m$-H (m = 1, 2...), attached to the C_5 atom in the glucose residue, and/or of $(CH_2CH_2O)_m$-H which substituted the second and/or third hydroxyl groups. According to the methylation/GC-MS analysis of HEC, the number of the glucosidic residues which have no substituent or have one substituent were less in the HEC of MS = 2.5 than in the HEC of MS = 1.7. Therefore, the experimental observation described above is quite reasonable.

The fact that ΔH^2 of HEC of MS = 2.5 is larger than the ΔH^2 of HEC of MS = 1.7 at lower temperatures suggests that the molecular packing is denser in HEC of MS = 2.5 than in HEC of MS = 1.7.

Temperature dependence of the real and imaginary parts of the complex Young's modulus c' and c'' for HEC films is shown in Figures 5 and 6. Temperature could not be raised above 100°C for viscoelastic measurements because HEC films became fragile at higher temperatures. The value of c'' showed three peaks: at −130°C, at −80°C ~ −60°C, and at 0°C ~ 50°C (Figure 6). The peak at −130°C was attributed to the rotational motion of CH_2OR, where R=H or $(CH_2CH_2O)_m$-H (m = 1, 2, ...), attached to C_5 atoms and/or $(CH_2CH_2O)_m$-H which substituted the second and/or third hydroxyl groups, which was also observed in dielectric measurements (23). The peak at temperatures from −80°C to −60°C was attributed to the commencement of the motion of frozen bound water, which was observed as a shoulder at the same temperature range in dielectric measurements (23). The peak in the temperature range from 0°C to 50°C was attributed to the melting of ice, which was not observed in dielectric measurements because it was hidden by the steep rise of the signal. At this temperature range, c' began to decrease rapidly, and the value of c' for a humid film became smaller than that for a dehydrated film. This can be explained by a plasticizing effect of water molecules at this temperature range. The value of c' of HEC of a humid film showed a larger value than that of a dried film at lower temperatures. This is also due to the structural stabilization by water molecules, and the rotational motion of CH_2OR, where R=H or $(CH_2CH_2O)_m$-H (m = 1, 2, ...), attached to C_5 atoms and/or $(CH_2CH_2O)_m$-H which substituted the second and/or third hydroxyl groups, is hindered. This corresponds well to the fact that the peak of c'' at about −130°C is more pronounced in a dried film than in a humid film. Similar behavior has been found in amylose and pullulan (10,11). This structural stabilization by a small amount of water has been found also in cellulose I (20), amylose (8,22), and collagen (18).

The temperature dependence of the real and imaginary part of the complex Young's modulus for HPC of different MS is shown in Figures 7 and 8. The film of HPC with MS = 4.1 was easily broken at higher temperatures than 80°C under the constant tension, and the measurements were carried out below 70°C. Here again, the imaginary part c'' showed a peak

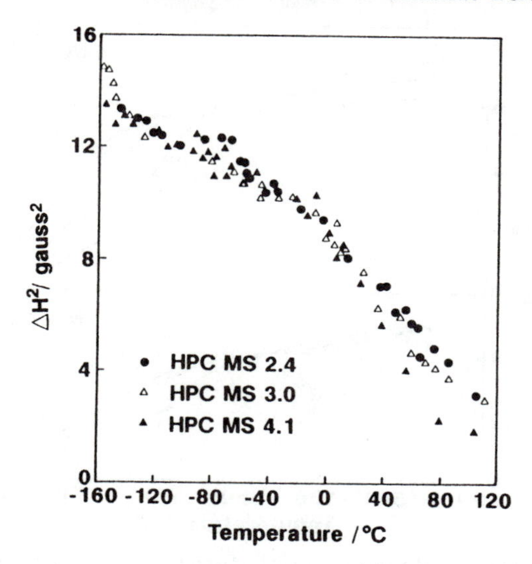

Figure 3. The temperature dependence of the second moment for HPC of different MS.

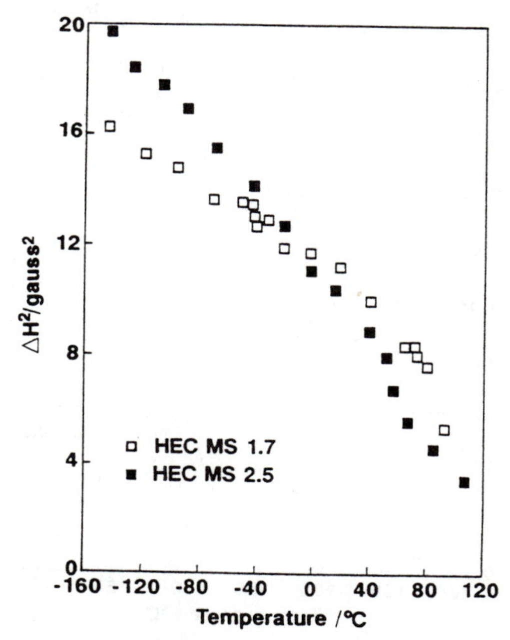

Figure 4. The temperature dependence of the second moment for HEC of different MS.

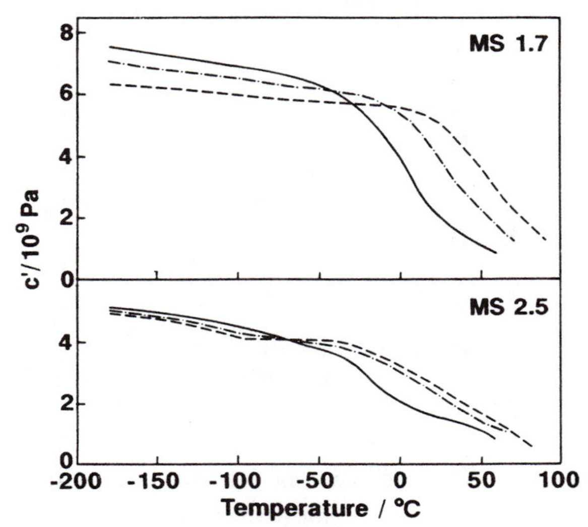

Figure 5. Temperature dependence of storage modulus c′ for HEC of different MS at various moisture levels. ——, non-dried; —·—·—, heated at 60°C for 40 min; — — — —, heated at 80°C for 40 min.
(Reproduced with permission from reference 23. Copyright 1991 Elsevier.)

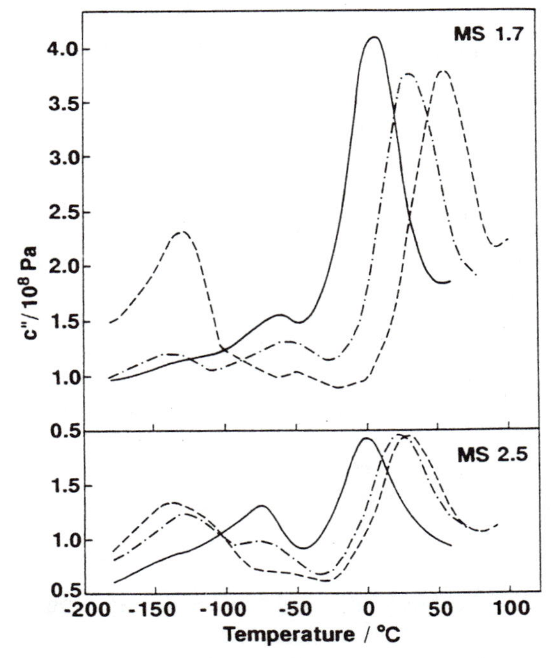

Figure 6. Temperature dependence of loss modulus c″ for HEC of different MS at various moisture levels. Symbols have the same meaning as in Figure 5. (Reproduced with permission from reference 23. Copyright 1991 Elsevier.)

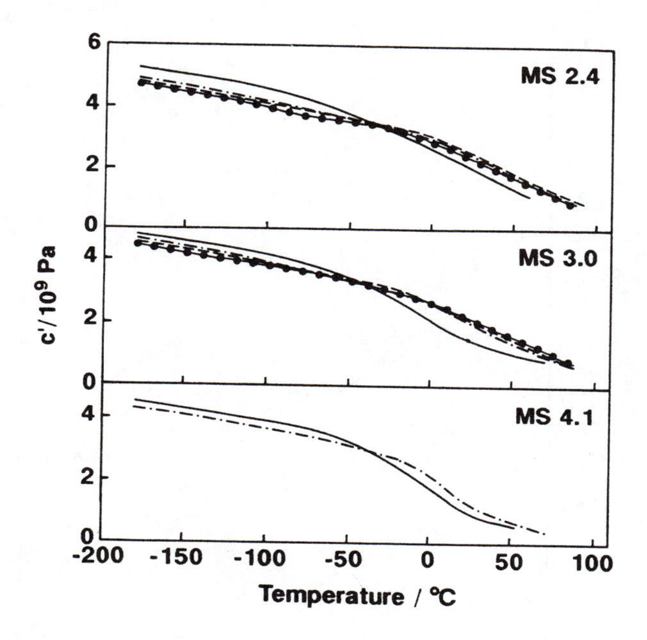

Figure 7. Temperature dependence of storage modulus c′ for HPC of different MS at various moisture levels. ——, heated at 100°C for 40 min. Other symbols represent the same meanings as in Figure 5.
(Reproduced with permission from reference 23. Copyright 1991 Elsevier.)

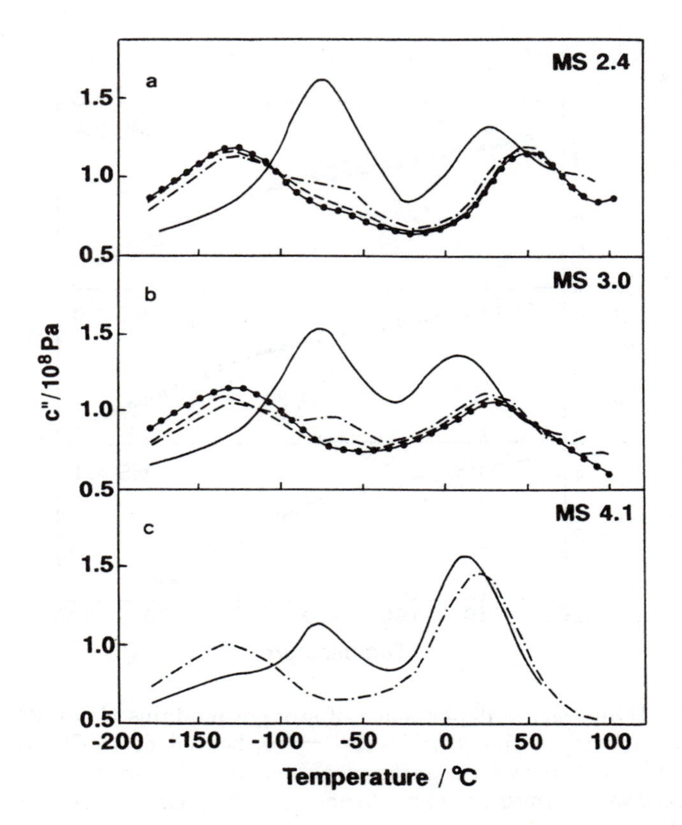

Figure 8. Temperature dependence of loss modulus c″ for HPC of different MS at various moisture levels. ——, heated at 100°C for 40 min. Other symbols represent the same meanings as in Figure 5.

at about −130°C (Figure 8). The peak temperature shifted to lower temperatures, and the peak height decreased with increasing MS just as in the dielectric case (23). The origin of this peak is thought to be a rotational motion of CH_2OR, where R=H or $(CH_2CH(CH_3)O)_m$-H (M = 1, 2, ...), attached to C_5 atoms and/or $(CH_2CH(CH_3)O)_m$-H which substituted the second and/or third hydroxyl groups in glucose residues. Since the peak temperature shifted to lower temperatures, and the peak height decreased with increasing MS, this peak must be caused mainly by hydroxymethyl groups (Table 2). The number of hydroxymethyl groups decreases with increasing MS as shown by GC-MS (Table 1). Thus, the peak height must decrease with increasing MS. The reason why the peak temperature shifted to lower temperatures is guessed as follows: The crystallinity will decrease with increasing MS, and then CH_2OR groups in the amorphous phase begin to move at lower temperatures than CH_2OR groups in the crystalline phase. The value of c′ at lower temperatures (around −150°C) decreased with increasing MS as shown in Figure 7. It may also be the case that the increased steric hindrance due to the increased MS would give rise to increased free volume for increased hydroxymethyl group mobility. In this situation, the mobility of hydroxymethyl groups increased, and therefore the peak temperature shifted to a lower temperature. These interpretations will be explored in the future.

Table 2. The temperature of mechanical loss peak and the storage modulus at −150°C for dried polysaccharide films

Polysaccharides	Temp. of c″ max/°C	c′ at −150°C/ 10^9 Pa
HEC MS 1.7[a]	−128	6.0
HEC MS 2.5[a]	−132	4.8
HPC MS 2.4[b]	−126	4.4
HPC MS 3.0[b]	−128	4.1
HPC MS 4.1[c]	−133	4.0
Amylose[d], Ref. 8	−75	8.4
Pullulan[d], Ref. 24	−100	7.8
Konjac Glucomannan[d], Ref. 25	−100	7.2

Drying temperature (a) 80°C, (b) 100°C, (c) 60°C, (d) 150°C. All samples were cooled rapidly to liquid nitrogen temperature, and then heated at 2°C/min.

A peak or shoulder of c″ for a slightly humid film of HPC (—·—·— and — — — — in Figures 8a, 8b, and 8c) at the temperature around −70°C is attributed to the commencement of motion of frozen bound water. It disappeared completely by dehydration (——) in Figures 8a and 8b.

A peak of c″ of HPC films in the temperature range from 10°C to 50°C may be attributed to the motion of longer side chains promoted by the plasticizing effect of water molecules just after the melting of ice (Figure

8). This peak is more pronounced in hydrated films than in dehydrated films. At the temperature range of this peak, c′ decreased most steeply.

The peak of mechanical loss at about −130°C was attributed to a rotational motion of both CH_2OR, where R=H or $(CH_2CH_2O)_m$-H (m = 1, 2, ...), attached at C_5 atoms and of $(CH_2CH_2O)_m$-H which substituted the second and/or third hydroxyl groups in the case of HEC, but in the case of HPC, the peak was attributed mainly to the rotational motion of a hydroxymethyl group. $(CH_2CH(CH_3)O)_m$-H groups are so long that they could not move so easily at this temperature range.

Acknowledgments

This work is supported by the Biorenaissance Programme, grant number BRP-91-III-B-1.

Literature Cited

1. Bolker, H. I. *Natural and Synthetic Polymers*; Marcel Dekker, Inc.: New York, 1974.
2. Davidson, R. L., Ed. *Handbook of Water-Soluble Gums and Resins*; McGraw-Hill Book Co.: New York, 1980.
3. Glicksman, M., Ed. *Food Hydrocolloids, Vol. III*; CRC Press: Florida, 1982.
4. Mikhailov, G. P.; Artyukhov, A. I.; Borisova, T. I. *Vysokomol. Soyed.* 1967, **A9**, 11, 2401.
5. Crofton, D. J.; Pethrick, R. A. *Polymer* 1982, **23**, 1609, 1615.
6. Sasaki, S.; Fukada, E. *J. Polym. Sci. Polym. Phys. Ed.* 1976, **14**, 565.
7. Bradley, S. A.; Carr, S. H. *J. Polym. Sci. Polym. Phys. Ed.* 1976, **14**, 111.
8. Nishinari, K.; Fukada, E. *J. Polym. Sci. Polym. Phys. Ed.* 1980, **18**, 1609.
9. Nishinari, K.; Tsutsumi, A. *J. Polym. Sci. Polym. Phys. Ed.* 1984, **22**, 95.
10. Nishinari, K.; Chatain, D.; Lacabanne, C. *J. Macromol. Sci.-Phys.* 1983, **B22**, 795.
11. Nishinari, K.; Shibuya, N.; Kainuma, K. *Makromol. Chem.* 1985, **186**, 433.
12. Lemieux, R. U.; Purves, C. B. *Can. J. Res.* 1947, **B25**, 485.
13. Cobler, J. G.; Samsel, E. P.; Beaver, G. H. *Talanta* 1962, **9**, 473.
14. Morgan, P. W. *Ind. Eng. Chem. Anal. Ed.* 1946, **18**, 500.
15. Hakomori, S. *J. Biochem.* 1964, **55**, 205.
16. Shibuya, N.; Iwasaki, T. *Phytochem.* 1985, **24**, 285.
17. Hiltner, A.; Anderson, J. M.; Baer, E. *J. Macromol. Sci. Phys.* 1973, **B8**, 431.
18. Andronikashvili, E. L.; Mrevlishvili, G. M.; Japaridze, G. SH.; Kvavadze, K. A. *Biopolymers* 1976, **15**, 1991.
19. Nakamura, K.; Hatakeyama, T.; Hatakeyama, H. *Tex. Res. J.* 1981, **51**, 607.

20. Hatakeyama, T.; Hatakeyama, H. In *Cellulose and Its Derivatives*; Kennedy, J. F.; Phillips, G. O.; Wedlock, D. J.; Williams, P. A., Eds.; Ellis Horwood Ltd., 1985; p 87.
21. Alince, B. Abstr., 10th Cellulose Conf., Syracuse, NY, May 1988; p 24.
22. Nishinari, K.; Chatain, D.; Lacabanne, C. *J. Macromol. Sci.* 1983, **B22**, 529.
23. Kim, K. Y.; Kim, N. H.; Nishinari, K. *Carbohydr. Polym.*, 1991, **16**, 189.
24. Nishinari, K.; Horiuchi, H.; Fukada, E. *Rep. Prog. Polym. Phys. Jap.* 1980, **23**, 759.
25. Kohyama, K.; Kim, K. Y.; Shibuya, N.; Nishinari, K.; Tsutsumi, A. *Carbohydr. Polym.*, in press.

RECEIVED December 16, 1991

Chapter 25

Internal Motions of Lignin

A Molecular Dynamics Study

Thomas Elder

School of Forestry, Alabama Agricultural Experiment Station, Auburn University, AL 36849

The available conformational space, and therefore flexibility, of a lignin oligomer has been examined using current methods in molecular dynamics. These techniques allow for the determination of chemical structure and energy as a function of time and temperature. Results indicate that the oligomeric structure under consideration begins with a somewhat coiled nature, which is retained throughout an extensive simulation. While the oligomer remains relatively compact, the structure appears to twist internally. This change may be needed to accommodate the increases in energy resulting from higher temperatures.

The flexibility of the lignin polymer represents a fundamental parameter exerting considerable influence on the macroscopic behavior of the bulk polymer. Changes in molecular mobility of lignin, and its derivatives, are frequently invoked in the interpretation of experimentally observed phenomena (1). Modifications in molecular mobility can be assessed by an exploration of the conformational space available, providing insights into the ranges of size and shape that may be assumed.

The impact of molecular level interactions on material properties has been described in terms of a number of hierarchical disciplines as shown in Figure 1. It is proposed by these authors (2) that an understanding of the interdependency of these organizational levels could lead to the *a priori* design of materials with specific properties.

While it has been proposed that conformation and flexibility at the molecular level may be among the factors that control material properties, there has been relatively little research in this area of lignin chemistry. In a recent article on the polymeric properties of lignin, Goring (3) discussed the minimal work that has been reported with respect to the conformation

0097–6156/92/0489–0370$06.00/0

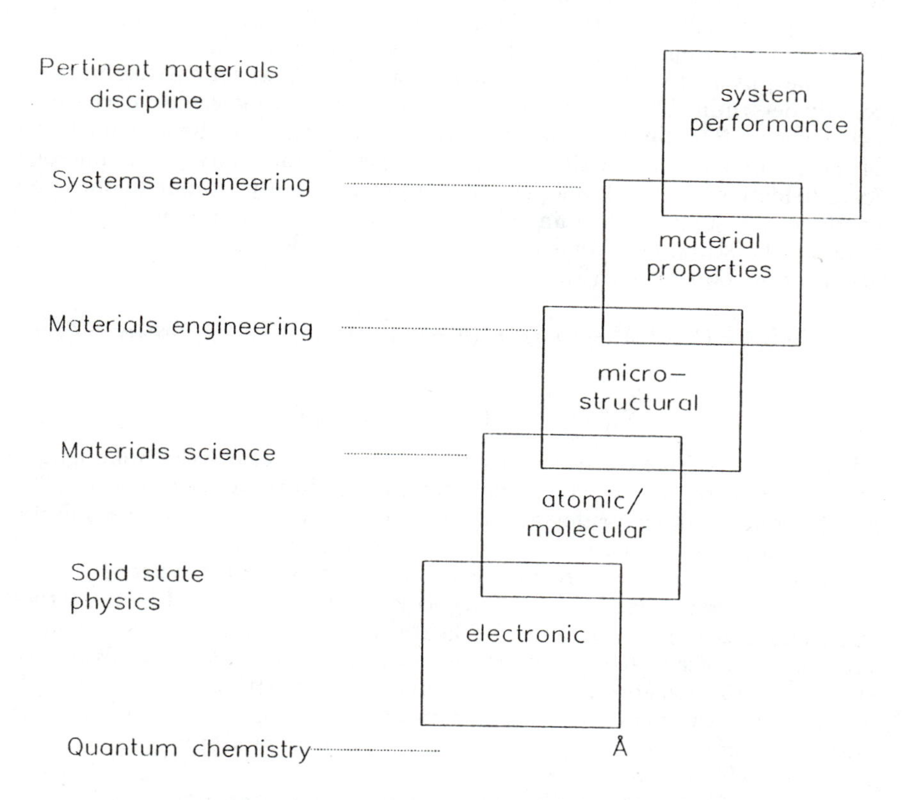

Figure 1. Hierarchy of models for materials by design. (Adapted from reference 2.)

of the macromolecule. Experimental efforts that were described include the determination of hydrodynamic properties and the information that may be obtained from size-exclusion chromatography. Furthermore, the secondary structure of proteins, and some possible comparisons with lignin, were mentioned. While protein structure has been explored by experimental methods (4), the application of computational chemistry and molecular modelling to these problems has become quite well-developed (5). The current paper will assess the conformation of a lignin oligomer and its ability to deform, by application of molecular modelling methods.

The physical structure and energy of large chemical entities may be readily determined by the application of force-field (or molecular mechanics) calculations. The fundamentals of these calculations and the minimization of energy are reviewed by Burkert and Allinger (6) and Clark (7). Numerous force-fields have been developed, but are conceptually similar in that the energy of a system is determined by summing the energy of bond stretching, bending, twisting, non-bonded interactions, and electrostatic interactions, as shown in Equation 1 (8).

$$V = (0.5)\Sigma k_b(b - b_0)^2 + (0.5)\Sigma k_\theta(\theta - \theta_0)^2 + (0.5)\Sigma k_\phi[1 + \cos(n\phi - \delta)] +$$

$$\Sigma[(A/r^{12}) - (C/r^6) + (q_1 q_2/Dr)] \quad (1)$$

where k_b is the force constant for a specific bond type, r is the bond length in the structure, and r_0 is a parameterized equilibrium bond length. The energy contributions for deviations in bond angles are calculated similarly. In some force-fields the dihedral term is approximated by a Hooke's Law form, but may take on more complex forms, as shown here, where k_ϕ is the torsional barrier, n is the periodicity, ϕ is the torsional bond in the structure, and δ is the parameterized dihedral angle. Lennard-Jones potential functions are often used to determine non-bonded interactions, and electrostatic contributions are assessed by coulombic interactions. In Equation 1, A and C are the Lennard-Jones parameters, r is the interatomic distance between atoms 1 and 2, q_1 and q_2 are the respective atomic charges, and D is the dielectric constant.

Energy functions and the corresponding structures may then be minimized by a number of non-linear techniques, such as steepest Descent, Newton-Raphson, BFGS (Broyden, Fletcher, Goldfarb, and Shanno), or conjugate gradients (6). At the present time, however, there are no minimization techniques that guarantee identification of the lowest energy structure (the global minimum). This can be particularly difficult when dealing with large polyatomic structures, since the potential energy surface is $3N - 6$ space, where N is the number of atoms. Several approaches to conformational searches, especially for large molecules, have been proposed (9).

The most straightforward of these is a systematic search, in which each dihedral angle is incrementally modified through a specific rotation. Only dihedral angles, rather than bond lengths or angles, are considered, since the former impart much larger structural changes and require less energy for distortion. The limitation of this approach lies in the fact that the number of conformations, for which energy calculations are performed, increases exponentially with the number of rotatable dihedral angles, rapidly resulting in a problem of intractable size given current computer technology.

As an alternative to a systematic search, the methods of molecular dynamics have been successfully applied to the conformational analysis and internal motion of proteins. Based on the potential energy and positional information from force-field calculations, the forces on individual atoms and their velocities may be determined. The application of Newtonian mechanics to velocities allows for the determination of positions as a function of time. The force (F), position (r), and potential energy (V) are described by the classical mechanical relationship shown in Equation 2.

$$F = -\partial V/\partial r \tag{2}$$

Based on the position x of a given atom at time t, the Verlet method (10) is used to calculate the position at Δt, as indicated in Equation 3, where v is the velocity.

$$x(t + \Delta t) = x(t) + v\Delta t \tag{3}$$

The velocity v is assumed to be equivalent to the instantaneous velocity at the midpoint in the time step (Equation 4).

$$v = v(t + \Delta t/2) \tag{4}$$

The midpoint velocity is estimated from the previous time step, and acceleration a, which is determined from the force acting on a particular atom (Equation 5).

$$v(t + \Delta t) = v(t - \Delta t) + a\Delta t \tag{5}$$

The new value for the velocity is substituted into Equation 3, and the process is repeated (5).

At current levels of computer technology, the total times that may be simulated are on the order of picoseconds (1 ps = 10^{-12} seconds), and the time steps within a simulation are in terms of femtoseconds (1 fs = 10^{-15} seconds). Larger time steps may be allowed by the use of the SHAKE algorithm that neglects high-frequency vibrations in bond lengths (5). Furthermore, the temperature is related to velocities by Equation 6,

$$3k_BT = \Sigma(mv \times v/N) \tag{6}$$

where k_B is Boltzmann's constant, m and v are the atomic mass and velocity, respectively, and N is the total number of atoms. Applications which hold temperature constant, allowing energy to fluctuate, correspond to the canonical ensemble from statistical thermodynamics.

Molecular dynamics methods have advantages over other minimization methods in that the latter cannot surmount energy barriers. In contrast, the energy imparted to a structure by atomic velocities may be sufficient to cross potential barriers, resulting in a more complete exploration of the possible conformations. Furthermore, the non-linear optimization calculations that have been described result in a single structure representing the calculated minimum, while dynamic methods are capable of showing a distribution of structures as a function of time, temperature, or energy.

Constant temperature calculations have been proposed as optimization tools through a technique described as "simulated annealing." In this approach, which has been applied to other minimization problems (11,12), the molecular system is initially held at an elevated temperature, such that virtually all conformations are accessible. The temperature is then slowly reduced, allowing preferred conformations to be assumed. This process may be repeated several times to identify groups of conformations, and the associated energies.

Methods

The lignin oligomer considered in this paper is adapted from a segment proposed by Glasser and co-workers in simulation studies, and is shown in Figure 2. The computer program used throughout this work was the integrated graphics and computational package "SYBYL" distributed by Tripos Associates. The lignin oligomer was input using standard fragments and geometric parameters within the program. Upon completion, a geometry optimization was performed for all atoms, using the Tripos 5.2 forcefield parameters (12), and conjugate gradient minimization. The resultant structure is shown as a stereo pair in Figure 3.

Beginning with this conformation, molecular dynamics calculations were performed using the simulated annealing technique. The system was rapidly heated to 1500°K, held for 200 fs, and then cooled to 0°K in 50°K increments, holding for 50 fs at each temperature. Upon reaching 0°K, the temperature was again rapidly increased to 1500°K. Six cycles of heating and cooling were performed (Figure 4). Structure and energy calculations were performed at 5 fs intervals throughout the molecular dynamics run.

The irregular nature of the lignin oligomer in question dictates the use of several structural descriptors for its characterization. These were: the vector connecting atom 4 to atom 195, taken as a measure of end-to-end distance; the angle formed by atoms 4-60-195; and the torsional angle formed by atoms 4-47-60-195. These atom numbers are as shown in Figure 1. In addition, the radius of gyration, the root mean square of distances from the center of mass, was determined.

Results and Discussion

Figure 5 indicates how the potential energy of the conformers changes as a function of temperature. As might be expected, this is a reasonably linear trend, with the distribution of energies increasing with temperature.

The next group of figures describes the variability in geometric parameters (interatomic distance, angle, torsional angle, and radius of gyration).

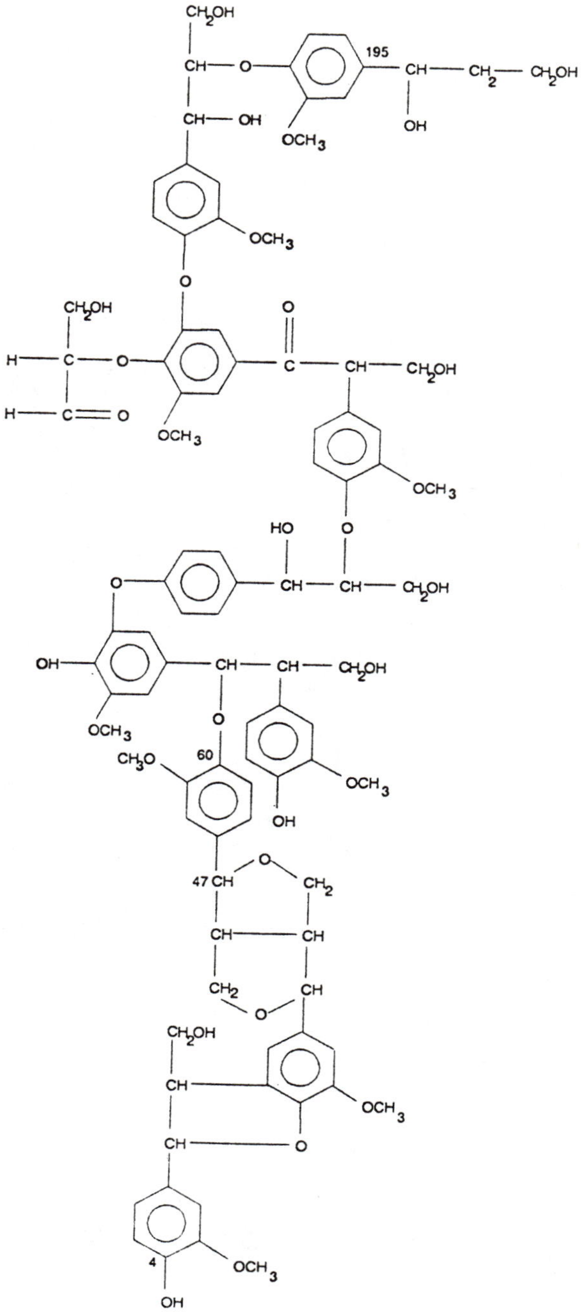

Figure 2. Lignin oligomer used in this study.

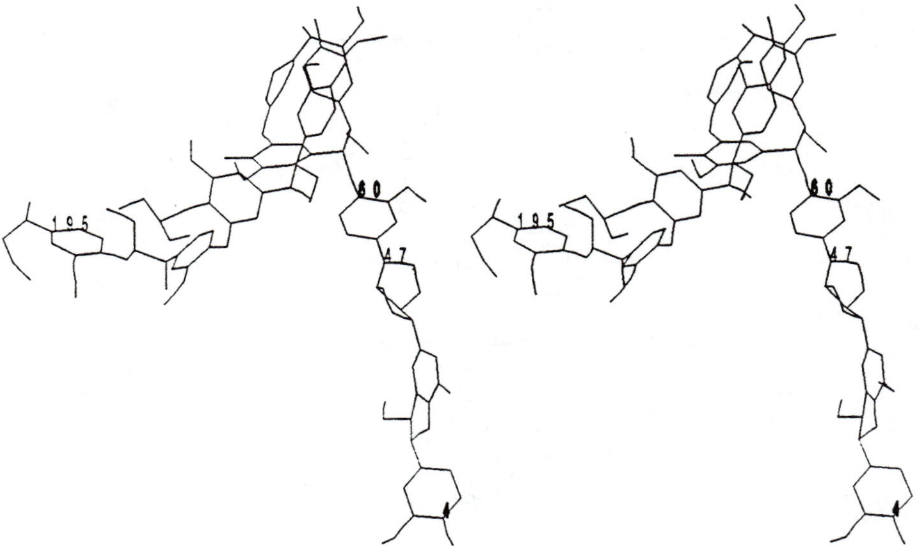

Figure 3. Stereo-pair for the minimized structure of the lignin oligomer.

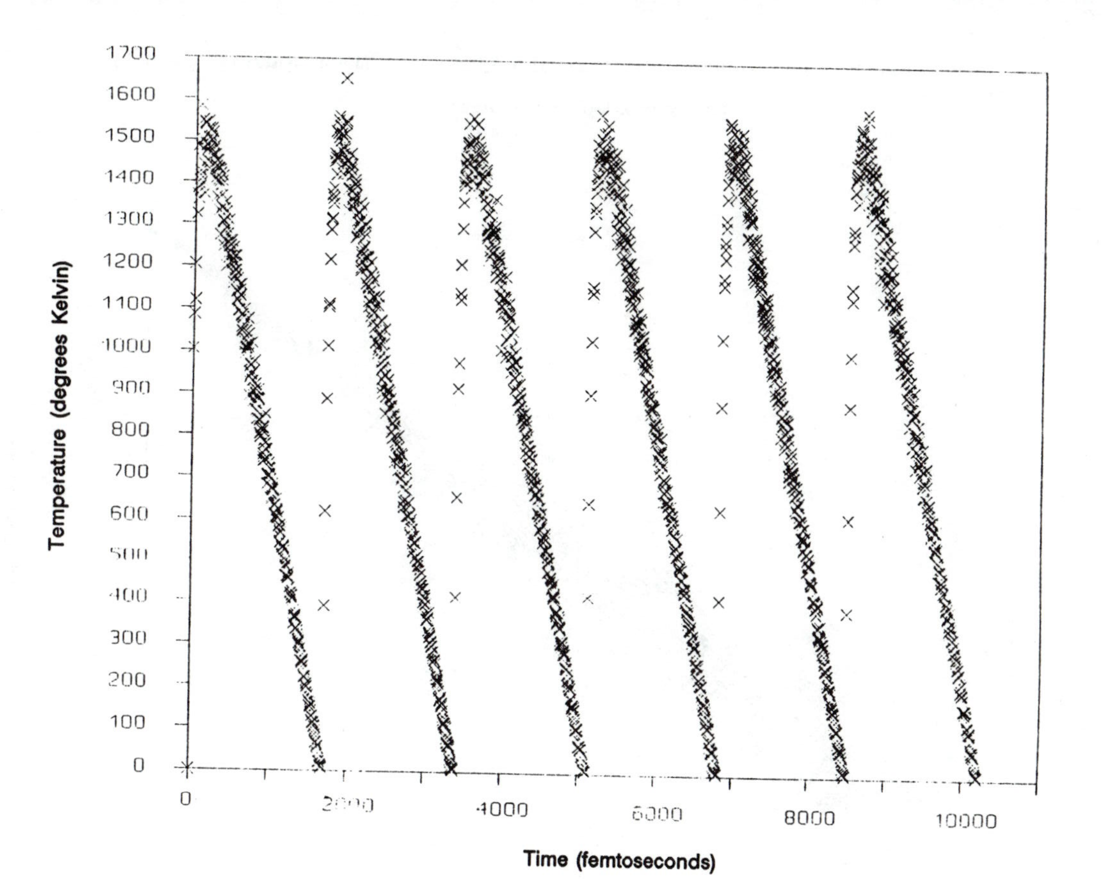

Figure 4. Time–temperature relationship for molecular dynamics calculations.

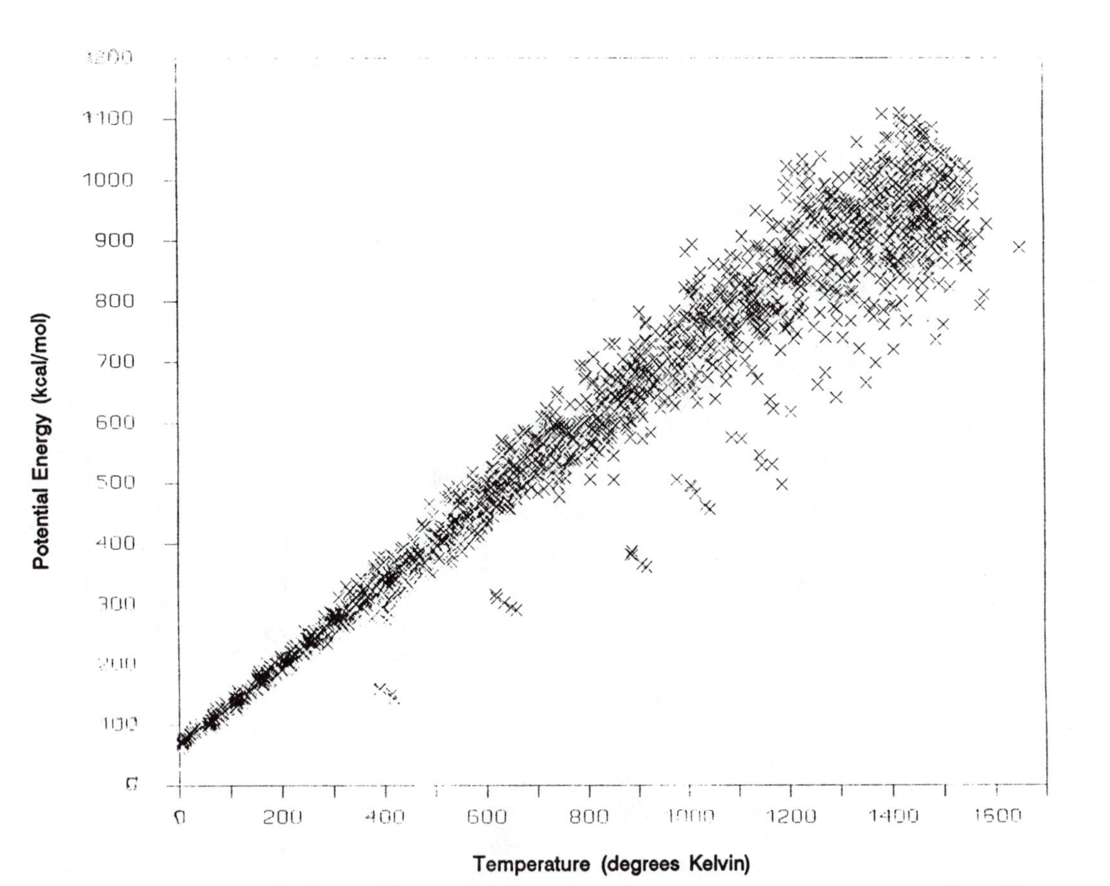

Figure 5. Potential energy–temperature relationship from molecular dynamics calculations.

These graphs include all data points for the cyclic time-temperature regime described in Figure 3. As a consequence, it will be observed that the cyclic behavior exhibited in geometry may also be discerned.

Figure 6 is a plot of the vector from atom 4 to atom 195 as a function of temperature, with each time cycle as noted. This distance, taken as a measurement of end-to-end distance, represents a fundamental factor in polymer behavior and flexibility. It can be seen that at elevated temperatures the oligomer exhibits wide variation. As the temperature is lowered and raised through the six time-temperature cycles, several patterns emerge, becoming relatively distinct even above 1000°K. As the temperature approaches 0°K, clusters of conformations can be seen within a narrow distance range. Perhaps more interesting is the behavior at the highest temperature. While the interatomic distance changes considerably, the range is still quite small (ca. 23-27Å). Across all temperatures, the 4-195 distance only changes by about 6Å from the lowest point to the highest point. From these data, and the initial structures shown in Figure 3, it appears that the oligomer has a somewhat coiled structure that is retained throughout the simulation. Consequently, the increases in energy associated with temperature must be accommodated through other structural changes rather than a simple extension of the polymer.

Figure 7, showing changes in the 4-60-195 angle, is somewhat similar, At low temperatures, this angle appears to be tightly constrained within a range of less than 10°. The exception to this generalization is the behavior exhibited during time cycle 4. It is also interesting to note that this cycle mirrors the 4-195 distance by having a major decrease at about 600°K. In contrast to the distance measurements, the angle varies across a larger range at the higher temperatures.

Figure 8 is a plot of the torsional angle formed by atoms 4-47-60-195. It was proposed in the discussion of the 4-195 distance that changes in energy and geometry must be accounted for through internal changes in structure, rather than simply the displacement of the ends away from each other. The torsional angle described by atoms 4-47-60-195 may help to account for the structural changes required to accommodate the increased energy at higher temperatures. It can be seen that at elevated temperatures, there is again a wide range of conformations, but even as the temperature is lowered, the overall range remains large. At the lowest temperatures, this torsional angle exhibits a range of about 60° and with the exceptions of cycles 4 and 6, do not converge to a common point. The flexibility shown in this term may be responsible for the structural changes in the oligomer.

The radius of gyration as a function of temperature is shown in Figure 9. At elevated temperatures, as usual, considerable variability can be seen. Similarly to the 4-195 distance results, the range over which this term varies does not markedly decrease as the temperature is lowered. It appears that the relatively compact nature of the oligomer is retained even at higher temperatures. Several compact regions of convergence can be seen as lower temperatures are approached. Furthermore, the depression exhibited during time cycle 4 in Figures 6 and 7, also appears for the radius of

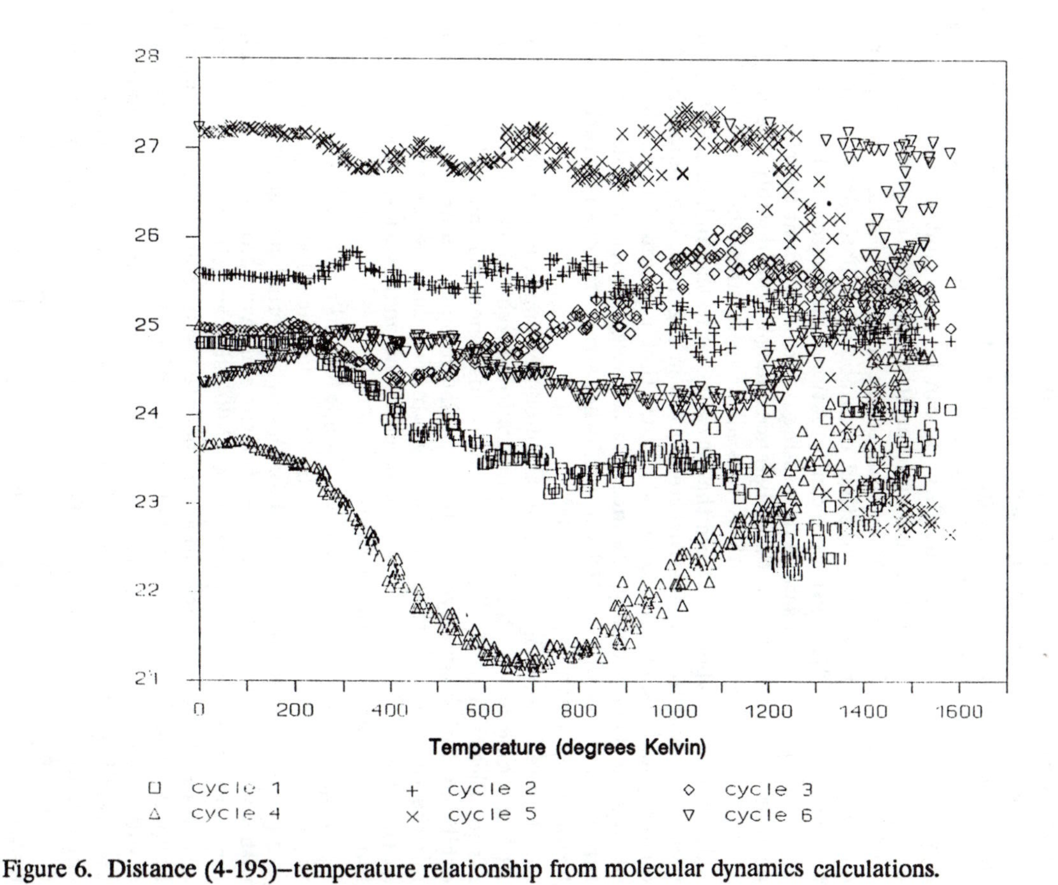

Figure 6. Distance (4-195)–temperature relationship from molecular dynamics calculations.

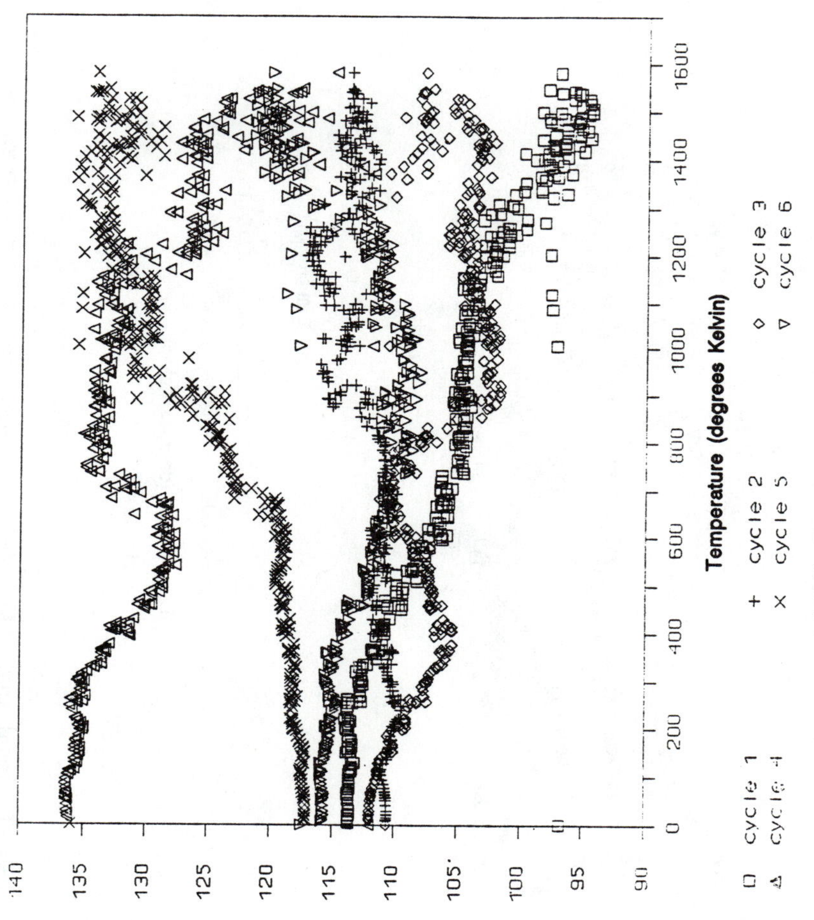

Figure 7. Angle (4-60-195)–temperature relationship from molecular dynamics calculations.

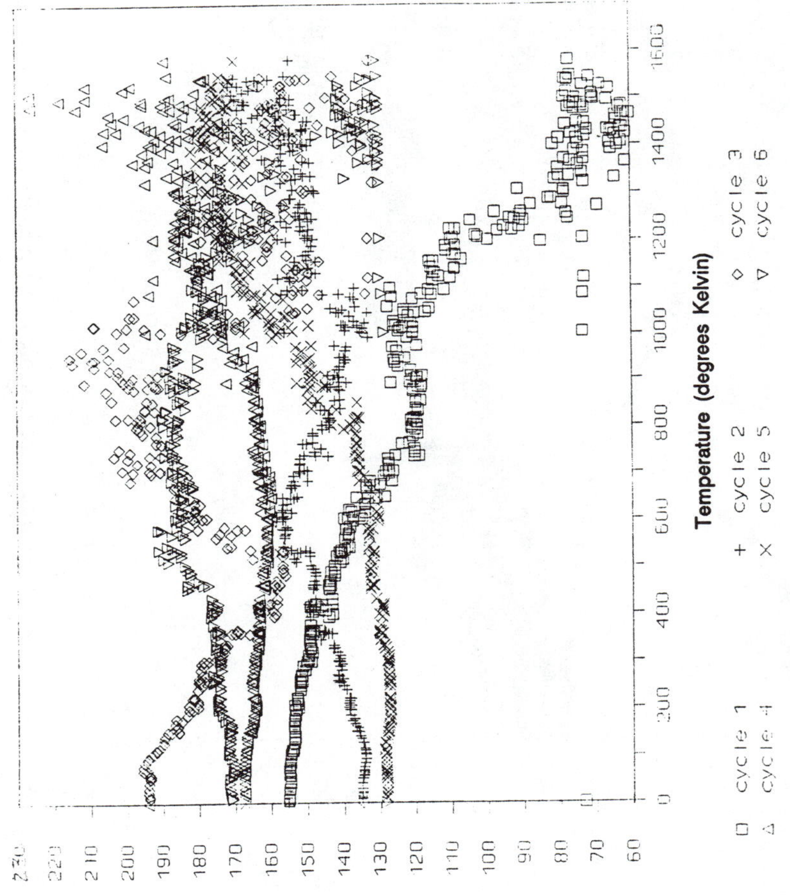

Figure 8. Torsional angle (4-47-60-195)-temperature relationship from molecular dynamics calculations.

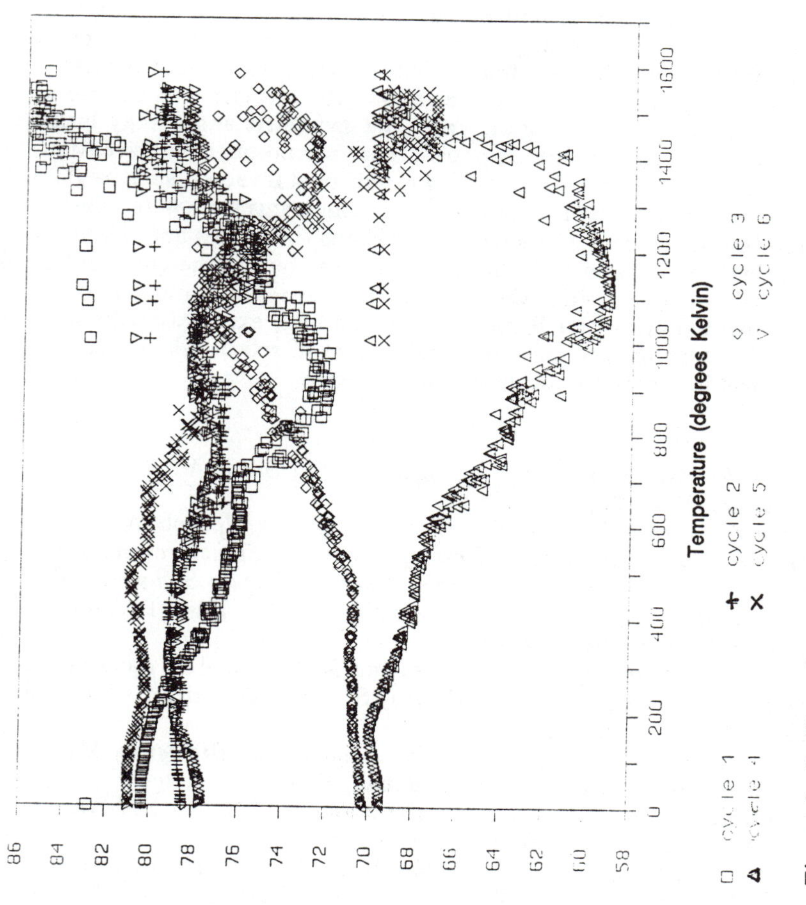

Figure 9. Radius of gyration-temperature relationship from molecular dynamics calculations.

gyration. It is interesting to note, however, that the temperature at which this decrease occurs is shifted to about 1100°K.

Conclusions

Based on end-to-end distance and radius of gyration data, it appears that the oligomer that has been examined retains a fairly compact form even at elevated temperatures. The selected torsional angle is somewhat more flexible, and is perhaps the parameter that allows the structure to deform at elevated temperatures and energies. From the geometric and energy terms that were examined in the current paper, there are no obvious discontinuities that might be interpreted in terms of experimental values, such as glass transition temperatures. This could be due to the terms that were selected to describe the structure of the oligomer. Perhaps the angles and distances selected are not sensitive to such changes. This may be an inherent problem in the analysis of lignin, since there are no simple repeating units in the polymer. The complexity of lignin notwithstanding, explorations will be continued to determine the existence and nature of relationships between molecular level interactions and material properties.

Literature Cited

1. Glasser, W. G.; Barnett, C. A.; Rials, T. G.; Saraf, V. P. *J. Appl. Polym. Sci.* 1984, **29**, 1815-1830.
2. Eberhardt, James J.; Hay, P. Jeffrey; Carpenter, Joseph A., Jr. In *Computer-Based Microscopic Description of the Structure and Properties of Materials; Proc. Materials Research Society Symposia*, Vol. 63; Broughton, J.; Krakow, W.; Pantelides, S. T., Eds.; Pittsburgh: Materials Research Society, 1985.
3. Goring, D. A. I. In *Lignin: Properties and Materials; ACS Symposium Series 397;* Glasser, W. G.; Sarkanen, S., Eds.; American Chemical Society: Washington, DC, 1989.
4. Cantor, C. R.; Schimmel, P. R. *The Behavior of Biological Macromolecules*; San Francisco: W. H. Freeman, 1980.
5. McCammon, J. A.; Harvey, S. C. *Dynamics of Proteins and Nucleic Acids*; Cambridge: Cambridge University Press, 1987.
6. Burkert, U.; Allinger, N. L. *Molecular Mechanics; American Chemical Society Monograph 177*; Washington, DC: American Chemical Society, 1982.
7. Clark, Tim. *A Handbook of Computational Chemistry: A Practical Guide to Chemical Structure and Energy Calculations*; New York: Wiley-Interscience, 1985.
8. Karplus, Martin. *Methods in Enzymology* 1986, **131**, 283-307.
9. Howard, A. E.; Kollman, P. A. *J. Medicinal Chem.* 1988, **31**, 1669-1675.
10. Verlet, L. *Phys. Rev.* 1967, **159**, 98.
11. Brünger, A. T. *J. Molecular Biol.* 1988, **203**, 803-816.
12. Kirkpatrick, S.; Gelatt, C. D.; Vecchi, M. P. *Science* 1983, **220**, 671-680.

RECEIVED February 10, 1992

Chapter 26

Thermal Behavior of Aromatic Polymers Derived from Phenols Related to Lignin

S. Hirose[1], H. Yoshida[3], T. Hatakeyama[2], and Hyoe Hatakeyama[1]

[1]Industrial Products Research Institute and [2]Research Institute for Polymers and Textiles, 1–1–4 Higashi, Tsukuba, Ibaraki 305, Japan
[3]Tokyo Metropolitan University, Fukazawa, Setagaya-ku, Tokyo 158, Japan

Two aromatic polyethers having phosphine oxide groups were synthesized from bisphenols having p-hydroxyphenyl groups. The p-hydroxyphenyl group is known to be a lignin-related structure. The molecular relaxation of amorphous polyethers having 2,2-diphenyl propane units (I) and those having 4,4′-biphenyl units (II) was studied by differential scanning calorimetry (DSC) and dynamic mechanical analysis (DMA). Enthalpy relaxation was observed for the samples annealed at temperatures below glass transition temperatures (T_g's) in DSC measurements. The value of relaxation time (τ) of polyether I was shorter than that of polyether II. The activation energies (E_a's) of enthalpy relaxation were 250 kJ/mol for polyether I and 410 kJ/mol for polyether II. Three relaxations, α-, β- and γ-relaxations, were observed at around 473K, 273K, and 173K, respectively, in DMA measurements. The E_a for each relaxation was calculated using the relationship between relaxation temperature and frequency. α-relaxation is attributed to long-range molecular relaxation of the main chain. β-relaxation is related to local mode relaxation of the main chain. γ-relaxation is assumed to be due to the rotation of phenyl groups in the side chain. It was concluded that the mobility of the main chain of polyethers I and II affects enthalpy relaxation and β-relaxation and that the 2,2-diphenyl propane units in polyether I are more mobile than the 4,4-biphenyl units in polyether II.

0097–6156/92/0489–0385$06.00/0

Aromatic polymers having phenylene groups in their main chain have been recognized as high-performance polymers due to their excellent thermal and mechanical properties. Recently, aromatic polyesters (1,2) were synthesized in our laboratory from bisphenols which were derived from phenols having core structures of lignin such as p-hydroxyphenyl, guaiacyl, and syringyl groups. We have studied these polymers with reference to the relationships between the chemical structures and thermal properties of the polymers. It has been found that the existence of methoxyl groups attached to the phenylene groups does not reduce the starting temperatures of thermal decomposition (T_d's) of polymers, and also that the glass transition temperatures (T_g's) of polymers with guaiacyl groups are lower than those of polymers with syringyl groups.

We also studied aromatic polyethers (3) having phosphine oxide groups which were synthesized from bisphenols having p-hydroxyphenyl groups. These polyethers were found to have T_g's higher than 450K and T_d's higher than 760K. Recently, attention has been paid to the relaxation process of aromatic polymers having high T_g's in the glassy state. This is because the physical properties of glassy polymers are markedly affected by the molecular relaxation at a temperature lower than T_g. At the same time, it has been suggested that the rate of enthalpy change is related to the local-mode relaxation of polymers, especially in the case of polymers having rigid phenylene groups in their main chain. In the present study, the relaxation process of polyethers having p-hydroxyphenylene groups, such as 2,2-diphenyl propane units and 4,4′-biphenyl units, was studied using differential scanning calorimetry (DSC) and dynamic mechanical analysis (DMA) in order to establish the relationship between chemical structure and relaxation behavior.

Experimental

Samples. Aromatic polyethers having phosphine oxide groups were prepared according to the procedure reported previously (3). The chemical structures of the polyethers are shown in Table I.

Measurements. A Perkin Elmer differential scanning calorimeter DSC II and a Seiko thermal analysis system SSC 5000 were used in the measurements for enthalpy relaxation. The sample weight was ca. 3 mg and the heating rate was 10K/min. The sample was sealed in an aluminum vessel and heated to a temperature higher than $T_g + 50$K, quenched to 300K and then reheated at 10K/min. The quenched samples were annealed at various temperatures and times in the DSC holder. Heat capacity (C_p) (4) and T_g were measured as reported previously (5). Dynamic mechanical analysis (DMA) was carried out using a Seiko Dynamic Mechanical Spectrometer SDM 5600 equipped with a Tension Module DMS 200. Films 20 mm in length, 3 mm in width, and 0.25 mm in thickness were used for DMA measurements. The initial stress was 3.4×10^5 Pa. The temperature was controlled from 120 to 520K. The measurements were carried out under a nitrogen atmosphere at a heating rate of 2 K/min and at a frequency of 1, 2, 5 and 10 Hz.

Table I. The chemical structures of the polyethers

Abbreviation	Chemical Structure
I	$-C_6H_4-P(=O)(C_6H_5)-C_6H_4-O-C_6H_4-C(CH_3)_2-C_6H_4-O-$
II	$-C_6H_4-P(=O)(C_6H_5)-C_6H_4-O-C_6H_4-C_6H_4-O-$

Results and Discussion

Figure 1 shows the C_p curves of polyethers I and II. A jump in T_g is clearly observed. Figure 2 shows representative DSC curves of the quenched and annealed samples of polyether I. The endothermic peak shown in Figure 2 increased with increasing annealing time. The enthalpy of the equilibrium state at a temperature below T_g was estimated by assuming that the C_p in the liquid state could be interpolated to $T_g - 50$K. Based on the above assumption, the excess enthalpy (ΔH_o) of the sample can be defined as follows:

$$\Delta H_o = \int_{T_a}^{T_g} C_{pl}(T_i)dT - \int_{T_a}^{T_g} C_{pg}(T_i)dT \qquad (1)$$

where C_{pl} is the C_p in the liquid state, C_{pg} is that at the glassy state of the sample immediately after quenching from the liquid state, $T_a = T_{g-a}$. In the present study, $a = 15$K was used.

Instead of equation 1, the equation $\Delta H_o = \Delta C_p \times a$ can be used (6), where $\Delta C_p = C_{pl} - C_{pg}$ at T_g. The enthalpy difference between the annealed glass and the quenched glass can be obtained from the experimental data as follows:

$$\Delta H_a = \int_{T_g-a}^{T_g+a} C_{pa}(T_i)dT - \int_{T_g-a}^{T_g+a} C_{pq}(T_i)dT \qquad (2)$$

where C_{pa} is the heat-capacity of the annealed glass and C_{pq} is that of quenched glass. The total excess enthalpy of an annealed sample, ΔH_t, can be obtained from equation 3:

$$\Delta H_t = \Delta H_o - \Delta H_a \qquad (3)$$

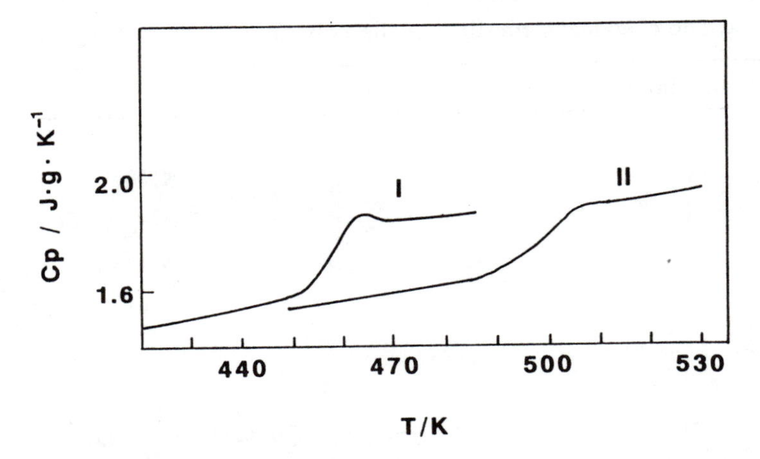

Figure 1. C_p curves of polyethers I and II.

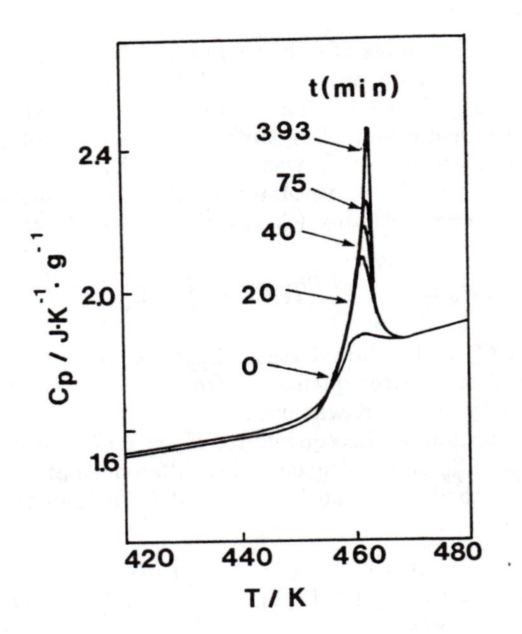

Figure 2. C_p curves of polyether I. T_a = 448K. The annealing times are shown in the figure.

When the annealing time increases, ΔH_a increases and ΔH_o decreases; that is, the state of the sample approaches the equilibrium state. The rate of this change and the relaxation time can also be calculated from equation 4:

$$\Delta H_t = \Delta H_q \exp(-t/\tau) \tag{4}$$

where ΔH_q is the enthalpy of quenched glass and t is the annealing time. The value of ΔH_q is almost the same as that of ΔH_o.

The C_p data indicated in Figures 1 and 2 were used in order to calculate ΔH_a and ΔH_t using equations 2 and 3. τ values can be obtained using equation 4. Figure 3 shows the relationship between τ and ΔH_t of polyether I.

Using the relationship shown in Figure 4, the relaxation time at $\Delta H_t/\Delta H_o = 0.5$ ($\tau_{1/2}$ value) can be evaluated for each annealing temperature. Figure 4 shows the relationship between $\tau_{1/2}$ and $(T_g - T_a)$ for two polymers. The activation energy (E_a) of enthalpy relaxation was also calculated from the relationship between $\tau_{1/2}$ and $1/T$ by assuming that the enthalpy relaxation will proceed according to Arhenius' kinetics equation $\tau_{1/2}{}^{-1} = A\exp(-E_a/RT)$. The plots of $\tau_{1/2}$ *vs.* $1/T$ are shown in Figure 5.

As shown in Figure 4, the $\tau_{1/2}$ values for polyether I are smaller than those for polyether II. The calculated E_a's were 250 kJ/mol for polyether I and 410 kJ/mol for polyether II. It has been suggested that the internal rotation of the main chain is related to the enthalpy relaxation of the amorphous chain, since molecular rearrangement of long-range order will not occur at a temperature below T_g. The difference in $\tau_{1/2}$ and in E_a values between polyether I and II suggests that the molecular relaxation is accelerated in the presence of 2,2-diphenylpropane units in the main chain.

Figures 6 and 7 show the changes of dynamic Young's modulus E′ and dynamic loss tanδ as a function of temperature. These results were obtained by the measurements at 1 Hz. The dynamic modulus E′ of polyether I decreased slightly at around 170 and 270K and markedly at around 470K. The change in E′ for polyether II was almost the same as that for polyether I. The tanδ curves show two small peaks and a large peak. The Greek characters α, β, and γ indicate each tanδ peak from the high temperature side.

Table II indicates the temperatures of tanδ peaks measured at 1 Hz and the calculated activation energies (E_a's). The calculated E_a's of the enthalpy relaxation are also listed in Table II. Concerning α-relaxation, E′ markedly decreases in the temperature range of the α-tanδ peak. At the same time, E_a of α-relaxation shows a high value of ca. 800 kJ/mol. In addition to this, glass transition was observed in this temperature region in DSC measurements. Therefore, it is reasonable to consider that the α-relaxation is glass transition which is related to the long-range molecular motion of the main chain.

γ-relaxation was observed in the tanδ curve at around 170K for polyethers I and II. The calculated E_a's are ca. 35 kJ/mol as shown in Table II. It has been reported (7,8) that polymers having phenyl groups in

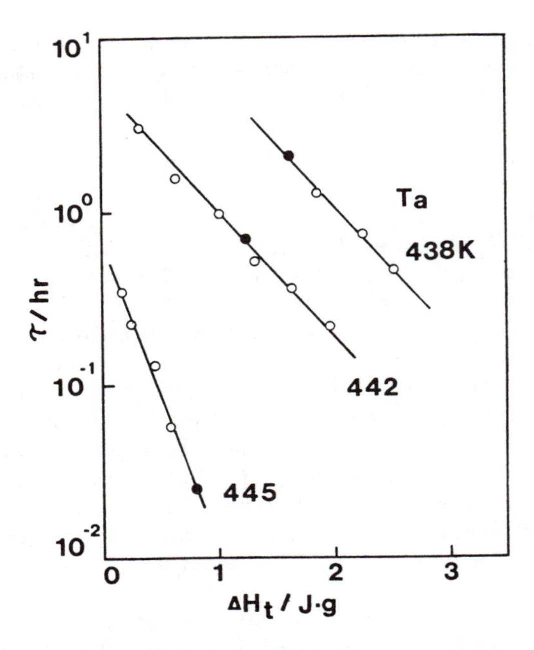

Figure 3. Relationship between τ and H_t of polyether I.

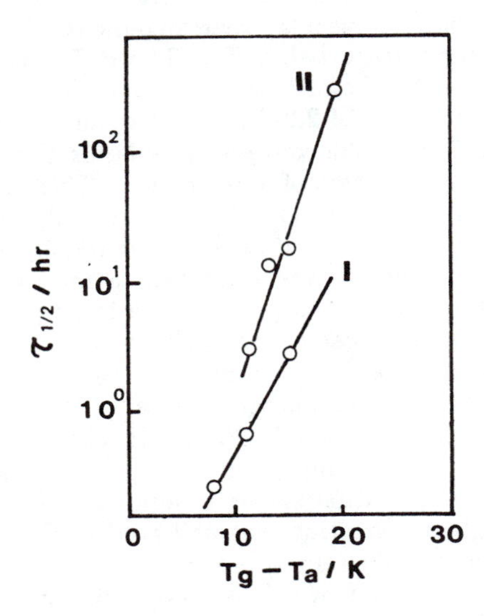

Figure 4. Relationship between relaxation time $(\tau_{1/2})$ and $(T_g - T_a)$ of polyethers I and II.

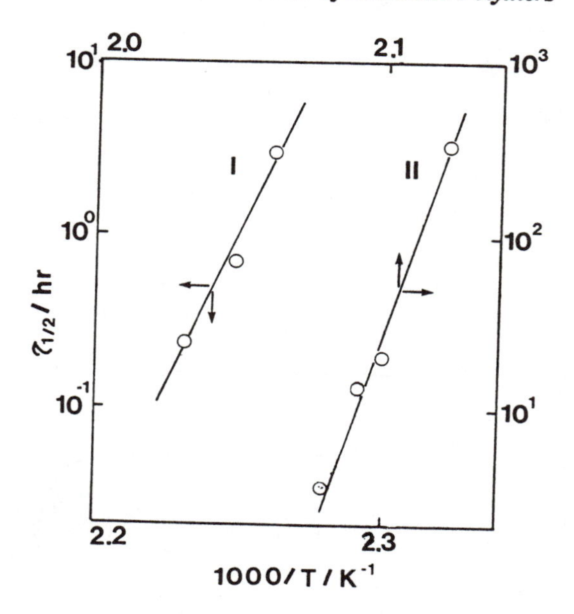

Figure 5. Relationship between relaxation time ($\tau_{1/2}$ and 1/T of polyethers I and II.

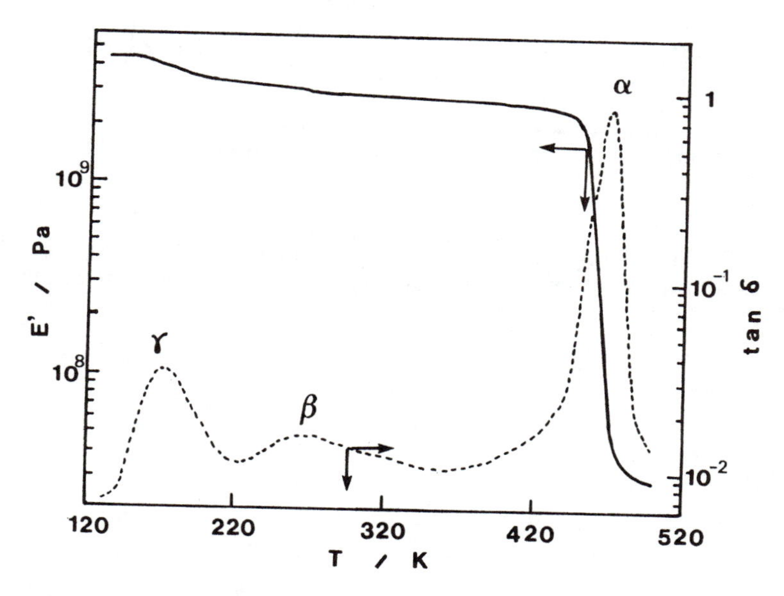

Figure 6. The changes of dynamic Young's modulus E′ and dynamic loss tanδ as a function of temperature of polyether I. The curves were obtained at 1 Hz.

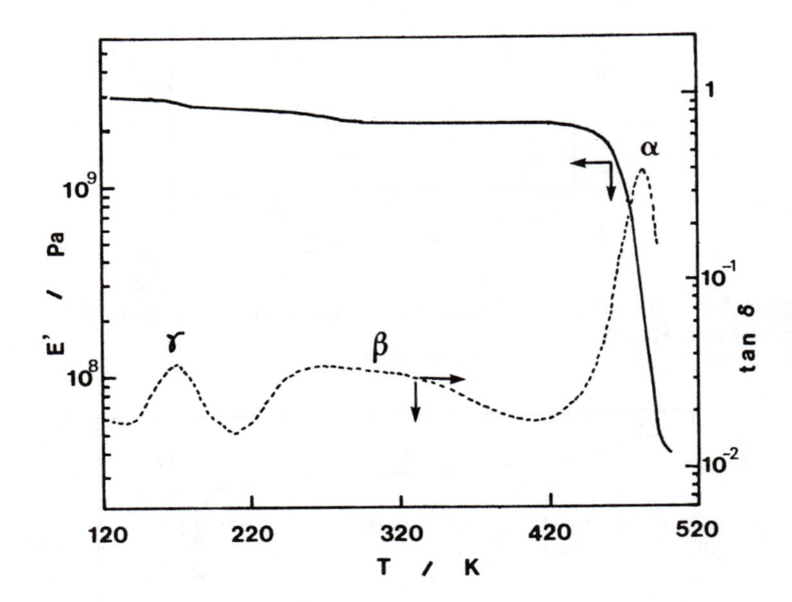

Figure 7. The changes of dynamic Young's modulus E′ and dynamic loss tanδ as a function of temperature of polyether II. The curves were obtained at 1 Hz.

Table II. The tanδ peak temperatures and calculated activation energies (E_a's)

Sample	Relaxation	Temperature, K	E_a, kJ/mol
	α	472	785
I	β	275	100
	γ	172	37
	Enthalpy	442-448	250
	α	483	800
II	β	273	90
	γ	157	33
	Enthalpy	472-480	410

the side chains, such as polystyrene and its copolymers, show γ-relaxation having similar E_a's and at similar temperatures to those of polyethers I and II. For example, polystyrenes were found to show γ-relaxation having E_a of 9 kcal/mol (ca. 36 kJ/mol) (7). This relaxation can be attributed to the restricted rotation of phenyl groups in the side chain (6). Therefore, it seems reasonable to consider that the observed γ-relaxation for polyethers I and II will also be related to the restricted rotation of phenyl groups in the side chains of polyethers I and II. E_a's of β-relaxation were 100 kJ/mol for polyether I and 90 kJ/mol for polyether II. It has been reported that the relaxation having similar values of E_a's to those for polyethers I and II are observed at temperatures below T_g for polymers having phenylene groups in their main chain, such as polyethylene terephthalate, polycarbonate (9-11). These studies have suggested that the local mode relaxation of the main chain can be attributed to β-relaxation (9-11). Accordingly, it can be said that β-relaxation which was observed for polyethers I and II is also related to the local mode relaxation of the main chain. The temperature range of β-relaxation for polyethers I and II was found to be different, although the E_a values were almost the same. β-relaxation of I occurred from 220 to 370K and that of II from 220 to 420K. The temperature range of β-relaxation for polyether I is wider than that for polyether II. This fact suggests that the distribution of β-relaxation of polyether II is larger than that of polyether I. It was also found that the peak intensity of β-relaxation for polyether II was larger than that for polyether I. This suggests that the units which are related to β-relaxation for polyether I are more mobile than those of polyether II.

Enthalpy relaxation was observed at temperatures higher than those for β-relaxation. Therefore, it can be said that the molecular motion units for β-relaxation are included in the enthalpy relaxation process. Furthermore, in the case of polyether I, the highest temperature at which β-relaxation is observed is ca. 80K lower than temperatures for the occurrence of enthalpy relaxation. This temperature difference is approximately 50K for polyether II. These results suggest that the molecular units related to β-relaxation in polyether I are more mobile than those in polyether II. It has been suggested that enthalpy relaxation of polymers having phenylene groups in their main chain is related to the internal rotation of three or four repeating units in the main chain (12). E_a's of enthalpy relaxation for polyethers I and II were 250 and 410 kJ/mol, respectively. This indicates that diphenyl propane units of polyether I are more mobile than biphenyl units of polyether II. These results obtained for enthalpy relaxation suggest that the molecular unit which is related to β-relaxation contributes also to the enthalpy relaxation process.

Literature Cited

1. Hirose, S.; Hatakeyama, H.; Hatakeyama, T. *J. Soc. Fibre Sci. Technol. Japan* 1985, **41**, T-433.
2. Hirose, S.; Hatakeyama, H.; Hatakeyama, T. In *Wood Processing and Utilization*; Kennedy, J. F.; Phillips, G. O.; Williams, P. A., Eds.; Chichester: Ellis Horwood Ltd., 1989; Ch. 22.
3. Hirose, S.; Nakamura, K.; Hatakeyama, T.; Hatakeyama, H. *J. Soc. Fibre Sci. Technol. Japan* 1987, **43**, 595.
4. Hatakeyama, T.; Kanetuna, H.; Ichihara, S. *Thermochimica Acta* 1989, **146**, 311.
5. Nakamura, S.; Todoki, M.; Nakamura, K.; Kanetuna, H. *Thermochimica Acta* 1988, **136**, 163.
6. Yoshida, H. *Netsu Sokutei* 1986, **13**, 191.
7. Yano, O.; Wada, Y. *J. Polym. Sci.* 1971, **A2, 9**, 669.
8. Illers, K. H.; Jenkel, E. *J. Polym. Sci.* 1959, **41**, 528.
9. Wada, Y. *Physical Properties of Macromolecules (Koubunshi No Kotai Bussei)*; Tokyo: Baihukan Publisher, 1971; p 381.
10. Coburn, J. C.; Boyd, R. H. *Macromolecules* 1986, **19**, 2238.
11. Aoki, Y.; Brittain, J. O. *J. Appl. Polym. Sci.* 1976, **20**, 2879.
12. Yoshida, H.; Kobayashi, Y. *J. Macromol. Sci.* 1982, **B21**, 565.

RECEIVED February 10, 1992

Author Index

Affiliation Index

Subject Index

Production: Peggy D. Smith
Indexing: Deborah H. Steiner
Acquisition: Barbara C. Tansill
Cover design: Sue Schafer

Printed and bound by Maple Press, York, PA